Factory Planning Manual

Michael Schenk · Siegfried Wirth · Egon Müller

Factory Planning Manual

Situation-Driven Production Facility Planning

Springer

Prof. Michael Schenk
Otto von Guericke University, Magdeburg
Institute of Logistics and
Material Handling Systems
Chair of Logistics Systems
Universitätsplatz 2
D-39106 Magdeburg
Director of the Fraunhofer Institute for
Factory Operation and Automation IFF
Sandtorstr. 22
D-39106 Magdeburg
michael.schenk@iff.fraunhofer.de

Prof. Siegfried Wirth
Chemnitz University of Technology
Institute of Industrial Management
and Factory Systems
IREGIA e.V.
D-09107 Chemnitz
s.wirth@mb.tu-chemnitz.de

Prof. Egon Müller
Chemnitz University of Technology
Institute of Industrial Management
and Factory Systems
Professorship of Factory Planning
and Factory Operation
D-09107 Chemnitz
egon.mueller@mb.tu-chemnitz.de

Translator: Michael Lorenz, Magdeburg
Translation: TECH GmbH, Magdeburg, www.t-ech.de

ISBN 978-3-642-03634-7 e-ISBN 978-3-642-03635-4
DOI 10.1007/978-3-642-03635-4
Springer Heidelberg Dordrecht London New York

Library of Congress Control Number: 2009939127

Cover design: WMXDesign GmbH, Heidelberg

Printed on acid-free paper

Springer is part of Springer Science+Business Media (www.springer.com)

Preface

The central purpose of this book is to impart knowledge, skills and practical implementation methods for the planning and operation of adaptable production facilities and factories.

It addresses planning methods and procedures for various types of production facility up to and including entire factories, and is aimed at practicing factory planners and students alike. The book provides facts and demonstrates practical processes using case studies for the purposes of illustration, so that ultimately skills can be acquired that make independent practical implementation and application possible. It is based on up-to-the-minute practical experience and universally applicable knowledge of the planning and technological design of adaptable production facilities (manufacturing and assembly) and factories.

In comparison to existing, thematically-similar reference books, what is innovative about this manual is that it provides the impulse for a more flexible planning approach for the efficient design of adaptable production facilities using responsive, unconventional planning and organizational solutions. The book aims to provide a way of integrating systematic and situation-driven planning methods in a meaningful way. Situation-driven planning is becoming increasingly important to production facilities in these fast-moving times of change, in particular in terms of resource and energy efficiency. Existing technical and organizational course of action in terms of resources (both human and technical) need to be selected for the specific case at hand, and changes (to workshops, products, processes and equipment) need to be managed. Project managers are responsible for assigning subcontracts, coordinating services and combining them in a single project. To this end, the questionnaires and checklists contained in the book and the discussion of potential for change in the case of key planning activities are particularly useful. The book's appendix expands upon investment appraisal methods, main building and production parameters, supply and disposal systems and the planning and control of information systems.

The book is structured so that it conveys an overview of *engineering services* for production facility planning.

The book's *scope of application* is focused on production facilities for manufacturing parts, assemblies and finished products in the following sectors of industry:

- mechanical engineering and plant construction, electrical and electronic equipment engineering
- process engineering (textiles, clothing, printing and packaging)
- automobile industry and supply industry
- ICT, automation and environmental technology

The main emphasis of the book is on businesses employing make-to-order, small batch and series production for manufacturing processes with differing mechanization and automation solutions and the following types of investment: new, expansion, rationalization and replacement investments.

The manual is intended to be of use to production facility planners, equipment suppliers and operators.

It assumes basic manufacturing and business knowledge and is designed as a self-help guide for:

- interested practitioners,
- students, for acquisition and consolidation of knowledge,
- additional and further education for specialists,
- specialists and managers from industry, services and business, primarily in the fields of production engineering, industrial engineering, production management, construction, architecture and logistics.

This manual is a practical addition to existing reference and text books, and in particular “Fabrikplanung und Fabrikbetrieb” *(Factory Planning and Factory Operation)* published by Springer (Schenk & Wirth 2004) and “Montage in den industriellen Produktionsstätten” *(Assembly in Industrial Production Facilities)* (Lotter & Wiendahl 2006), as well as the teaching materials of various educational and training establishments. Papers on *Factory Project Design* (Helbing 2007), “Facility Design and Engineering” (Hanna & Konz 2004) and “Changeable Manufacturing” (Wiendahl 2007) are also included. To aid understanding, these books, and in particular “Fabrikplanung und Fabrikbetrieb” *(Factory Planning and Factory Operation)*, are recommended to help readers acquire a basic knowledge of the subject.

Both within and beyond the European Union’s borders, manufacturing industry is subject to a multitude of country-specific standards, directives and regulations. In order to guarantee the uniformity, transparency and comprehensibility of the examples, we have referred primarily to EU standards (EN, ISO and DIN), VDI/VDE guidelines and German domestic regulations. Country-specific stipulations must therefore be accommodated and adapted separately.

The authors would particularly like to thank the staff of the Institute of Industrial Management and Factory Systems, department of Factory Planning and Factory Operation and of the Fraunhofer Institute for Factory Operation and Automation (IFF), as well as the Institute of Logistics and Material Handling Systems (ILM), department of Logistics Systems and the companies and institutions for their kind cooperation in supplying academic papers.

Special thanks go to Dr. Elke Glistau (Otto von Guericke University, Magdeburg), Gert Kobylka Dipl. Ing. and Kerstin Rockstroh (Chemnitz University of Technology) for their consistently professional assistance.

Magdeburg and Chemnitz, May 2009
Michael Schenk Siegfried Wirth Egon Müller

Contents

List of abbreviations

0 + 5 + X model	Planning model 0 = Project definition (01 to 05) 5 = Project development (5.1 to 5.5) x = Project implementation (x6 to x10)
2D	2 dimensional
3D	3 dimensional
ABC analysis	NB: does not stand for activity based costing (process costs), but for a quantity and value analysis for the categorization of any given item in the 3 groups A, B and C. Commonly used for procured goods, customers and warehouse stock.
ABC-XYZ characteristic	Categorization of any given object (e.g. customers, orders, parts, assemblies, replacement parts, products) into 3 groups according to quantity and value (A, B, C) and demand behavior (X, Y, Z) criteria. This produces a 9 field matrix of possible combinations (e.g. group AX) for which generic strategies can be derived.
AC	Alternating current
AGBG	German Act Governing General Terms and Conditions (**G**esetz zur Regelung des Rechts der **A**llgemeinen **G**eschäfts**b**edingungen)
AMM	Auxiliary manufacturing material
APS	Advanced Planning System
AR	Augmented Reality
ArbStättV	German regulations governing the workplace (**Arb**eits**stätt**en**v**erordnung or Workplace Ordinance)
ASR	German workplace guidelines (**A**rbeitsstätten**r**ichtlinien)
AWSA	Assembly workstation area
BAT	Biological Tolerance values (**B**iologischer **A**rbeitsstoff-**T**oleranz-Wert)
BGB	German Civil Code (**B**ürgerliches **G**esetz**b**uch)
BMV	German Federal Ministry of Transport (**B**undes**m**inisterium für **V**erkehr)
BSC	Balanced Scorecard
CAD	Computer Aided Design
CAFM	Computer Aided Facility Management system
CAP, CAPP	Computer Aided Planning systems, Computer Aided Process Planning system
CIM	Computer Integrated Manufacturing
CL	Client
CNC	Computerized Numerical Control
CRM	Customer Relationship Management systems
CO	Contractor

CONWIP	Constant Work in Process
CS	Components and parts suppliers
DBS	Database System
DC	Direct current
DIN	German Institute for Standardization (**D**eutsches **I**nstitut für **N**ormung)
DNC	Distributed (or Direct) Numerical Control
EKA	Exposure Equivalent for Carcinogenic Substances (**E**xpositionsäquivalent für **k**rebserzeugende **A**rbeitsstoffe)
ERP	Enterprise Resource Planning system
FAQ	Frequently asked questions
FEM	European Federation of Materials Handling (Federation Européenne de la Manutention)
FMC	Flexible manufacturing cell
FTT	(Jigs &) Fixtures, tools and testing equipment
GeWo	German Trade Commerce and Industry Regulation Act (**Ge**werbe**o**rdnung)
GI	Goods inwards
GO	Goods outwards
HGB	German Commercial Code (**H**andels**g**esetz**b**uch)
ICP	In-cyclical parallelism
ISO	International Organization for Standardization
JIS	Just in sequence
JIT	Just in time
LAN	Local Area Network
LTUs	Conveyor
MAP	Manufacturing Automation Protocol
MDA	Machine data acquisition
MEK	Maximum Emissions Concentration (**M**aximale **E**missions-**K**onzentration)
MES	Manufacturing Execution Systems
MIK	Maximum Immissions Concentration (**M**aximale **I**mmissions-**K**onzentration)
MMO	Multiple machine operation
MSS	Management Support System
NC	Numerical control
NS	Number of shifts
OEM	Original Equipment Manufacturer
OP	Operation
OPC	OLE (object linking and embedding) for Process Control
OSACA	Open system architecture for controls within automation systems
OSI	Open System Interconnection

PC	Processing center
PCC	Production control center
PDA	Production data acquisition
PDAs	Personal Digital Assistants
PDCA cycle	Deming's Plan – Do – Check – Act cycle
PDM	Production data management
PerP	Performance program
PLC	Programmable logic controller
PMS	Project Management Systems
PP	Production program
QFD	Quality Function Deployment
QMS	Quality Management Systems
R&D	Research and development
R&D Tools	Research and development tools
RefP	Reference period
RP	Replacement parts
SS	System supplier
STEP	Standard for the exchange of product model data
SUB	Subcontractor
SWOT	Strengths, Weaknesses, Opportunities and Threats
TBS	Technical building systems
TCP/IP	Transmission Control Protocol/Internet Protocol
TGL	German technical quality and supply standards (**T**echnische **G**üte- und **L**iefer-bedingungen)
THS	Transport, handling, storage
TOP	Technical and Office Protocol
3-phase AC	Three-phase alternating current
TPT	Throughput time
TR/TG	Technical rules / Technical guidelines
TRep	Type representatives
TRK	Technical reference concentration (**T**echnische **R**icht**k**onzentration)
TÜV	German Technical Inspection Association (**T**echnischer **Ü**berwachungs**v**erein)
UDM	Universal lathe
UVV	German accident prevention regulations (**U**nfall**v**erhütungs**v**orschriften)
VBG	German employers' liability insurance association (**V**erwaltungs-**B**erufs**g**enossenschaft)
VDI	Association of German Engineers (**V**erein **D**eutscher **I**ngenieure)
VDE	German Association for Electrical, Electronic & Information Technologies (**V**erband **d**er **E**lektrotechnik Elektronik Informationstechnik)

VR	Virtual Reality
WHG	German Federal Water Act (**W**asser**h**aushalts**g**esetz)
WLAN	Wireless Local Area Network
WMS	Workflow management system
WP	Wearing parts
WTR	Working time requirement

1 Introduction

1.1 Business Enterprises, Workshops and Factories

a) *Business enterprises* are commercial operations that earn profits by charging prices that exceed their costs. A business enterprise develops products and provides services.

Manufacturing enterprises manufacture products. Procurement, manufacturing and sales and distribution departments together with the necessary production facilities and factories are needed to develop products.

Production facilities, workshops and factories are basic tools used by businesses to add value. Figure 1.1 illustrates the production facility's position in the enterprise environment.

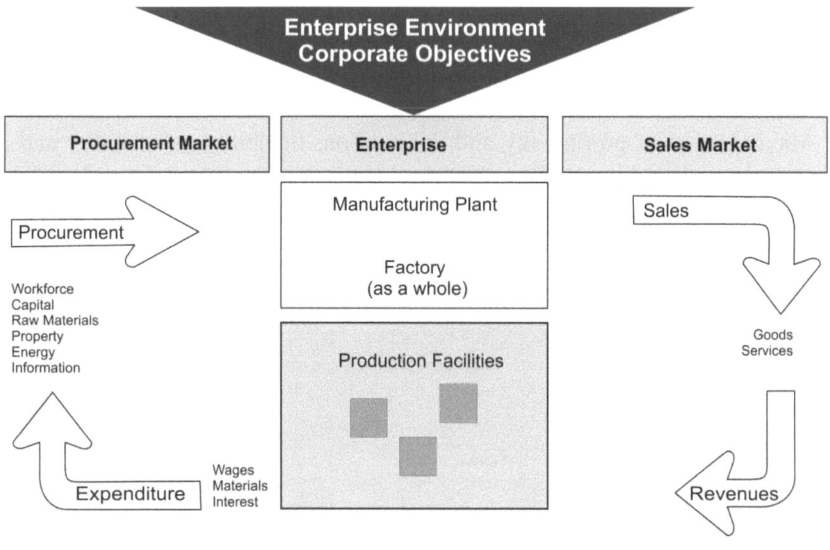

Fig. 1.1 Production facilities as part of an enterprise (Schmigalla 1995)

Manufacturing enterprises are subject to constant *changes* that are influenced by innovation, policy, the environment and the economy. Changes demand flexibility and adaptability on the part of manufacturing enterprises as well as flexible planning. Figure 1.2 shows factors that have an influence on change.

M. Schenk et al., *Factory Planning Manual*, DOI 10.1007/978-3-642-03635-4_1,

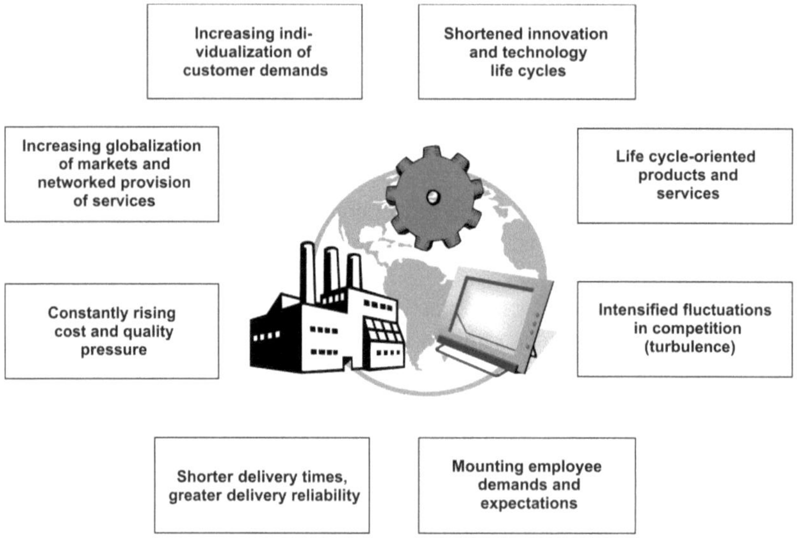

Fig. 1.2 Significant changes for manufacturing enterprises

Maximization of profitability and, in addition, flexibility, adaptability and attractiveness will continue to be corporate objectives in the future (fig. 1.3).

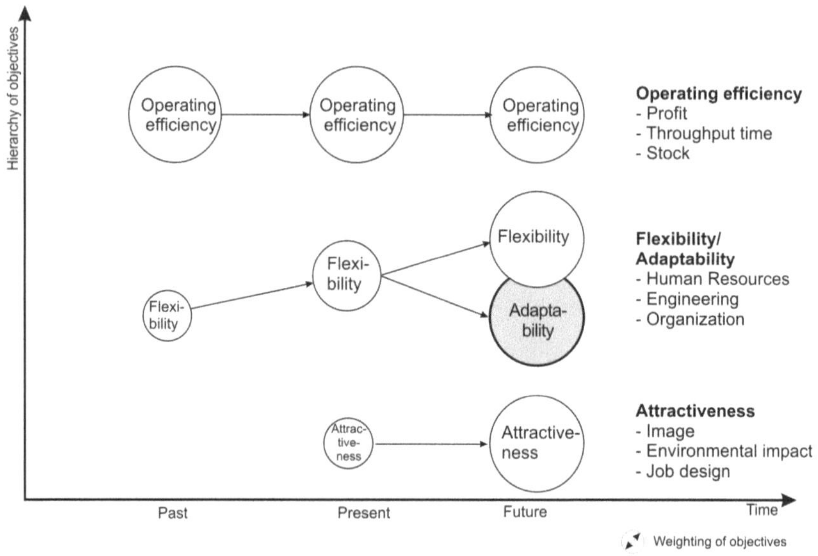

Fig. 1.3 Future objectives (Günther 2006)

Flexibility and adaptability apply both to the enterprise - including its production facilities - and the processes for planning efficient organizational solutions.

Finally, the corporate units responsible for production scheduling and execution in particular must ensure through planning that the technical order process is efficient for the production facility's operator (cf. fig. 1.4) (Spur 1994).

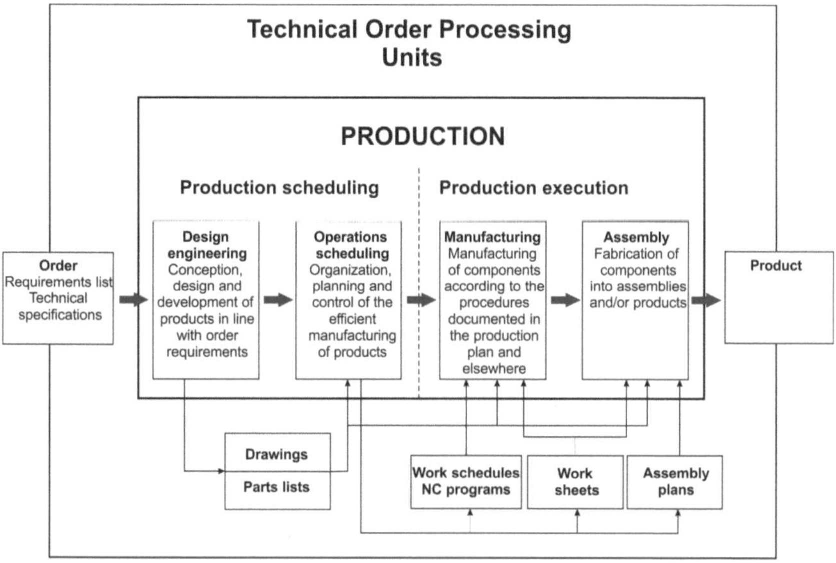

Fig. 1.4 Technical order process

b) *Workshops* – the technical and organizational part and economic unit of a business enterprise (company). They have the job of producing and commercializing goods and services to satisfy customer demands. They can be broken down into structural and functional units as shown below.

Structural units:

- division, department (production, management)

  → section

  → group

  → workstation

  These are technical and organizational elements of a factory, which are (hierarchically) structured according to management and leadership principles.

- Cost center

  This is a part of a plant to which costs incurred are allocated and recorded separately.

Functional units:

parts manufacture, assembly and logistics (transport, handling and storage)

Figure 1.5 shows a *basic model of a workshop* with the task assigned, input and output variables, the scope of work of both human resources and equipment, plus the working environment and workflow.

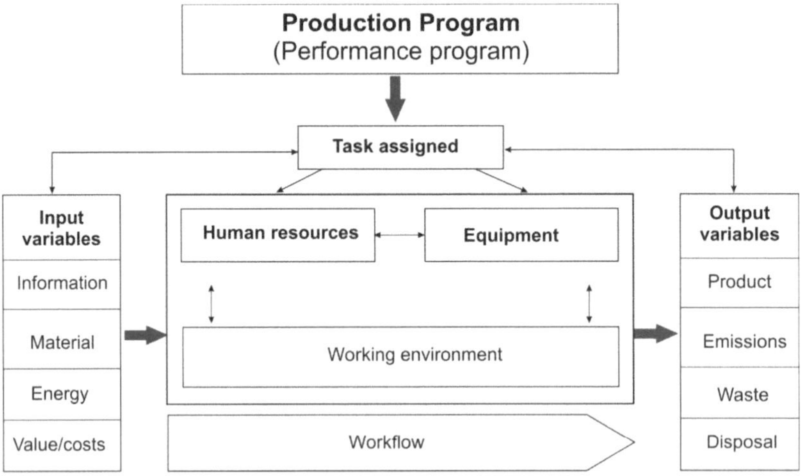

Fig. 1.5 The basic workshop model (Warnecke 1992)

c) *Factories* - are industrial operations that pursue profit-making and cooperative goals. They have organizational areas for which various processes (functions) and facilities must be planned and carried out. Production and operating facilities and workstations are all constituent parts of a factory. Figure 1.6 illustrates the various organizational areas for personnel/workforce, equipment and technical systems in a factory. (See also chapter 6).

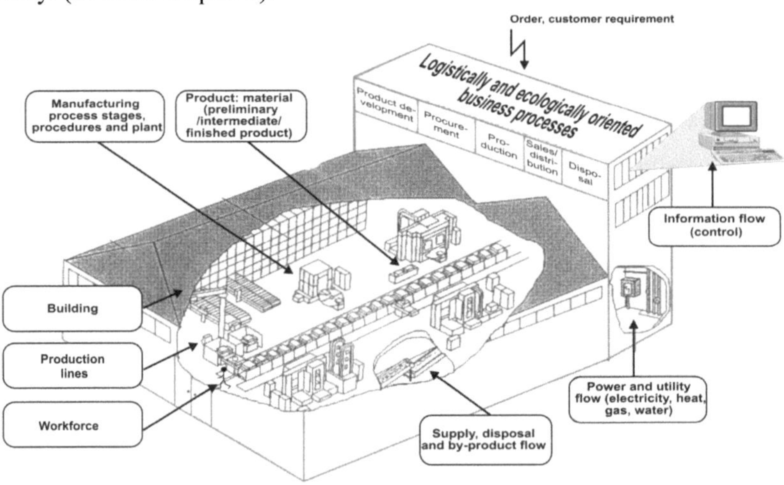

Fig. 1.6 The organizational areas and facilities of a manufacturing site

General elements of a factory/production facility include (cf. also ch. 6):

Personnel/workforce

- number, gender
- qualifications, skills

Machinery and equipment:

- manufacturing and assembly equipment: machinery/workstation including fixtures/auxiliary equipment and tools
- logistics facilities: transport, handling, storage and order picking facilities including auxiliary warehouse and transport equipment
- quality assurance equipment: measuring and testing equipment, jigs and fixtures/auxiliary equipment
- control, information and communication systems
- safety, emissions and interference suppression systems
- supply and disposal systems for utilities, power; raw materials and auxiliary materials; waste and residual materials

Technical systems (in conjunction with their structures):

- structural equipment: supporting structures, foundations, pillars, beams, roof structure
- envelope: facades, roofs including windows, doors, gates
- interior: flooring, ceilings, dividing walls, openings
- building systems: heating, ventilation, air conditioning, sanitary facilities
- supply and disposal systems for utilities: power, gas, water (drinking and industrial water), electricity, raw materials, auxiliary, waste and residual materials

Operating materials:

- liquid materials (fluids, media): water, oils and greases, coolants, acids and bases, solvents, cleaners, polishing materials and abrasives, fuels, paints, biological materials
- gaseous materials: technical gases, technical fuel gases, gas mixtures, steam
- solid materials: fuel, paper and cardboard, glass, administrative equipment

Compatibility, deconfigurability and reconfigurability, mobility, modularity and universality all characterize adaptable equipment, plant and production systems (Wiendahl 2005, Spur 2007). Figure 1.7 illustrates common change scenarios that give rise to changes in personnel and technical resources.

Different types of factories (factory types) can be categorized according to their different characteristics. Figure 1.8 lists the characteristics and various attributes, which are mainly based on technical/organizational and economic aspects. They can be combined to produce a multitude of factory types with their various production facilities. Figure 1.8 shows an example of a factory type/production facility.

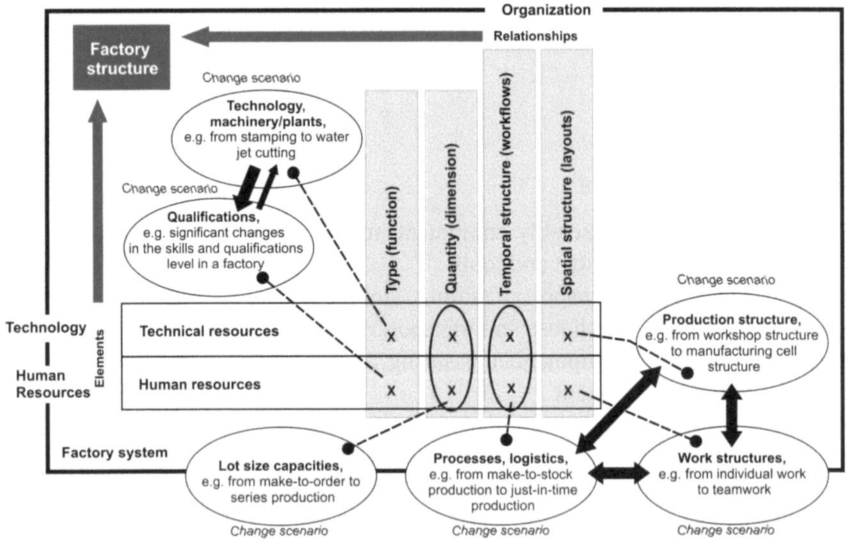

Fig. 1.7 Selected change scenarios (Hildebrand 2005, p. 19)

Characteristic	Attributes					
Enterprise size	Small enterprise	Small and medium-sized enterprise	Large enterprise			
Product size	Micro	Small	Medium	Large		
Production process	Continuous	Batch	Combined			
Investment period	Short-term	Medium-term	Long-term			
Location strategy	Local	Regional	National	Global		
Location changes	Mobile	Permanent	Combined			
Factory orientation	Process	Product	Workforce	Buildings		
Type of production	Make-to-order production	Small batch	Medium batch	Mass		
Value-added stages	Marketing	Research/ development	Procurement	Production	Sales & Marketing	Service
Operator models	Buy	Rent	Lease			
Production stages	Single part	Structural elements, components	Assembly, system assembly	Units (vehicles)	Plant systems	
Networking	Autonomous factory	Networked factory	Competence networks	Virtual factory		
Use	Reuse	Further use	Recycling	Disposal		

Example: standardized factory

Fig. 1.8 Morphology for determining factory types – extract (Schenk & Wirth 2004, p. 18) (cf. 5.2.1.2)

Note: factories and production facilities are unique entities. Every factory is different in terms of its human resources, products, processes, systems, function, dimensions, structure, layout, profitability and corporate philosophy. The ability of production facilities to adapt is becoming a top priority for modern enterprises and is a perpetual task for management.

1.2 Product, Processes and Plants

In an enterprise, the job of a production facility is to perform competitively on the market by producing (material) goods as products. The interrelationship between the product, processes and plant thus deserves attention (cf. fig. 1.9).

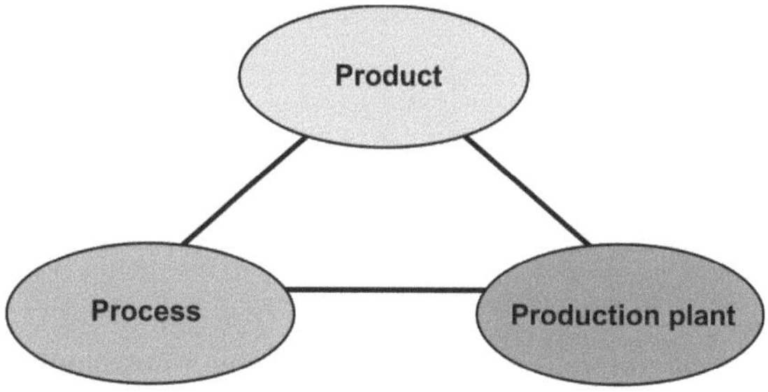

Fig. 1.9 Relationship structure between product, production process and plant

Note: the product desired by the customer determines the process (in units of individuals, equipment and organization) and the process determines the plant (equipment, facilities and items) that individuals operate, control and supervise.

Changing one component results in changes to other components. This applies to production facilities as well as to an entire factory, as in figure 1.10.

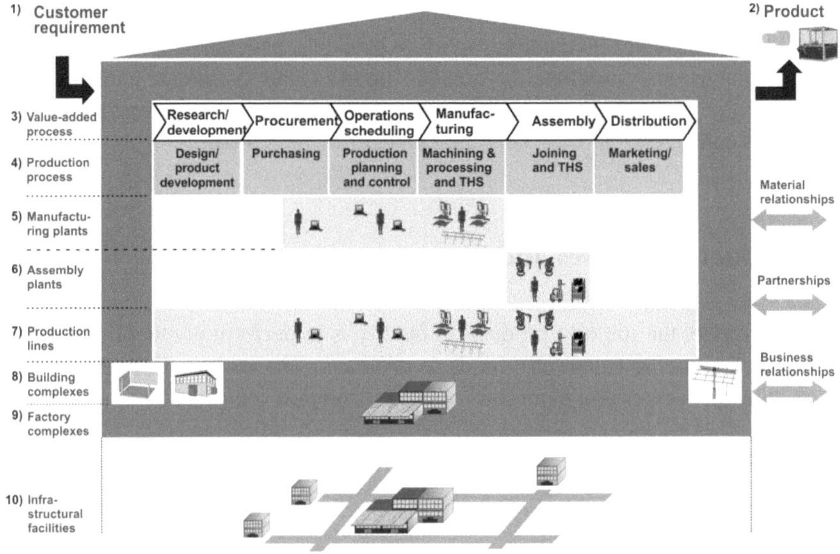

Fig. 1.10 A factory/production facility's product, processes and plants (Schenk, Wirth 2004, p. 18)

Customer requirement (1) – a concept devised by a customer or group of customers that is fulfilled by competitive products in the form of physical goods and services with high customer value.

Customer order – order placed by the customer for physical goods and services.

Product (2) – The outcome of operating processes that serves to satisfy customer requirements. (Product, assembly, component, repetitive parts, raw materials)

Value-added process (3) – The set of all commercial activities that are carried out to meet a customer requirement; it is implemented by value-added units in the value-added chain (research, development, procurement, operations scheduling, manufacturing, assembly, distribution).

Production process (4) – All processes involved in the production of goods and services in a combination of human resources (workforce), technology (object being worked upon and equipment) and organization. It encompasses design engineering/development, purchasing, production planning and control, machining and processing and THS, assembly and THS, and distribution, sales and service.

Manufacturing plant (5) – The production of individual parts by means of machining and processing equipment and systems, including transport, handling and storage equipment, with purchasing, production planning and control (and limited design engineering and sales).

Assembly system (6) – Creation of component assemblies (system components, products) using joining and assembly equipment (systems) including transport, handling and storage equipment, with purchasing, production planning and control (and limited design engineering and sales).

Production line (7) – The integration of machinery and plants for different technological manufacturing (manufacturing lines) and assembly (assembly lines) processes, including transport, handling and storage equipment, with design engineering, purchasing, production planning and control, distribution and sales.

Building complex (8) – A building with its geometric and load parameters including technical building systems to house production facilities or parts thereof. The building is the operational repository of technological processes and technical systems, the site where goods are produced, and a key element adapted to the environment with infrastructure connecting to the site. Technical building systems (TBS) include building, supply and disposal systems, e.g. water, waste water, gas, heating, ventilation, power, IT, safety and automated systems.

Factory complexes (9) – Buildings with production lines and connections to infrastructure.

Infrastructure (10) – The site and factory's supply and disposal systems (power, water, gas, transport routes, etc.) installed in the location.

1.3 Structure of Production Facilities

The structure (Wirth 2000) is subdivided into:

a) hierarchical organization levels (fig. 1.11)

This corresponds to the structure of the workshop from the division through the section (department), to the group and workstation for manufacturing and assembly operations.

> **Note:** the manufacturing workstation is the smallest unit. A manufacturing group consists of several manufacturing workstations, a manufacturing section consists of manufacturing workstations and manufacturing workstation groups, and so on. The interfaces between them are formed by storage areas connected by flow systems.

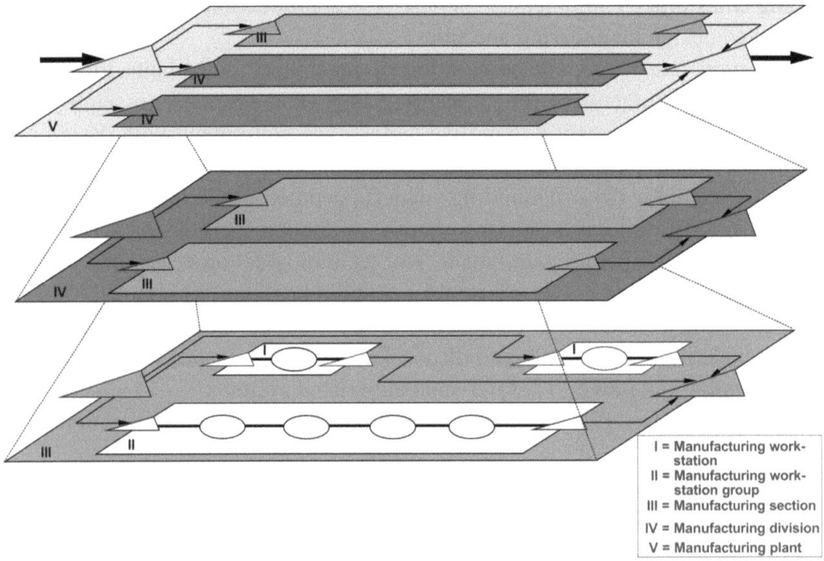

Fig. 1.11 Hierarchical organization of a production facility

b) Peripheral areas (fig. 1.12)

These are based on the main parts manufacturing and assembly processes. Varying degrees of interconnection (direct or indirect) with the product being manufactured (production program) yields three peripheral areas.

First periphery - Systems that are *directly* connected to the product and thus directly connected to the main process (connected to the object being worked on), e.g. quality control, warehouse, control.
Second periphery - Systems that are *not* connected to the product but directly connected to the main process systems (connected to equipment), e.g. maintenance, auxiliary materials
Third periphery - Systems that are *independent* of the main process and its systems. These include social and management facilities (dependent on the workforce), e.g. sanitary facilities, administrative services.

Note: planning always proceeds from the center (main process) and then in sequence from the first periphery to the second and third.

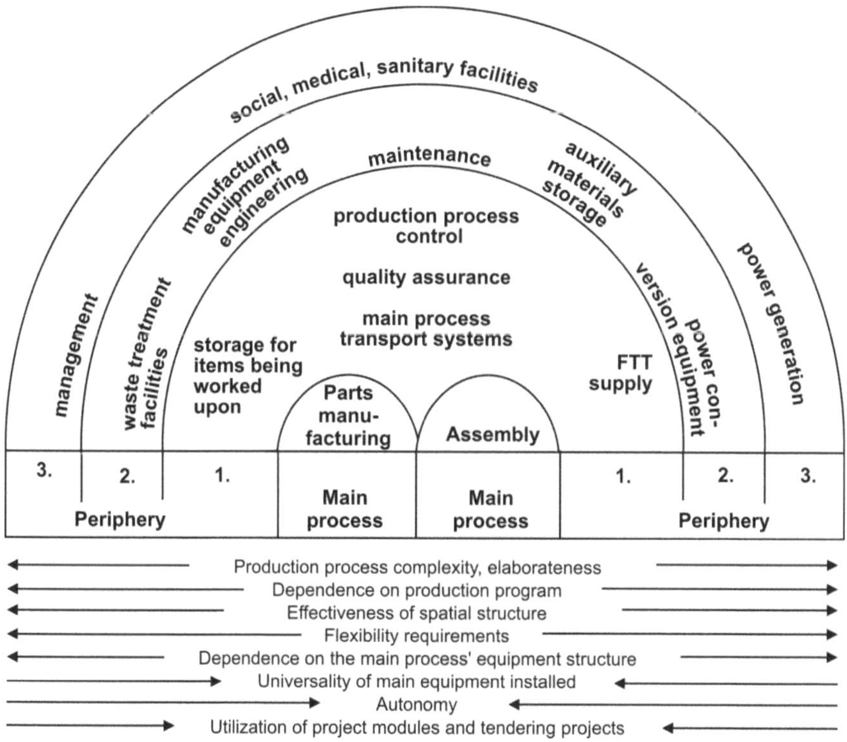

Fig. 1.12 Peripheral areas of the main production processes

c) Functional Organization (cf. fig. 1.13)

In a production facility different processes occur that need to be planned for different process elements. Process elements might be material, information and energy. Since process functions are also called flow functions, they are known as material, information and energy flows. Thus the flow object is the material, information or energy as well as their related systems and facilities (cf. ch. 6.2 and 6.3.6.4).

Material flow systems	Energy flow systems	Information flow systems
Product/material flow systems - Parts (unfinished/finished parts) - Units (assemblies) Finished products - Purchase parts and standard parts - THS equipment **FTT flow systems** - Jigs & fixtures - Tools - Testing equipment		**Production scheduling information flow systems** - Information processing in the management units (organization, planning/controlling) - Procurement/processing of external management information - Information processing in design engineering and operations scheduling
Supply and disposal/building flow systems		**Production execution information flow systems**
- Auxiliary manufacturing materials - Waste (turnings & chips, parts scrap) - Air (fresh air/exhaust air) - Water (drinking and fresh water/ wastewater)	- Electrical energy (power units, heating, IT) - Compressed air / hydraulic system - Technical gases - Indoor air (air conditioning) - Steam, hot water (heating)	- Information processing in production planning and control - Information processing to control machinery - Information processing to control and monitor processes - Information processing to capture operating data

Fig. 1.13 Material, energy and information flow systems

Material flow: within material flow, the flow of unfinished parts through to the finished product takes top priority in terms of planning. The Sankey diagram in figure 1.14 shows the product flow interrelationships on a machinery production line. The material flow also includes operating materials such as fluids (liquids, powders) and wastes.

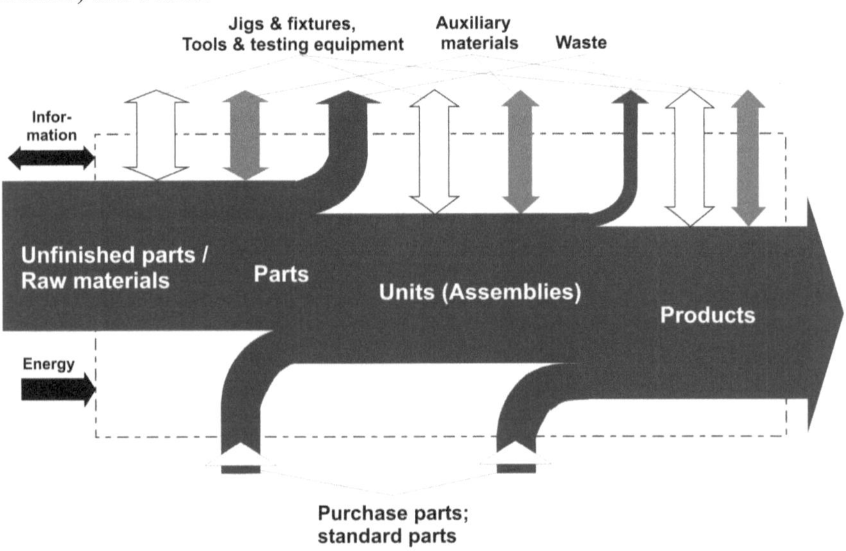

Fig. 1.14 Product flow on a machinery production line (Sankey diagram)

Energy flow: every process requires energy to fulfill its function. The forms of energy can vary. Therefore power equipment for electricity (DC, AC and three-phase), liquid (water, wastewater, oil) and gaseous media (steam, technical gases, compressed air) can all be required.

Information flow: information (data) with the pertinent IT equipment (e.g. computers, memory, cable) is needed to prepare for and execute production. This applies to the planning, scheduling, coordination, communication and technical control of plants.

Other additional flows include:

Personnel (work) flow: every process takes place under the supervision and with the interaction of workers. The workflow stipulates the allocation of labor in the process flow.

Capital and cost (value) flow: the value-added process is assessed using costs. The costs and value of a product change with the manufacturing process. The value of a product increases and the value of the equipment (plant) decreases during the production process (cf. ch. 6.2).

Note: material, energy, information, personnel and capital (value) flows, plus their systems and equipment and their connection to one another must be planned and implemented for every production facility. First of all, product and material flows must be planned with the workflow followed by the information, energy and value flows. Changes in a flow have a sustained effect on the other flows and their equipment. Flexible processes and systems/equipment improve adaptability and thus competitiveness. Every flow must be organized to be resource and - in particular - energy efficient.

1.4 Demands Placed on the Production Facility

The organization of production facilities relates to both "physical products" and to "services" and consequently represents a "hybrid product" for the customer whereby physical products and services blend with one another. The functions that are associated with this represent a package composed of a tailored combination of physical products and services geared towards the customer (Bundnek 2007).

In relation to physical products, when it comes to selling consumer and industrial goods, the marketing of services as an additional area of business is ever increasing, and offers strong potential for engineering services.

a) Physical products result from the specified product, process, system, human resource and administration-related requirements and their technical solutions.

Table 1.1 Demands placed on the production facility

Product and production process-related requirements	Human resource-related requirements
Product technology	Social and sanitary protection
Production technology	Occupational safety, workspace climate
Security of supply	Air conditioning
Security of disposal	Occupational ergonomics/usability
Climate	Color scheme
Flow reliability	Minimal noise, immission control
Connections/interfaces	Illumination/daylight/lighting
Flexibility/adaptability	Collision protection, protection from harmful interference
	Level of protection, protection against fire
	Controllability and manageability
Plant-related requirements	**Management-related requirements**
Accessibility and freedom of movement	Administrative tasks and equipment
Access, openings	Functionality of management
Universal use	Communication, IT equipment (EDP)
Upgradability	Office space
Configurability, modularity	
Utilization of space and spatial geometry	

These requirements, which are based on Helbing (2007), should be reviewed on a case by case basis.

b) *Services* result from the environment of the physical goods produced by the consumer and industrial goods industry. They rank among industry-oriented, product-related services. Table 1.2 summarizes production-related services for the fields of mechanical engineering and plant construction for the entire product life cycle from planning, commissioning and operation through to maintenance, reuse and disposal (Naumann 2008).

Table 1.2 Production-related services

Planning				Commissioning	
Analyses and Studies		**Planning tasks**		**Contracts (no services)**	
Raw materials inspections	x	Factory planning		Sale of products	x
Technical testing, analysis	x	Project management (schedule and cost control)	x	Renting	
Troubleshooting/needs assessment	x	Procurement	x	Sale of use/lease contract	x
Site inspections	x	Documentation	x	Sale of service / operator model	x
Market research / market studies		Network management	x	Replacement parts contracts (defined lead times)	x
Value analyses		Solving of interface problems	x	Maintenance contracts	x
Profitability analyses / return on capital studies	x	Organizational development		Service contracts	x

Feasibility studies	x	Processing of approval procedures		Patent and license agreements	
Organizational analysis		Loan brokerage	x	Expertise agreements	
Environmental impact investigations		Financing		Management contracts	
Pre-competitive product development (industrial research and development)	x	Cost estimates	x	Rental machines to bridge the gap until delivery	
Process analyses	x	Cost estimation support		Machine insurance	
Time studies, time management	x	Development of technology	x	**Training courses**	
Risk analysis, securing of CE mark	x	Material flow planning, process design (simulation)	x	User/operator training	x
Consulting		Planning of technological concepts	x	Computer-based training	
Technology consulting	x	Development of factory logistics concepts	x	Online user training	
Technical consulting	x	Product development (general framework specifications and requirements specification)	x	Technology training	x
Operating resources consulting		Technical planning	x	Maintenance training	x
Environmental consulting		Construction	x		
Legal advice		Drive dimensioning project planning	x	**Transport**	
Advice relating to tools		Factory, layout planning	x	Transport organization	x
Organizational consulting		Ergonomic workstation design, industrial engineering (data calculation)	x	Transport insurance	x
Financing advice (R&D funds)	x	Control/safety concepts	x	**Ramp-up management**	
Provision of advice and support in the design of a quality and environmental management system	x	Concepts relating to the safety of personnel		Building works / facilities	x
Manufacturing in the strictest sense		Simulation of workpiece throughput (process suitability of the machines)		Assembly	x
Sample production	x	Tests using virtual reality		Adaptation to existing plant (updating)	x
Manufacturing to bridge the gap until delivery	x	3D ergonomic simulation		Production scheduling	x
Assembly	x	Software planning	x	Commissioning	x
Development of CNC/PLC and MDA/PDA program	x	**Replacement/wearing parts (RP/WP)**		Test pieces	x
Development of measuring station program	x	Hotline	x	Troubleshooting	x
Help for machinery operation		Supply with own RP/WP	x	Process security	x
On-site production support	x	Ordering of original RP/WP	x	Production of pre-launch and pilot batches	x
Remote machinery and plant operation consulting	x	Replacement part service (24h)		Introduction of change management	x
Production-related training courses	x	Marketing of external RP/WP	x	**Plant improvement**	
FAQs (answers to frequently asked questions)		Consignment (buffer/supply) stock – RP/WP	x	Updating/modernization of machinery	x
Hotline/teleservice	x	Replacement part management (documentation, logistics, stock control, statistics, determination of requirements)		Processing technical inquiries	x
Online self-service (Helpware); Online manuals		**EDP services**		Troubleshooting	x
Animated multimedia documentation		Downloads of software (e.g. simulation/diagnosis software, maintenance tools)	x	Machinery and process diagnosis	x
Tooling (tool making)		Updating/Upgrading	x	Investigating idle time	
NC parts programming	x	Adaptation programming / modification	x	Project-related technology /process consulting and op-	x

				timization	
Recruiting services		**Customer communication**		Remote optimization of plant and processes	
Data management		Complaint management	x	Retrofitting and upgrading	x
MDA: Machine, (manufacturing, process) data acquisition, storage, processing and evaluation	x	Receipt of claims and complaints	x	Investigating energy savings	
PDA: Parts (product) data acquisition, storage, processing and evaluation (quality inspection, parts traceability)	x	Data management	x	Safety, risk and hazard analysis	x
ODA: Order (operating) data acquisition, storage, processing and evaluation (production statistician)	x	Provision of product documentation (manuals)	x	Machine relocation	x
Maintenance		Recording of complaints customer suggestions/problems	x	**Reuse**	
Cleaning of machinery	x	Reference customer visit	x	Development of de-integration plans	
Preventative maintenance	x	User groups (customer exchange of experience)	x	Acceptance of returned machinery, equipment, used parts	x
Surveying	x	Non order-related training		Automatic reuse	
Remote diagnosis/teleservice (mobile maintenance)	x	Product-related symposia		Trade in used machinery	
Breakdown management	x	Publication of interesting findings	x	Sale of used parts, equipment and machinery	
Repair/servicing	x	Newsletter (e.g. case studies, tips, news)		Brokerage of used machinery	
Servicing	x	Customer magazine		**Reconditioning**	
General overhauling	x			Disassembly	x
Manufacturer-independent repair of competitor's products				Large-scale inspection and plant refurbishment	
				Revamping	x
				Conversion / refitting	x
				Overhauling, retrofitting	
				see also system development.	
				Disposal	
				Withdrawal from service	
				Organization of decommissioning process	
				Scrapping of old equipment	
				Recycling of materials	
x = preferred engineering services				Waste management	

Note: production facility design integrates physical products with services and represents a "hybrid product" for the customer.

2 Systematic and Situation-Driven Planning Methods

2.1 Planning Project

Planning production facilities means envisioning production in advance. This necessitates using instruments that efficiently design the planning process. A systematic, methodical approach is influenced by situation-driven decisions. It serves the development of a (planning) project through internal and/or external planning activities.

Project design denotes a creative design activity that utilizes preprepared technical building blocks/modules (components, assemblies, individual systems, etc.) and organizational solutions to design, dimension, structure and configure a user friendly technical unit (device, machine, plant, building, production facility, etc.). The result is a planning project. Figure 2.1 shows the features of a planning project.

A planning project involves the development of and is a prerequisite for the construction of production facilities in preliminary and execution planning.

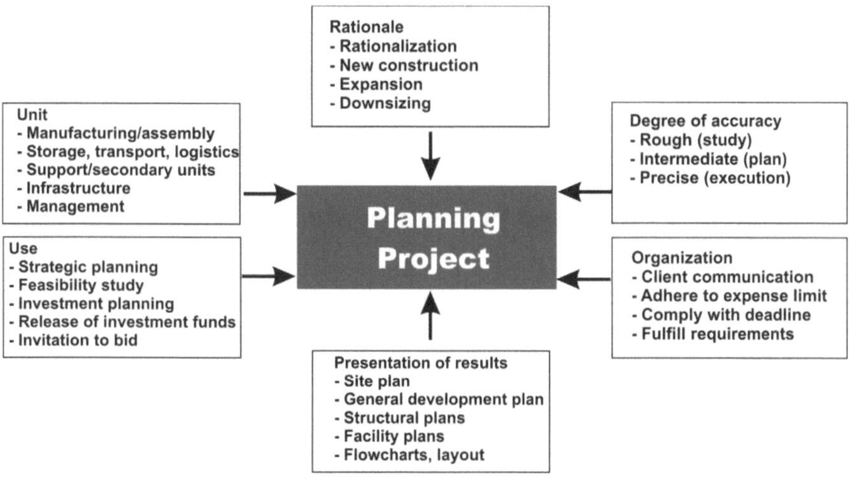

Fig. 2.1 Features of a planning project

Technical Disciplines Involved

The following professional disciplines - among others – that are part of a planning project must be managed during the planning and implementation process:

M. Schenk et al., *Factory Planning Manual*, DOI 10.1007/978-3-642-03635-4_2,

Table 2.1 Production-related technical disciplines

Project engineer and project manager	→	Production facilities and factory design Functional design, budget, deadlines, quality
Architects	→	Building design
Specialist engineers	→	Structural analysis, heating, sanitary facilities, electrical systems, etc.
Production engineers	→	Machinery, equipment, jigs and fixtures, tools
Logisticians	→	Transport, handling, storage
IT engineers	→	Planning, control and automated systems
Design engineers	→	Product specifications
Business managers	→	Target costs, operating efficiency, budget
Ergonomists	→	Working time and remuneration systems, ergonomics
Psychologists	→	Conflict management, motivational techniques
Suppliers	→	Trades, technical building systems (TBS)
Authorities	→	Permits, approvals
Experts	→	Reports and surveys
Attorneys	→	Contracts

The point of departure for all planning is the customer order as the basis for verification of performance agreed upon by the client and the contractor in the form of technical and requirements specifications (in accordance with DIN 69905). This results in the planning and project order that includes the planning basis for products (production programs), quantities, times, production processes, resources (workforce, plant, floor space, personnel), investments (costs, turnover and profit) and legal aspects.

2.2 Planning Process and Procedural Models

A planning project can be developed systematically and/or situation-driven on the basis of various planning process and procedural model views.

a) Systematic Planning Processes

(1) Production facility and factory life cycle design planning phases and stages (fig. 2.2).

Planning activities span a production facility's entire life cycle from development/planning through setup, execution and operation to phase-out. Three planning stages are always implemented within the individual phases.

The following reflections concentrate on "planning/project design" (the planning project) and setup or "execution planning" (the implementation project).

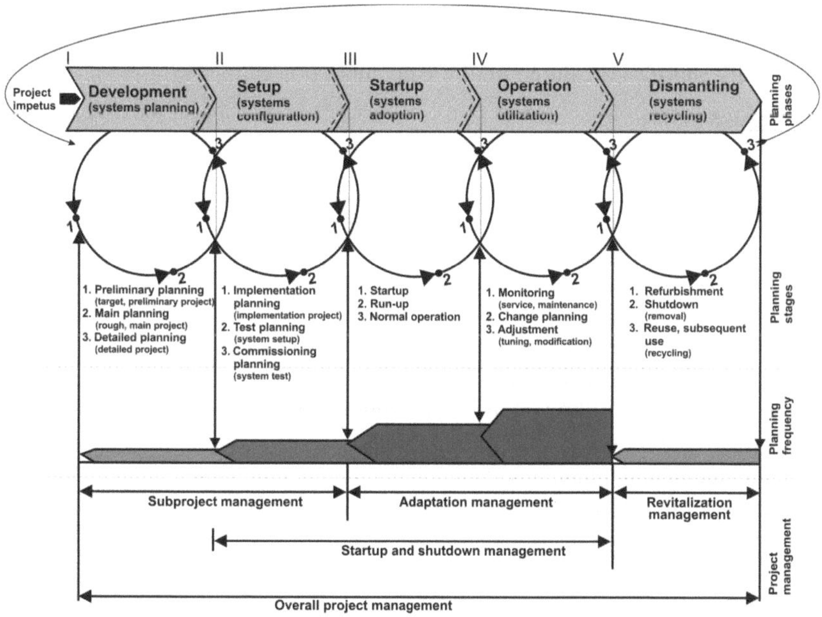

Fig. 2.2 Production facility and factory life cycle design planning phases and stages (Schenk, Wirth 2004)

(2) Views of the planning process based on planning levels, stages and steps (fig. 2.3)

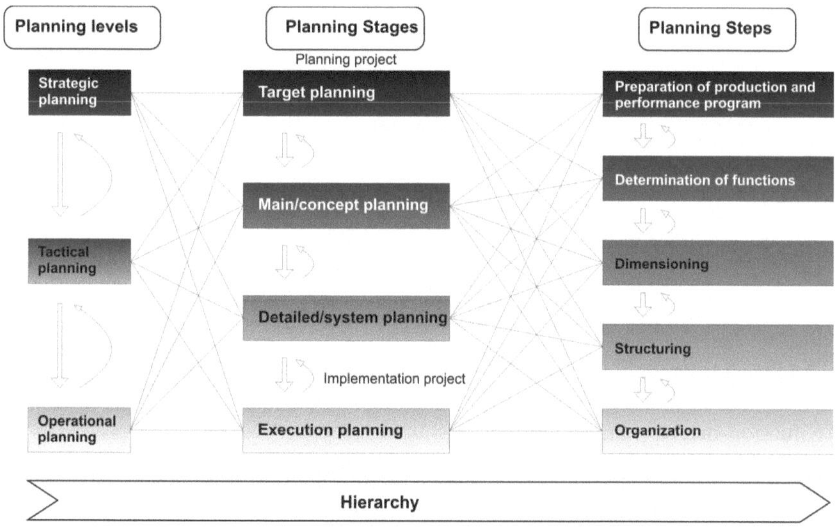

Fig. 2.3 Views of the planning processes

It is clear that there is a multitude of interconnections that have repercussions for one another. The proportion of "operational" planning is increasing and this substantially shortens the duration of the planning process.

b) Planning Models

(1) Systematic planning models

- Rough sequence of the planning of processes and manufacturing plants (fig. 2.4)
- Project design takes place in the following sequence (cf. fig. 1.12): main process → 1st periphery → 2nd periphery → 3rd periphery and must be implemented in the required planning stages (e.g. target planning).
- Complex planning model ("0 + 5 + X"), see point 3.
  This consists of the following planning complexes:
  I Project definition
  II Project development
  III Project implementation

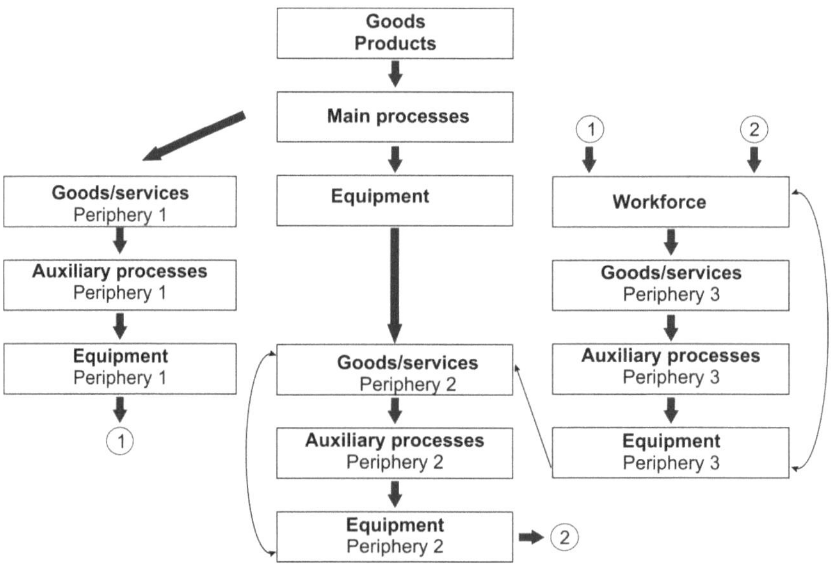

Fig. 2.4 Rough sequence of the planning of processes and manufacturing plants

(2) Situation-driven planning model

Situation-driven planning differs from systematic (algorithm based) planning in that as a result of operational decisions (e.g. changes to target, data, product, technology, requirement, time, profitability or quality specifications) all or part of the planning process and sequences have to be changed. This pertains to the planning levels, stages and steps, the process planning sequences and resources (workforce, production equipment) as well as the complex planning model covering project definition, project development and project implementation.

The planning model is adapted operationally and redeveloped based on the given situation and uses selected systematic planning components and activities. Figure 2.5 shows an example of a situation-driven planning model for small business enterprises employing make-to-order and series production methods.

Operational planning and project design is extremely important for short-term adaptation and change processes (Kirchner 2003). For this purpose, evaluating typical change characteristics in accordance with table 2.2 is recommended. (Wirth, Schenk 2001; Wiendahl, Hernandes 2002; Westkämper 2002)

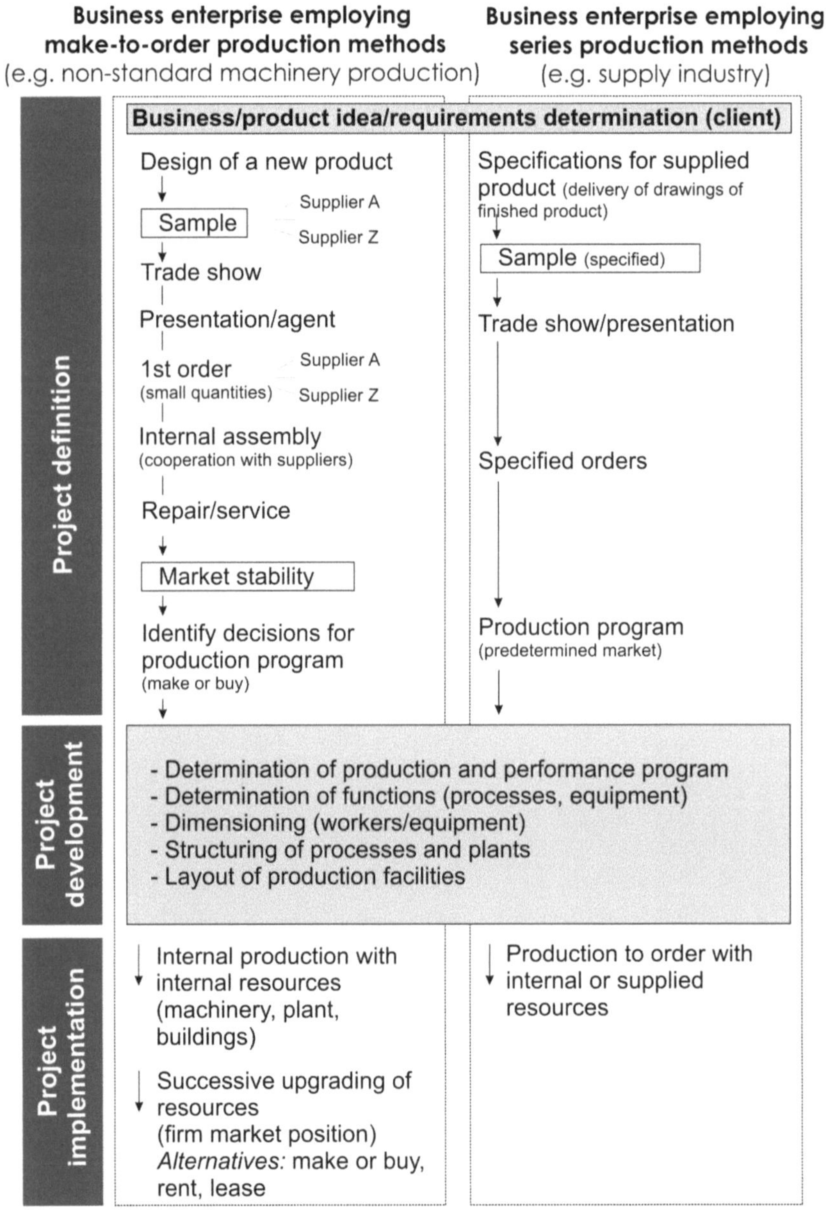

Fig. 2.5 Situation-driven planning model (example)

Table 2.2 Evaluation of Change Characteristics

Evaluation of Structures	Approach	Results
Technological flexibility	Forecast changed customer product requirements and consequently derive technological requirements for every function	• Tolerance range of potential technological modifications per product group • Identification of resources that limit technological flexibility
Capacity-related flexibility	Forecast changed customer requirements and consequently estimate required capacities (max./min.)	• Tolerance range of capacity per product group [hours] or [unit] • Identification of resources that limit capacity-related flexibility
Structural flexibility	Forecast changed customer requirements and consequently derive requirements placed upon structural units	• Tolerance range of requirements placed upon structures • Identification of resources and structures that limit structural flexibility
Logistical flexibility	Forecast changed customer requirements and consequently derive the logistics service to be provided	• Tolerance range for fulfillment of logistics requirements • Identification of system or process elements that limit logistical flexibility
Variability	Forecast changed customer requirements and consequently derive requirements for upgrades, modifications and extensions	• Evaluation of structures based on their suitability for modification • Designation of modifications and extensions
Mobility	Forecast changed customer requirements and consequently derive mobility requirements	• Evaluate and actively facilitate or enhance mobility • Identification of the resources to be implemented or reorganization of structures
Agility/vitality	Forecast changed customer requirements and consequently derive requirements for performance relative to dynamic and sudden changes	• Evaluate agility • Make provision for navigation instruments • Derive organizational and qualifying measures

Customized solutions frequently demand the making of immediate decisions and dealing with unforeseen events on the basis of incomplete data and facts. This has a negative impact on the accuracy of the results of project development and implementation.

Typical situation-driven planning case studies are discussed below.

2.3 Planning Tools and Methods of Evaluation

a) Planning Methods and Tools

(1) Planning and project design methods

- indicator-based project design – indicators for different project planning activities
- model-based project design – two or three-dimensional models for layout design
- building block-based project design – reusable solutions in the item and method areas (also type and module project planning)
- catalog-based project design – project planning catalogs for objects (production resources, plant, facilities)
- computer-aided project design – computers with planning and project design software

(2) Computer-aided planning tools

Computer-aided project design tools are used in every phase, stage and step of project planning, especially to produce drawings, models, workflows (simulations) and layouts based on relevant data and software (VR and multimedia applications). Table 2.3 shows potential tool applications.

In the case of virtual reality (VR), in the digital factory/production facility, all the manufacturing and logistics processes that characterize the actual factory are reproduced in a computer model using databases. An additional step is planning using augmented reality (AR). This is achieved by enriching an actual environment with virtual objects so that the relationship with reality never becomes obscured and the user does not become completely immersed in an artificial world (Haller 2006). The aim of AR is to effectively supplement human perception with virtual information and thus enhance a real factory's productivity.

Advantages of AR include: increased planning certainty. The plants are realistic. Planning errors are detected faster without having to reproduce every detail. Shortened planning time. Digital plant data does not have to be updated in such precise detail. Furthermore, savings are generated by avoiding errors caused by planning based on incomplete or erroneous data. Reduced planning costs due to reduction in the time and work needed for modeling. Wider-reaching analyses can be performed without great effort whenever planning assumptions are modified. Improved communication. Presenting virtual systems in a familiar environment makes it easier to understand complex processes. Establishing a common knowledge base facilitates the integration of employees on-site. (Schreiber and Doil 2005)

Table 2.3 Planning tool evaluation (Günther 2005)

A – Drawings/templates
B – CAD systems
C – VR systems
D – Model-based tools
E – Simulation
F – Production system planning tools

● - applicable, necessary
◘ - partly applicable, partly necessary
○ - not applicable, not necessary

Criteria		A		B		C		D			E		F			
		Technical drawings	Templates	AutoCAD, Microstation	Visio	Tucan, worldup	Mantra 4D	Build-it planning table	Fastdesign, FactoryCAD	Tarakos	Automod, ED	emPlant	Layout Planner, em Planner	Falop, Spiral, Planopt	Malaga, Factotum, Matflow	LCAD
Procedure	Initial procedure	○	○	○	○	○	○	○	○	○	○	○	○	●	●	○
	Corrective procedure	○	○	○	○	○	○	○	○	○	○	○	○	◘	○	○
Type	Analog	●	●	○	○	○	○	○	○	○	○	○	○	○	○	○
	Digital	○	○	●	●	●	●	●	●	●	●	●	●	●	●	●
Type of model	Formal model	○	○	○	○	○	○	●	●	◘	●	●	●	●	●	○
	Analog model	●	●	◘	●	○	○	○	○	○	○	○	●	◘	○	○
	Pictorial model	●	●	●	●	●	●	●	●	●	●	●	●	◘	●	●
Depiction	2D	●	●	●	●	●	●	●	●	●	●	●	●	●	●	●
	3D	◘	○	●	○	●	●	●	●	●	●	●	●	○	●	○
	VR	○	○	○	○	●	●	○	○	●	○	○	○	○	○	○
Operation	Multi-user operation	●	●	○	○	●	●	●	○	●	○	○	○	○	○	○
	Apportioned planning	○	○	○	○	○	○	○	○	○	○	○	●	○	○	○
	Intuitive handling	○	●	◘	◘	◘	◘	●	◘	◘	◘	◘	◘	◘	◘	◘
	Specialist knowledge	○	○	●	◘	◘	◘	○	●	◘	●	●	●	●	●	◘
	Planning support	○	○	○	○	○	○	○	◘	○	○	○	◘	◘	◘	◘
Compatibility	CAD	○	○	●	●	◘	●	◘	●	◘	○	○	●	○	●	○
	Database	○	○	◘	◘	◘	●	◘	●	○	●	●	●	○	●	○
Modeling	Time and effort required for development	●	●	◘	◘	◘	◘	◘	◘	◘	◘	◘	◘	◘	◘	◘
	Time and effort required to make changes	●	●	◘	◘	◘	◘	◘	◘	◘	◘	◘	◘	◘	◘	◘
	Model library	○	○	○	○	○	◘	●	●	●	●	●	●	○	○	◘
Scope	Rough layout planning	●	●	●	●	●	●	●	●	◘	●	●	●	●	●	●
	Detailed layout planning	●	◘	●	●	◘	●	◘	●	●	◘	◘	●	○	●	●
Evaluation	Flow intensities	◘	○	○	○	○	○	◘	●	◘	●	●	◘	◘	●	○

b) Evaluation Methods

The results of planning and project design activities must be evaluated so that decisions can be made about the manner in which the work will progress. Evaluation methods exist to do this, as summarized in table 2.4.

Table 2.4 Evaluation Methods (Illes, 2007)

Classification aspects			Methods
Evaluation of argumentative aspects (ARGUMENTS)			• Advantages/disadvantages
			• Strengths/weaknesses
			• Opportunities/risks
			• Delphi method
			• SWOT matrix
			• S curve
			• Experience curve
Evaluation of time-dependent aspects (TIME)			• Environmental development trend forecast
			• Exponential equalization
	1 characteristic		• Scenario technique (best case, worst case, trend extrapolation)
Evaluation of inter-relationships (CLASSES, TYPES)	2 to 3 characteristics		• Pareto methods
		simultaneously	• Portfolio
	more than 2 characteristics	sequentially	• Cluster analysis
		time-independent	• Hierarchical class formation (grouping)
			• Morphological boxes
			• KO process
Evaluation of objectives (TARGET-ACTUAL)			• Checklists
			• Benchmarking
			• Target fulfillment level
			• Balanced Scorecard (BSC)
			• Quality Function Deployment (QFD)
		Benefits / Cost	• Measurement of customer preferences
			• Utility value analysis
			• Work expenditure analysis
			• Usefulness analysis
		Dynamic profitability assessment	• Benefit/cost analysis
			• Revenue/effort analysis
			• Ecopoint evaluation
			• Net present value method
			• Internal rate of return method
Evaluation of quantifiable aspects (BENEFITS, COSTS)			• Capital recovery method
			• Dynamic payback period

	Static profitability assessment	• Costing principle
		• Determination of the optimum replacement time
		• Useful life
		• Cost comparison method
	Business evaluation	• Accounting rate of return method
		• Rate of profit method
		• Break-even analysis
	Evaluation of environmental aspects	• Portfolio effect analysis
		• Revenue calculation
		• Budget calculation
		• Future earning capacity value method
		• Capitalized earnings value method
		• Asset value method
		• Excess profit payment
		• Waste balance
		• Raw materials balance
		• Pollutant balance

Special attention is paid below to the application of individual evaluation methods for selected project planning activities.

Note: modern planning and project design require a combined application of systematic and situation-based approaches and methods. Systematic planning methods and models are predominantly employed for "standard projects." In the case of "situation-driven projects" operational planning takes precedence, for which problems are solved on the basis of a specifically-developed planning model. A relevant planning model should be devised for every planning and project task, and this will be implemented with or without modification. The systematic project design activities and functions discussed below should be selected for a planning model on the basis of the given situation. The order in which they are implemented is based on a "model" or "orientation path" that must be developed.

References (Chapters 1 and 2)

Bonnemeier S, Ihl C, Reichwald R (2007) Wertschaffung und Wertaneignung bei hybriden Produkten. Arbeitsbericht Nr 03, BWL, München

Günther U (2005) Methodik zur Struktur- und Layoutplanung wandlungsfähiger Produktionssysteme. IBF, TU Chemnitz

Haller M, Billinghurst M, Thomas B (2006) Emerging Technologies of Augmented Reality: Interfaces and Design. Idea Group Publishing, Nov 2006, 300 p.

Hanna SR, Konz S (2004) Facility Design and Engineering. Holcomb Hathaway, Scottsdale

Helbing K (2007) Fabrikprojektierung. book manuscript, Rostock

Hildebrand T, Mäding K, Günther U (2005) Plug + Produce – Gestaltungsstrategien für die wandlungsfähige Fabrik. IBF, TU Chemnitz

Illes B, Glistau E, Coello Machado N (2007) Qualitätsmanagement. Miskolc, 978-963-887738-1-4, p. 193-194
Kirchner S, Winkler R, Westkämper E (2003) Unternehmensstudie zur Wandlungsfähigkeit von Unternehmen. wt Werkstattstechnik online, 93/4, p. 254 - 260
Lotter B, Wiendahl HP (2006) Montage in der industriellen Produktion. Springer, Berlin
Naumann A (2008) Gestaltungsmöglichkeiten für Dienstleistungen. Dipl-Arb, TU Chemnitz
Schenk M, Wirth S (2004) Fabrikplanung und Fabrikbetrieb. Springer, Berlin
Schmigalla H (1995) Fabrikplanung, Begriffe und Zusammenhänge. Hanser, München
Schreiber W, Doil F (2005) Augmented Reality Einschnittstellen zwischen der realen und virtuellen Fabrik. 8. IFF Wissenschaftstage, Magdeburg
Spur G (1994) (Hrsg) Fabrikbetrieb. Hanser, München
Spur G (2007) Wandlungsfähigkeit von Produktionssystemen. ZWF, ZH 3, S 96-97
Warnecke HJ (1992) Der Produktionsbetrieb. Springer, Berlin, p. 18
Wiendahl HP, Nofen JH, Kußmann et al (2005) Planung modularer Fabriken. Hanser, München
Wiendahl HP et al (2007) Changeable Manufacturing - Classification, Design and Operation. Annals of the CIRP, Vol. 56/2/2007
Wiendahl HP, Hernandez R (2002) Fabrikplanung im Mittelpunkt. Herausforderung Wandlungsfähigkeit. In: wt Werkstattstechnik (online) 92 (2002) 4, p. 133-138
Wirth S (2000) Werkstätten- und Produktionssystemprojektierung. TU Chemnitz, IBF
Wirth S, Schenk M (2001) Wandlungsfähige Produktionsfabriken. BVL – Berichtsband 1, Huss, München, p. 487-498

3 The 0 + 5 + X Planning Model

The 0 + 5 + X planning model (fig. 3.1) is the outcome of a multitude of projects in different sectors together with many years of practical experience.

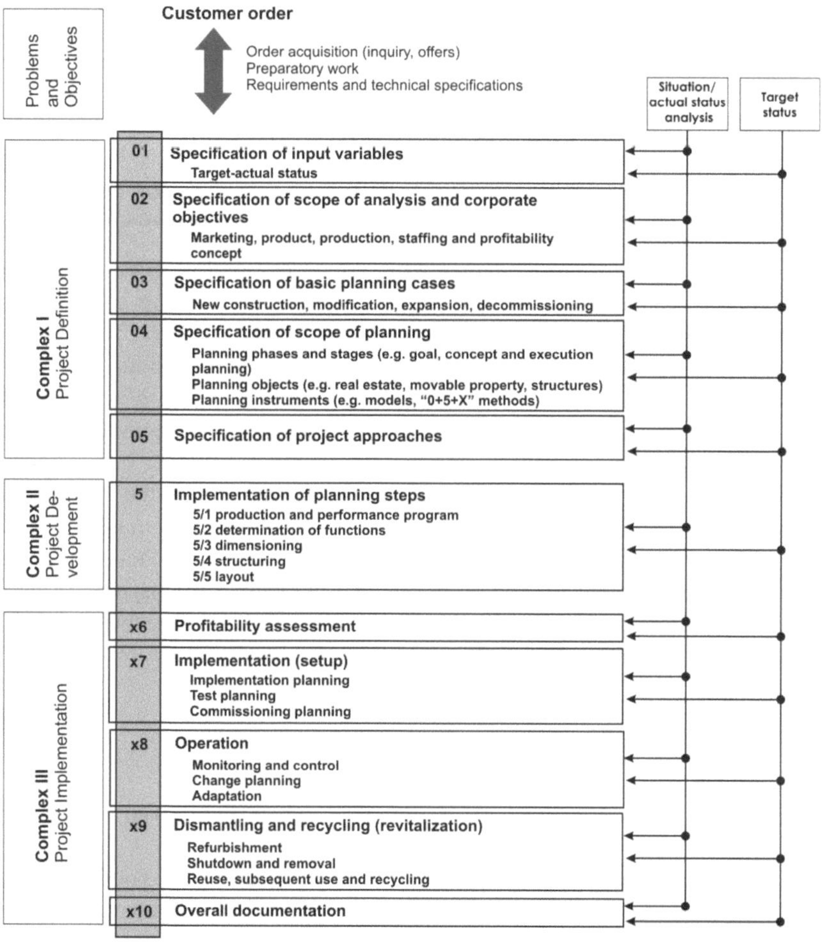

Fig. 3.1 Complexes of the 0 + 5 + X planning model (abbreviated version)

It contains complexes that cover a range of areas from the processing of the problem and task definition to acquisition of the customer order and the project definition, project development and the project implementation and execution. The results of the complexes can be developed either individually or in their totality as engineering planning services for customers, clients or investors. At the same time, the complexes define fundamental decision-making levels for complet-

M. Schenk et al., *Factory Planning Manual*, DOI 10.1007/978-3-642-03635-4_3,

ing or continuing the planning work from the task definition to the realization of the production facility.

The customer order is a prerequisite for project development and implementation. It must lead to the project definition (complex I) agreed upon between the customer and the contractor (manufacturer/supplier).

The planning project should be implemented in three stages (complexes) based on a planning model.

Complex I: five specifications (01 – 05) to qualify the *definition of the project.*

Complex II: five project design steps (5/1 – 5/5) for the *development of the project.*

Complex III: five specifications (x6 – x10) for the *implementation of the project.*

The three complexes are implemented in a loop or spiral process that reevaluates every upstream and downstream project design activity whenever changes are made and verifies their effect on the overall project.

In this manual checklists and comments such as "NB" and "note" support the implementation of a planning project employing the *0 + 5 + X planning model.* The checklists are used to provide a logical sequence for the selection and review of project design activities and are assigned to complexes I - III.

The planning model is only a rough guideline beginning with the identification of objectives and the design of a solution. A methodical approach must provide for adaptability in the planning process and the design of production facilities.

The approach is demonstrated by means of a complex planning project of systematic production facility planning for the production of worm gears (cf. 3.2.5).

3.1 Project Definition (Complex I)

This includes the five specifications in figure 3.2, which are derived from the customer order.

The customer order is the outcome of intensive collaboration, consultation and influence exerted by the potential planner as contractor - CO - (planning engineer, supplier) with the potential client - CL - (investor, operator). Compiling details of the problems and objectives relevant to the project necessitates preparatory work (e.g. project studies and analyses) on new developments and specifications.

Examples of preparatory work relevant to development include research findings (products, processes/technologies, machinery and plant), patents, inventions, technical developments, forecasts, innovations, reference designs, technologies, plants and projects, conceptual designs, reports and inspections.

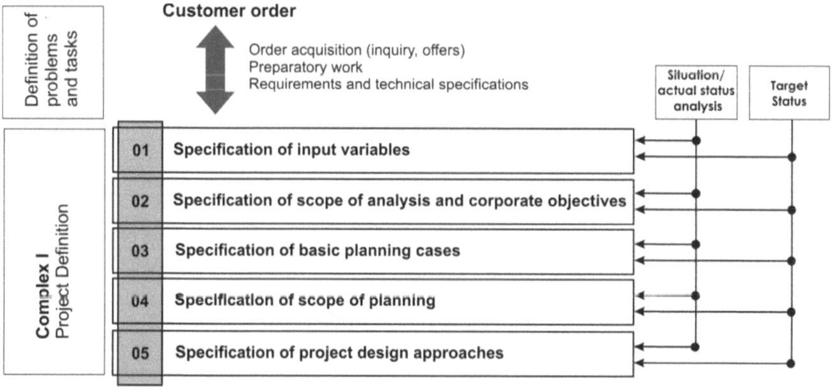

Fig. 3.2 Complex I of the 0 + 5 + X planning model: project definition

Examples of preparatory work relevant to specifications are production and sales programs, products (design, features), technology, quantities, times, quality, costs, restrictions and permits, partners (general contractors, subcontractors, contractors for equipment, facilities, construction and trades); deadlines, work and decision stages and investments.

In order to prepare the content of the project definition (complex I), the questionnaire (checklist) in Table 3.1 should be completed.

Table 3.1 Checklist – 0 + 5 + X Planning Model - Customer Order/Project Definition (Complex I)

❒	0	**Customer Order**			
		❍	*Requirements specification:* Definition of customer's task (problem) – what has to be done and why to achieve the scope of service demanded		
		❍	*Technical specifications:* Project engineer/supplier's implementation concept – how and by what means is customer's task (problem) to be resolved?		
❒	01	**Project definition**			
			Specification of input variables		
			a) Actual status		
		❍	Data available	❍	No data available
			b) Target status		
		❍	Data available	❍	No data available
		❍	Redevelopment	❍	New design/plan
❒	02	**Analysis Phases/Corporate Objectives**			
			a) Corporate objectives		
		❍	Market position	❍	Return on capital
		❍	Finance	❍	Social
		❍	Prestige		
			b) Concepts		
		❍	Sales, marketing	❍	Product
		❍	Production	❍	Personnel
		❍	Profitability		

❐	03	**Basic Cases**			
		❍	New building	❍	Remodeling/rationalization
		❍	Extension	❍	Decommissioning
		❍	Revitalization		
❐	04	**Planning Phases, Objects and Instruments**			
			a) Planning phase		
		❍	Development/planning	❍	Set-up
		❍	Start-up	❍	Operation
		❍	Dismantling		
		❍	Main/conceptual planning	❍	Objective planning
		❍	Execution planning	❍	Detail/system planning
			b) Planning objects		
		❍	Personnel	❍	Structure
		❍	Equipment/plants	❍	Property/building
			c) Planning instruments		
		❍	Models	❍	Methods
		❍	Tools	❍	Theories
❐	05	**Project design principle**			
		❍	Overall	❍	Stages
		❍	Top down	❍	Bottom up
		❍	Variants	❍	Flexibility
		❍	Improvement	❍	Situation
		❍	Synergy	❍	Profitability

The purpose of order acquisition is to negotiate the most important requirements and specifications to which the production facility must respond, as well as the scope of supply and services. A detailed request for quotation or offer (technical and commercial part) can then be drawn up based on these preliminary negotiations and preparations. The following documents have proved valuable in this respect:

Requirement specifications (DIN 69905): client's (CL) definition – what has to be done and why to fulfill the scope of performance required?

Technical specifications: project engineer/supplier's (CO) implementation concept – how and by what means is customer's task to be accomplished (task definition and solution)?

Table 3.2 shows the functions of the requirement and technical specifications.

Table 3.2 Functions of the requirement and technical specifications

Functions of the requirement specifications	Functions of the technical specifications
Description of the required scope of performance including the basic conditions (from the client's perspective)	Implementation concept for the required scope of performance (prepared by the contractor)
Description of requirements regarding the scope	Implementation of the requirements in the

of performance (→ user requirements)	form of technical and organizational solutions
Basis of tenders and requests for quotation	Auxiliary equipment (objectives, specifications) in support of project execution
Basis for offers, orders and contracts	Basis for project design, implementation and acceptance

The feasibility in terms of content, time and financing of the scope of performance set out in the requirement specification is coordinated and reviewed by the CO (planner, supplier) in conjunction with the CL, and the technical specification is drawn up by the CO as its implementation concept.

The implementation concept includes the scope of performance required, the implementation of the requirements in economically efficient supply chains with technical solutions, the auxiliary equipment in support of project execution, the coordinated implementation of planning activities, stages and steps and decision and control measures with specific target dates.

The cycle: together with the CL and CO's joint specifications in the project definition stage (complex I), the request for quotation/order to the CO by the CL, the drawing up of the offer by the CO, the commissioning of the scope of services by the CL and the joint conclusion of the contract all form the basis for the project development (complex II) and project implementation (complex III). This process must be managed and this is known as project management.

In principle there are two separate stages to defining the content of the requirement and technical specifications:

a) Project development (complex II)
for the complex production facility planning as the planning-side basis for the execution/implementation project. This includes, among other things, all essential resources and items (machinery/plant/facilities/trades) based on non-binding (indicative) offers. The equipment list equates to a "bill of materials" for the entire project.

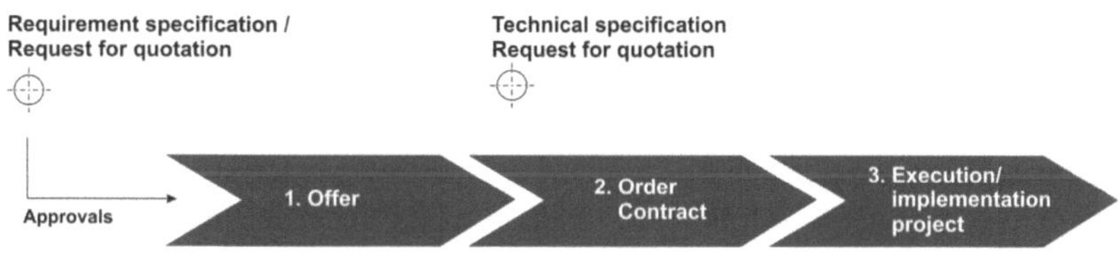

Fig. 3.3 Project development procedure (complex II)

b) Project implementation (complex III)
has the objective of implementing all resources and items (machinery/plant/facilities/trades) based on "binding" orders and contracts for all items in the equipment list. For this reason from this point forward binding quotations and agreements must be obtained / concluded for every single item right up until the

material and technical execution. The approaches to and special features of planning and execution are discussed separately in "Project Implementation (complex III)".

The project definition comprises important specifications regarding the customer's (general) aims. A network should be set up between the customer and project team and the participating partners to evaluate short-term changes and their consequences and derive inferences for the subsequent work. In the process, decisions need to be made in the following five areas of project design activity:

(01) Specification of input variables

The input variables are configured according to the target status. In the case of existing production facilities that are to be "redeveloped," the actual status needs to be analyzed (actual status analyses). To do this, existing inventory data (documents, drawings, products, production quantities, technologies, processes, location and structure of the production and logistics systems and building plans) should be used. A comparison between the target and actual status should make any shortfalls apparent. This applies for the client's specifications for construction, technology, demand progression, level of technology, floor space/rooms, times, restrictions and profitability.

NB: Input variables

a) General input variables

- corporate objectives
- production program, products, product groups (quantity, quality, times, costs) taking market developments into account
- materials (commodities, raw materials, semi-finished products), origin, quality and quantity for core and by-products
- production technologies and processes
- resources and capacities for core, auxiliary and subordinate processes
- manufacturing information: manufacturing/assembly, lot sizes per unit of time; make-to-order, small batch and series production, manufacturing stages, vertical integration
- logistical processes
- plant and equipment information: main systems, mechanization and automation, procurement possibilities
- logistical organization and technology
- working time, remuneration systems, workforce capabilities
- organization (structure, processes, functional units)
- profitability including target costs (target costing, target pricing), performance and revenue statistics, capital spending
- health, labor and environmental protection including resource consumption
- timetable for planning and execution

- basic conditions (selected laws, ordinances, etc.)

b) Input variables in the case of existing production facilities

The assessment of the situation and potential based on existing data constitute a necessary prerequisite.

Situation analyses incorporate all data about the product, process and plants including building elements and employees, and a check of all inventory data. The assessment of potential incorporates the determination of capabilities together with an evaluation of strengths, weaknesses and opportunities.

c) Evaluation by assessing potential

Objectives of the assessment of potential

- creation of initial data for the subsequent planning process
- obtaining of detailed knowledge about the planning object, discovery of technical, operational and organizational weaknesses

Fundamentals

- clear definition of aims, contents and scope of the analysis
- use of suitable employees
- application of special methods and auxiliary equipment
- determination of basic conditions

Approach

- specification of areas to be examined when assessing potential
- specification of investigative effort and the methods employed to carry out the production facility analysis

Information obtained from

- product quantity analysis (e.g. ABC analysis)
- manufacturing process analysis
- material, energy and information flow analysis
- order processing analysis
- plant and equipment analysis
- cost analysis
- workflow and organizational structure analysis

Note: the outcome of a well-founded assessment of potential forms the basis for ascertaining the planning task, providing a detailed definition of the initial position and targets and evaluating solution principles. Input variables must be examined for possible changes and adaptability. Target and actual data need to be determined for production and order program planning.

(02) Specification of the scope of analysis and corporate objectives

This is based on the anticipated value-added and supply chains and corporate objectives. Decisions must be made regarding the basic corporate objectives in accordance with fig. 3.4 and planning concepts.

NB: For the corporate objectives take note of the aim categories.

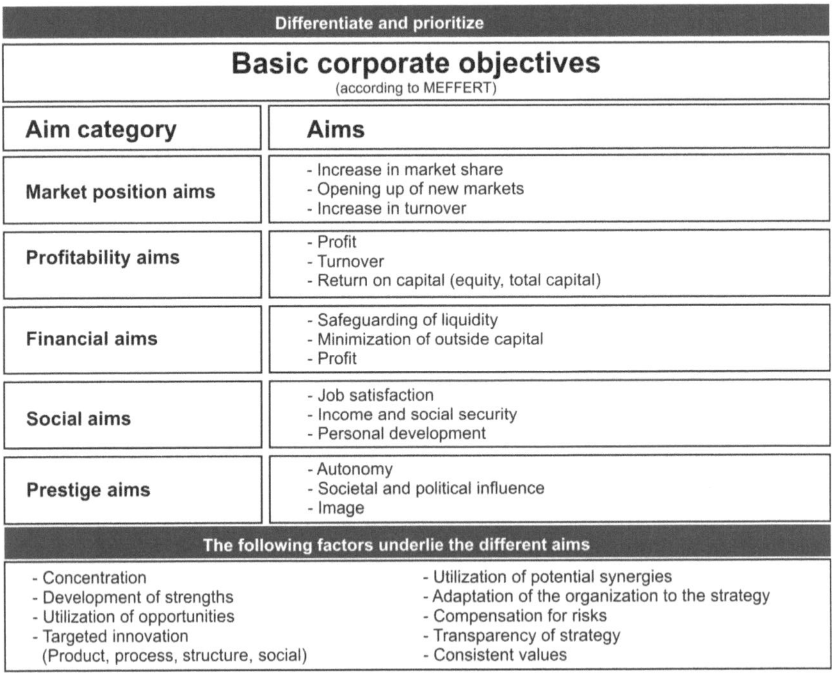

Fig. 3.4 Strategic principles and aims of the manufacturing enterprise (adapted from Meffert and Wiendahl)

NB: for the planning concepts

a) Sales and marketing concept

Table 3.3 Questions regarding the sales and marketing concept

Enterprise related questions, including about	Sales market related questions, including about
the product's share of turnover and profit the fixed and variable costs associated with the	the absolute and relative market share of the product

product the product's position in its life cycle	the (uptake) volume and the market's opportunities/threats
the requisite manufacturing and marketing capability	the market position
the enterprise's product-related strengths and weaknesses	

b) Product concept
This contains all information on economically-marketable products (quantity, form, costs). Sub-concepts need to be developed in accordance with the product life cycle. Sub-concepts include: product development, use and recycling.

c) Production concept
This is based directly on the product concept and determines the product manufacturing processes (technologies) in terms of their technical, technological and organizational form. Table 3.4 shows the decision-making principles.

Table 3.4 Principles of product development

Influences	Examples		
Product	• Quality • Quantity per unit of time • Availability	⇒	need-/customer-based
Business enterprise/ Profitability	Manufacturing costs	⇒	profit oriented
	Process reliability, operating and occupational safety	⇒	compliant with the law
	Employees (number, skills)	⇒	• need-based • socially acceptable • profitable
	Flexibility (adaptation of processes to product life cycle, product changes)	⇒	market/sales-based
Environmental precautions	Resources (materials, energy, information)	⇒	economical, minimum-cost, resource and energy efficient
	Emissions	⇒	environmentally compatible
	Waste	⇒	environmentally friendly/ recycling-based

d) Personnel concept
Sub-sectors of personnel planning are

- personnel requirements, future personnel requirements and workforce
- recruitment, planned/future recruitment in the event of shortages
- personnel deployment, optimum allocation of staff to jobs (a) medium to long term and (b) short term/operational
- personnel development (qualitative personnel planning), qualification of workforce in accordance with job requirements as a significant contribution to supply of personnel

e) Profitability concept (assessment)

All of an enterprise's business processes influence its profitability in that they either drive value creation directly and/or indirectly, and consequently result in costs.

The profitability concept should be regarded as the definition of concrete return on investment targets and financial goals to be achieved by the enterprise as well as the corresponding basic implementation measures (cost reduction, investment and pricing, among others).

Each assessment and manipulation of the profitability of a business enterprise is based on monetary values. These values obtain their regulating effect from the markets. The corporate accounting department connects the monetary values to in-house processes. Key and in-process tasks (bookkeeping, financial accounting) include

- cost accounting – pricing (process-oriented cost accounting and pricing)
- investments - financing
- financial accounting – benchmarking (financial accounting process and benchmarking)

Note: variants of possible changes should be taken into account in the concepts. In the process, the targets, benefits, costs and risk also need to be evaluated (cf. evaluation methods tab. 2.3)

(03) Specification of basic planning cases

These form the basis of the reason for formulating a project. Decisions must be made regarding the project function based on the four basic cases.

NB: Selection of basic planning cases

Basic case A – New development of production facilities

- significant preliminary planning period in terms of time and content
- global specifications for production program and development
- determination of optimal location incl. infrastructural integration
- targeting of optimal process solutions based on high degrees of freedom

Basic case B – Reconfiguration of existing production facilities

- aims are rationalization or modernization of existing production complexes
- relatively precise specifications for production program and development
- continuous adaptation of the production complexes to production program changes (market) or to cost-effective process and plant innovations.

Basic case C – Expansion of existing production facilities

- leads to increased floor space and room utilization in the existing location

- can be combined with identification of location for additional capacities (cf. basic case A)
- can call the existing location into question and lead to relocation to a new site

Basic case D – Decommissioning (revitalization) of existing production facilities

- leads to reduction in floor space and room utilization in the existing location
- can lead to complete closure of the site
- floor space that is freed up can be assigned to new uses

Note: flexibility and adaptability in particular must be taken into account for each of the basic cases.

(04) Specification of the planning phases, objects and instruments

In the case of the planning of production facilities, the planning phases must be related to the life cycle, planning objects and planning instruments. The scope of consideration of the planning cube thus consists of x-y-z axes, by means of which the different planning tasks are to be developed (see fig. 3.5).

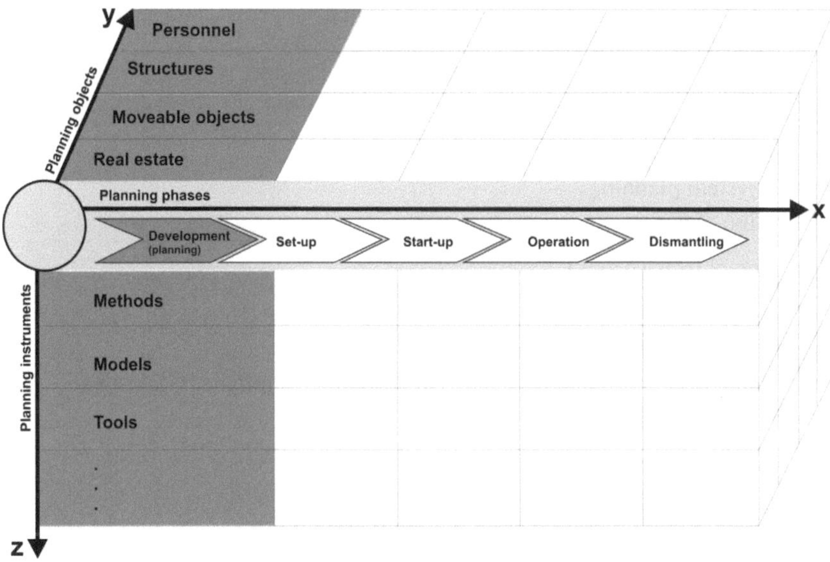

Fig. 3.5 Scope of planning consideration (Schenk, Wirth 2004, p. 107)

Each planning task is different and complex. It must be executed so that the planning phase, object and instrument are given equal consideration and aligned to the practical realities.

The planning phases form a closed loop. There is positive feedback between the individual phases, and the ways in which they interconnect and interact must be taken into account. In practice only the development (I) in the form of the planning phase, setup (II) in the form of the execution phase and startup (III) and operation (IV) in the form of the commissioning phase are integrated. Work contents and results that are generated by service providers (planning agencies) and in-house planning departments are:

Planning phase
Studies, concepts, analyses, projects (variants), task definitions, function descriptions, layout depictions, functional models

Execution phase
Assessment and selection of offers, construction sequence plans, relocation plans, project management plans (trades and schedule flows), capacity allocations, coordination processes), commissioning stages

Commissioning phase
Commissioning and acceptance protocols, quality and protection certificates, proof of performance in the startup phase

NB: Selection of planning phases, objects and instruments

a) Planning phases (x axis)

- development/planning
- target planning
- core/conceptual planning
- detail/system planning
- execution planning

b) Planning objects (y axis)

- personnel (number, skills, motivation)
- structures (workstation, section, division)
- moveable objects (machinery, plant, workstations, equipment, logistics)
- property (buildings, infrastructure, land)

c) Planning instruments (z axis)

- theories
- models
- methods/procedures
- tools and instruments

Note: the scope of planning consideration allows for a multitude of possible combinations, so that for each planning task specific planning methods and the corresponding models and tools may be selected and utilized.

(05) Specification of project design principles

Principles and rules that are based on experience have become established for managing projects. Decisions must be made as to which specific principles will apply to the project development.

NB: selection of principles (Rockstroh 1973 ff, Kettner 1984, Schmigalla 1995, Schenk/Wirth 2004, Helbing 2007)

1. Totality Principle

Global design of the planning process (phases and stages), interdisciplinary composition of teams, participative forms of procedure and cooperation. First analyze, then systematize.

2. Phase Principle

For each phase in the life cycle of products, processes, plants and buildings, it is necessary to execute a separate project (e.g. for planning, implementation, commissioning).

3. Stage principle

Progressive execution of definable, logically-ordered planning stages and steps within the project design process according to the TOP DOWN and BOTTOM UP principles. The approach that proceeds from the aggregate to the individual is also designated as the TOP DOWN principle, or as the analytical approach. The inverse is the BOTTOM UP principle, or synthetic approach. Fig. 3.6 shows both options.

Analytical approach	Synthetic approach
- Planning "from the outside in" or "top down" - starts with the whole and gradually works towards detailed solutions - follows the order used in construction: "plan the whole and then the details"	- Planning "from the inside out" or "bottom up" - starts from the smallest element (e.g. workstation) and by combining (or synthesizing) the elements calculates the next largest area until the overall result - the main focus is on production-related requirements

Fig. 3.6 Fundamental approaches in planning

4. Variant principle

The generation of variables and variant decisions leads to preferred solutions based on evaluations (vary and optimize).

5. Profitability principle

Applies to the planning (project development) as well as to the execution and operation of the production plants taking time factors into account.

6. Project constancy principle
Safeguarding of the project's aims, adherence to schedules and costs as well as contract-driven service provision.

7. Order principle
Prerequisite for systematic work and the application of existing solutions, building blocks and modules.

8. Flexibility principle
Flexibility and adaptability in planning, execution and operation of the objects.

9. Improvement principle
Continuous process of constantly monitoring the project for problems that might hinder the timely, cost-effective and quality-compliant completion of the task, and their systematic logging and resolution. Fig. 3.7 shows the permanent process of refinement as a PDCA cycle.

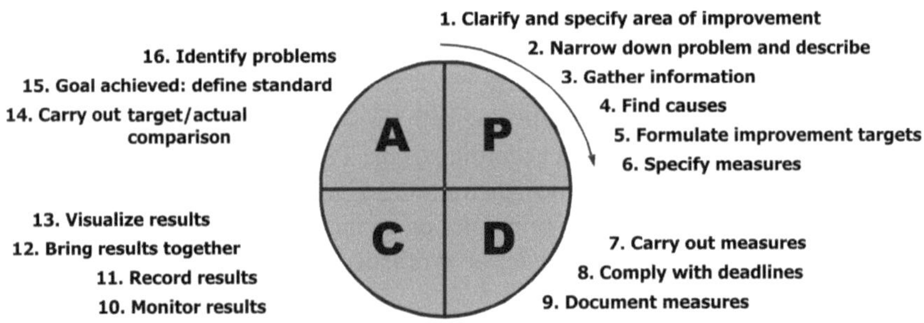

Fig. 3.7 Plan-do-check-act (PDCA) cycle

10. Situation principle
Planning, execution and operation subject to situational changes in terms of content and/or time.

11. Synergy principle
Cooperation through joint and efficient utilization of resources for value creation.

NB: Rules

From the general to the detailed	→ First as a whole, then subdivided → First centralized, then decentralized → First general, then individual → First aggregated, then detailed → First global, then in concrete terms → First general verification, then detailed verification → First the meters, then the millimeters

From the individual to the typical	→ First quality, then quantity → First analyze, then systematize → First individual processes, then standard processes
From the individual to the integrated	→ First individual, then integrated solutions → First integrate functions, then automate
From the ideal to the real	→ First legality, then profitability
From trust to control	→ First control, then delegation
From the refinement process to the decision process	→ First determine processes, then make decisions

Note: the planning of production facilities and plants is a process of refinement with different decision-making stages. The Customer Order/Project Definition checklist (fig. 3.1) is recommended to help you determine the focal areas for project definition. It is subject to a process of continuous improvement. From the principles listed, the "improvement principle" (PDCA cycle) is of primary importance for situation-driven planning (fig. 3.7)

References

DIN (1997) DIN 69905 – Projektwirtschaft, Projektabwicklung, Begriffe. Berlin

Helbing K (2007) Fabrikprojektierung. Book manuscript, Rostock

Kettner H, Schmidt J, Greim HR (1984) Leitfaden der systematischen Fabrikplanung. Hanser, München

Rockstroh W (1973) Technologische Betriebsprojektierung. Technik, Berlin, Band 1 - 4

Schenk M, Wirth S (1994) Fabrikplanung und Fabrikbetrieb. Methoden für die wandlungsfähige und vernetze Fabrik. Springer, Berlin

Schmigalla H (1995) Fabrikplanung, Begriffe und Zusammenhänge. Hanser, München

3.2 Project Development (complex II)

In order to develop a project (planning activities 5/1 to 5/5) it is necessary to complete the project design steps based on the project definition. Fig. 3.8 (Wirth 2000) shows the project design steps and what they entail.

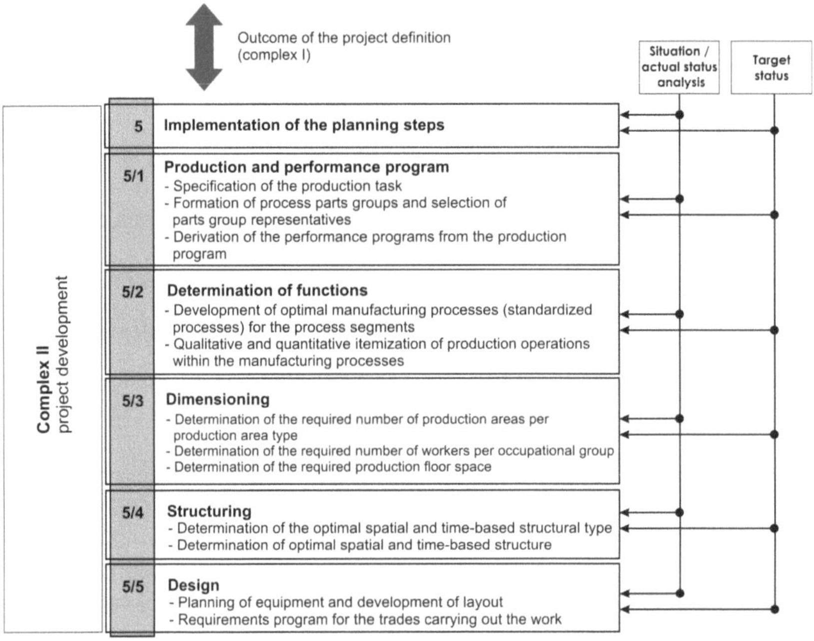

Fig. 3.8 0 + 5 + X Planning Model - Complex II - Project Development

Steps 5/1 to 5/5 are executed systematically and consecutively in a closed loop process with feedback to the preceding project design steps. Table 3.5 contains a questionnaire (checklist) for conducting a "broad brush" assessment of project development activities that must be analyzed.

Tangible and intangible products (services) can be developed by implementing processes. Tangible products are developed using production processes, i.e. manufacturing processes for the production of parts and assemblies in combination with THS processes.

Table 3.5 Checklist – 0 + 5 + X Planning Model - Project Development (complex II)

❐	5/0	**Project definition**					
			Production task				
		❍	Data available	❍	yes	❍	no
			b) Product program				
		❍	Data available	❍	yes	❍	no
			c) Technologies				
		❍	Data available	❍	yes	❍	no
			d) Quantity				
		❍	Data available	❍	yes	❍	no
			e) Times				
		❍	Data available	❍	yes	❍	no
			f) Costs				
		❍	Data available	❍	yes	❍	no
			g) Quality requirements				

		❍	Data available	❍	yes	❍	no
❒	5/1	**Production and performance program**					
			a) Production program preparation				
		❍	Length of planning period	❍	Production types		
		❍	Production program types	❍	Definitive/ restricted	❍	Undifferentiated /extended
			b) Performance program				
		❍	Material flow	❍	Energy flow		
		❍	Information flow	❍	Value flow		
			c) Production program preparation				
		❍	Product structure	❍	Make or buy		
		❍	Type representative	❍	Variable PP	❍	Flexible PP
❒	5/2	**Functional and processes determination**					
			a) Processes				
		❍	Partly known	❍	Known	❍	New
			b) Production/Manufacturing structure				
		❍	Partly known	❍	Known	❍	New
			c) Manufacturing procedure				
		❍	Manufacture of parts	❍	Assembly	❍	Combination
			d) Manufacturing process				
		❍	Manufacturing process stages	❍	Manufacturing process operations		
		❍	Workflow scheme	❍	Functional scheme		
			e) Equipment				
		❍	Technical, functional				
		❍	Manufacturing/assembly cells	❍	Special purpose machinery	❍	Transfer line
		❍	Processing center	❍	Standard machinery	❍	Assembly area
		❍	General purpose machinery	❍	Manufacturing/assembly system		
		❍	Economic, ecological				
		❍	Operating costs	❍	Investment/acquisition costs		
		❍	Energy consumption	❍	Water consumption		
		❍	Recyclability	❍	Reusability		
❒	5/3	**Dimensioning**					
			Methods				
		❍	Static	❍	Dynamic		
			Times				
		❍	Time calculations	❍	Utilization times	❍	Order times
			a) Machinery and equipment				
		❍	Individual	❍	Several	❍	Plants
			b) Staff				
		❍	Multiple machine operation	❍	Individual operation	❍	Qualifications, skills
		❍	Group/team structure	❍	Number	❍	Evaluation
			c) Floor space				
		❍	Approximate floor space structure	❍	Detailed floor space structure	❍	Floor space assessment
		❍	Determination of total floor space	❍	Structured determination of floor space		
❒	5/4	**Structuring**					
			a) Technical and organizational structural hierarchies				
		❍	Supply-chain network	❍	Building structure	❍	Area structure
		❍	Section	❍	Workstation	❍	Procurement
		❍	Production	❍	Sales logistics		

		b) Spatial structure				
	❍	Material	❍	Information	❍	Energy
	❍	Work	❍	Value flow		
		b) Time-related structure				
	❍	Sequential progression	❍	Parallel progression	❍	Combined progression
		d) Workflow principle				
	❍	JIT	❍	JIS	❍	KANBAN
	❍	CONWIP				
		e) Structural form				
	❍	Stationary production	❍	Workshop production	❍	Group production
	❍	Serial production	❍	Continuous flow production		
		f) Internal and external structuring				
	❍	Segmenting	❍	Customer order decoupling point		
	❍	Center organization				
	❍	Profit	❍	Cost	❍	Revenue
	❍	Service				
❐ 5/5	**Design**					
		Layout				
	❍	Rough layout	❍	Detailed layout		
	❍	Ideal layout	❍	Actual layout		
		a) Restrictions				
	❍	existing	❍	yes	❍	no
		b) Object alignment to the transport route				
	❍	parallel	❍	at right angles		
	❍	diagonal	❍	other		
		c) Dimensional layout				
	❍	Minimum clearances	❍	Safety clearances		
		d) Foundations				
	❍	Foundations	❍	Block foundations	❍	Vibrating foundations
	❍	Foundationless	❍	Fixed assembly	❍	Flexible assembly
		c) Ergonomic design				
	❍	Seated workstations	❍	Standing workstations		
	❍	Combined workstations	❍	Single operator workstations		
	❍	Multiple operator workstations				

3.2.1 Production and performance program

3.2.1.1 Production program (PP)

In the case of the production program, tangible and intangible (service) products can be developed through efficient process design. Tangible products are made using production processes for manufacturing parts and assemblies. The starting point is the determination of the production task as defined by the project definition via the five specifications.

The client (customer) specifies the framework within which the planner must operate. He also stipulates:

- the product (manufactured object), the processes and plants in combination with the specifications on floor space and capacities (technical and human resources)
- the costs and investment and
- the timetable, with decision-making stages.

What needs to be clarified is what should be produced, when, by whom, where, for what period of time, how and with what (Rockstroh 1982).

Note: in production program planning, an "idealized" task definition (job, order program), which is valid for a certain time period, is assumed. In general, the production program, resources, technical requirements, price ratios and timetables change during the planning and project implementation process. This requires not only empirical knowledge, but also regular evaluation of inventions, research results and expert knowledge as well as a permanent flexibility in the planning and production process. For example, the energy situation compels manufacturers to seek strict resource and energy efficiency and new products and technologies.

Fig. 3.9 shows the features of a production program.

A *production program* defines the production according to

- **material** (type, size),
- **quantity-based** (lot size, mass),
- **financial** (price, costs) and
- **time-related** (production period, planning period)

aspects.

Fig. 3.9 Production program features

Production programs are the outcome of corporate market and sales analyses. Production programs can be classified according to different aspects.

NB: Classification

a) according to the duration of the planning period:

- short-term operational planning with an effective duration of up to one year
- medium-term tactical planning with a planning horizon of 2 to 5 years
- long-term strategic planning with a planning horizon of 5 years plus

Short-term operational production program planning is usually done on the basis of customer orders, whereas long-term strategic planning requires sales forecasts to be made.

b) according to production type:

- make-to-order: each product is produced individually or only once (e.g. special purpose machinery, steel structures, special equipment).
- series production: this is characterized by the simultaneous or consecutive creation of several homogenous products (series). This also includes prototypes.
- batch production: different variations of similar products are manufactured in batches one after the other. As a rule, the batches differ from one another, for example, in terms of size or quality. The manufacturing sequence and equipment used are the same.
- varying batch production: this type of batch production is when a production process is not exactly the same, for example with regard to quality, raw materials or equipment used.
- mass production: homogenous, standardized products are manufactured in high lot sizes across an undefined period of time.

c) according to production program types:

Table 3.6 Production program types

Differentiation of production programs		Types	
Product types	Machinery	Tools	Jigs and fixtures
Complexity of the products	End products	Assemblies	Parts types
Manufacturing structure	Company, division	Section	Manufacturing workstation
Level of detail of PP	Definitive (detailed)	Definitive restricted (aggregated)	Undifferentiated (generalized)
Quantity-based certainty / product	Constant	Variable with upper and lower limits	Variable with lower limit
Financial certainty of the PP (turnover volume)	Constant	Variable with upper and lower limits	Variable with lower limit

d) according to level of detail:

Table 3.7 Production program types according to level of detail (adapted from Schmigalla 1995)

Definitive production programs (clearly defined production programs)	Definitive production program Detailed production program
	Definitive restricted production program Aggregated production program
Undifferentiated production programs (indeterminate production programs) Generalized production programs	Undifferentiated extended production program
	Undifferentiated production programs

Fig. 3.10 describes the production programs' level of detail with respect to planning.

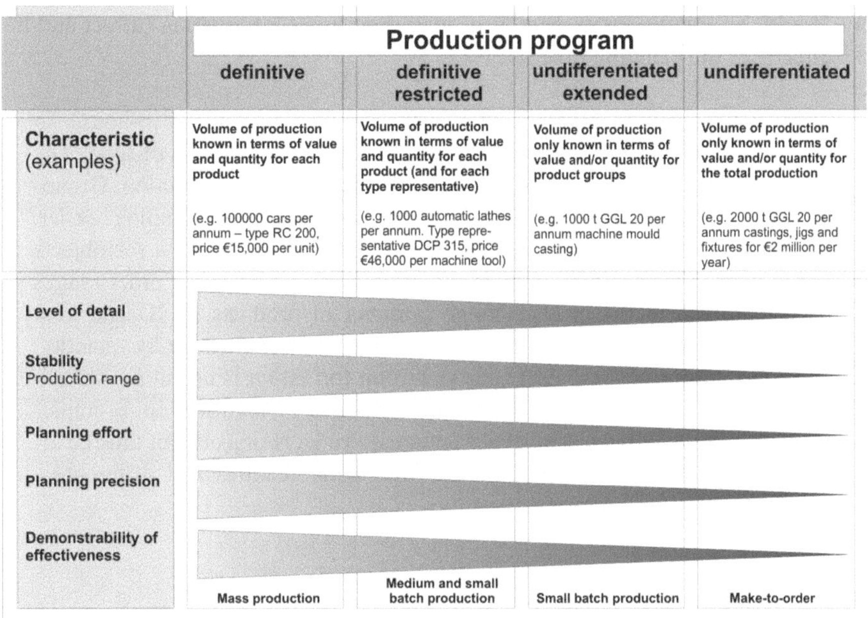

Fig. 3.10 Characteristics of program types with respect to level of detail

NB: initial situation

- In an enterprise, several types of product may be relevant (e.g. washing machines, refrigerators) or the company may make company-specific tools, jigs and fixtures in-house (possibly also in profit centers).

- At the start of the project, often only the production program (PP) for the end products is known; the PP for assemblies, part types and peripheral structural units must be derived from it.
- As a general rule, the complexity of the products in the PP increases with the hierarchy level of the functional units (exception: single station assembly).
- With increasing standardization of the production (from make to order, to series, to mass production) the level of detail of the PP increases.
- PP must be forecast. PPs with constant production quantities per product and a fixed volume of turnover for the entire PP carry the characteristic of probability (probable PP).
- If lot size limits are prescribed for individual products, with a constant turnover volume it is possible to plan with a variable PP, and by specifying turnover limits with an optimal PP.
- The transitions from detailed to aggregated and from aggregated to generalized PPs are fluid.
- The PP provides a large amount of important input information (direct and indirect) for the planning process.

Note: what needs to be clarified is what should be produced in-house, in what quantities, to what level of quality and over what period of time. The more accurate the information, the more precise the project solution. Grouping and classification procedures can be used to make planning easier. Groupings characterize features and quantities and weight them for objects or products. For example, items in the product range, customer order ranges by the breadth of the product range (number of products by ID and item numbers), depth of product range (proportion of overall range by quantity, value and turnover – e.g. ABC parts). Group formation is useful for simplification and transparency for standard strategies. Classification is implemented using a number of selected features. Products/objects can thus be allocated to a certain category that has the same features and differences, weighting and type representatives.

3.2.1.2 Performance program (PerP)

The performance program is the performance requirement to be met by the production plant that is derived directly from the production program. It contains the input information required for the detailed and execution planning for functionally varied (peripheral) areas/flows. Performance programs are defined for material, information and energy flows (e.g. the supply of auxiliary manufacturing materials, production planning and control, energy consumption in accordance with chapters 6.2, 6.3 and 6.4). In general the following information is compiled with a performance program, which at the same time is used to select the appropriate operating resources:

- description of the equipment, workstations, areas - "consumers" - to be operated or supplied and/or disposed of by the (peripheral) area being planned, with regard to: type, model, denomination, location; specific supply or disposal requirements according to type and quantity per unit of time or service (e.g. l/h auxiliary manufacturing materials; t/h scrap turnings and chips; m^3/h compressed air; kW electric power); with regard to qualitative requirements (e.g. nominal pressure and permissible moisture levels in the compressed air to be supplied); nominal voltage, type of voltage, permissible voltage fluctuation (e.g. the electrical energy to be supplied) as well as with regard to specific connection conditions (e.g. under- or over-floor connection for the disposal of turnings and chips) → consumer list
- determination of the quantity/quantities per unit of time/service to be yielded by the (peripheral) areas being planned (e.g. l/h auxiliary manufacturing materials; t/h scrap turnings; m^3/h compressed air; kW electric power); → determination of demand (e.g. electrical power KWh; water m^3/h; compressed air Nm^3/min)
- investigation of the actual status of equipment (that may already exist) with a view to its recyclability or reusability and the structural conditions with regard to any adjustments required
- recording of all restrictions applicable to the (peripheral) performance program being planned (laws, regulations – framework conditions of the overall planning project) plus the available financial means

Note: performance programs must be drawn up for all flows and their equipment, e.g. materials, information, energy, FTT and operating material flows (solid, liquid and gaseous substances).

3.2.1.3 Derivation of production and performance programs

The progress of planning requires further function-dependent production and performance programs to be derived from the production program and performance program for the end products (cf. fig. 3.11).

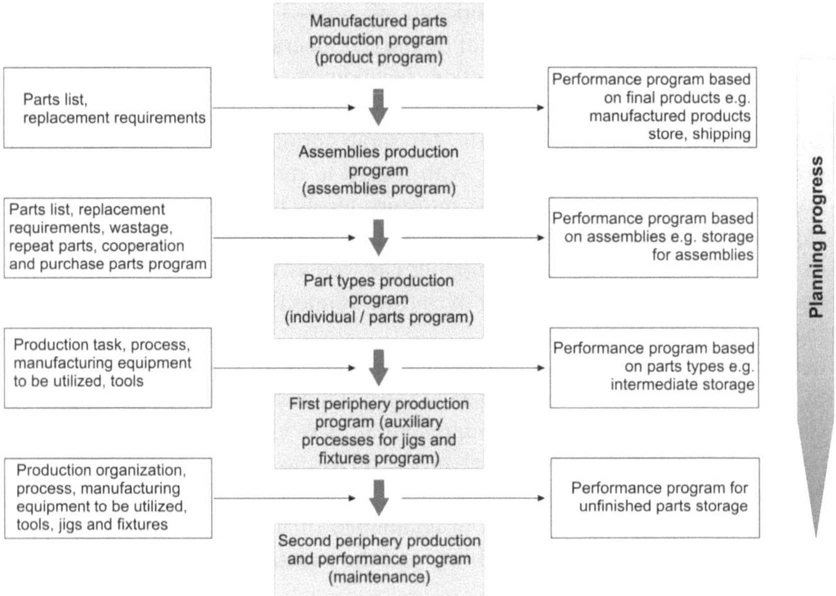

Fig. 3.11 Derivation of production and performance programs

The production program (PP) and the performance program (PerP) play a dominant role in planning. For peripheral structural units the PP/PerP can only be derived once the main process' structural units have been designed. As the planning progresses, the level of detail of the PP/PerP increases. The type and number of facilities and equipment as well as the structural units and their interconnections cannot be derived directly from the PP/PerP, but rather only via the processes required for its implementation. Relevant data (parts lists etc.) are required for planning.

a) Preparation of the production program

It is important to check whether there is an existing defined production program available or if a new one needs to be drawn up. In both cases, it is necessary to classify the products within the product structure. From this it is possible to derive the relationship of the end producer's product, e.g. component assembly or individual part, to the whole. There are various different ways of expressing the breakdown of the product structure (group and parts), as shown in fig. 3.12

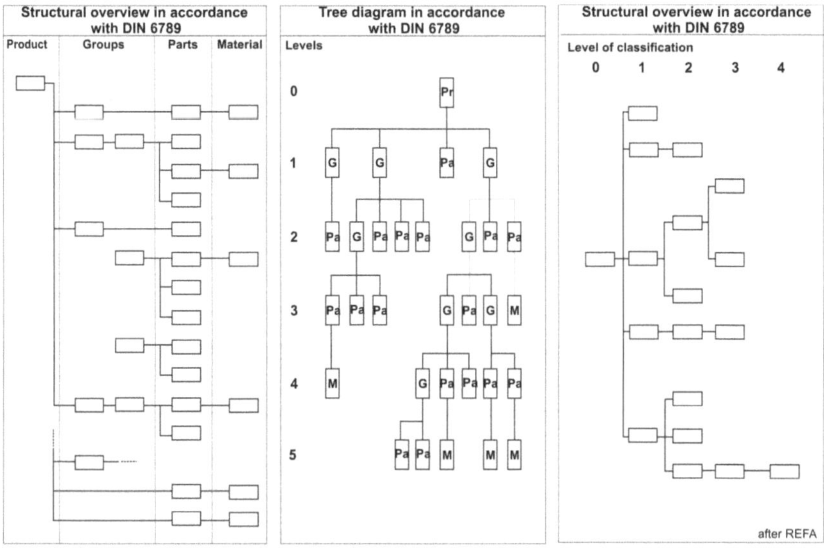

Fig. 3.12 Ways of expressing the product structure

The process of defining the production program on which the product is based begins by breaking down the product structure and with resource balancing in the enterprise in conjunction with the integration of a "make or buy" decision in the manner shown in fig. 3.13.

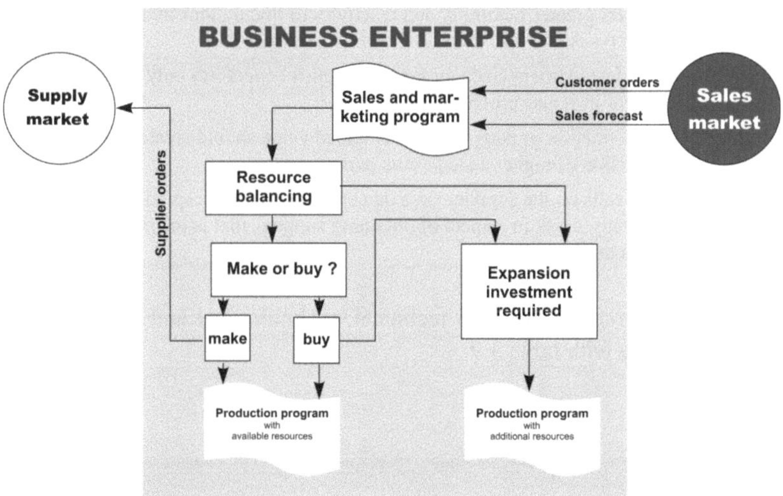

Fig. 3.13 Step-by-step definition of the production program (after Schmigalla 1995)

In practice the end product is made up of parts produced in-house and supplier parts. For workshop planning, only those products (parts) that will be produced in the workshop being planned come into consideration. A production program analysis based on production and financial criteria should be carried out for this purpose.

A longer term specification of the production program can generally be affected by three circumstances:

(1) As part of strategic enterprise planning, the production program is predetermined.
(2) As the basis for a specific project task, the production program is developed by the planning team in the course of the target planning phase and must be approved by the company's management.
(3) The production program is compiled on the basis of a make or buy decision.

The latter is done via a thorough assessment of the production program by answering the question: external procurement or in-house production? (cf. example)

(1) general logistical criteria of the make or buy decision (table 3.8) including by quantity (fig. 3.14) and costs and

Table 3.8 General criteria applying to a make or buy decision

	Reasons
1	Higher quality of the OEM parts due to supplier specialization
2	Cheaper supplier cost structure (technology or location advantages)
3	The supplier offers greater flexibility and reactivity to fluctuations in demand with regard to quantities and types
4	Investment and development costs for new parts must henceforth only be financed by the company itself for in-house produced parts.
5	Relocation of production of parts with lower added value should create space for concentration on production of higher added-value parts.
6	Demand-driven calls on the supplier save on costs arising from capital being tied up and reduce the company's risk in respect of obtaining supplies that is incurred when manufacturing parts in-house.

(2) detailed criteria according to technical, technological and economic aspects in accordance with table 3.9

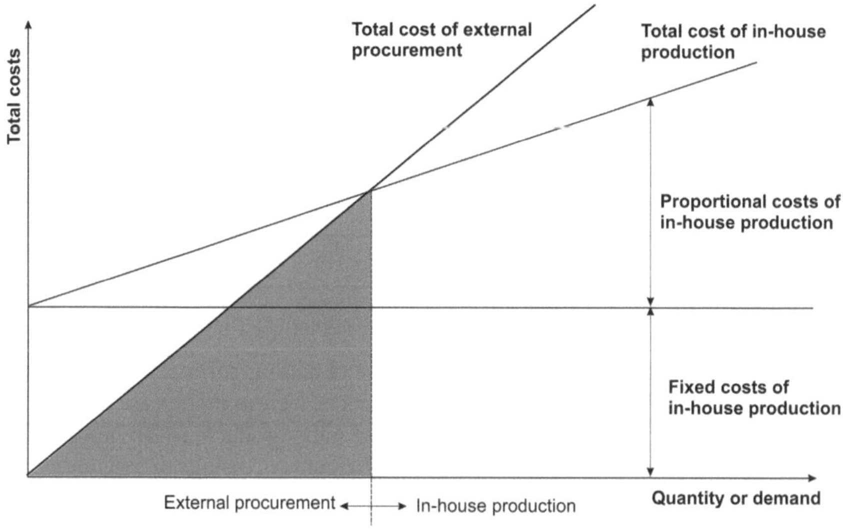

Fig. 3.14 Make or buy decision

Table 3.9 Criteria for the make or buy decision (H. BUSCH, German Logistics Congress '84)

Technical, technological, logistical and economic criteria for the make or buy decision		
	In-house production, if	**External procurement, if**
Readiness to invest	exists	does not exist
Short-term/long-term liquidity	plentiful	not plentiful
Personnel	available	not available and cannot be recruited in the short term
Manufacturing facilities (free capacity)	available	not available and cannot be made available in the short term
Manufacturing floor space	available, not planned over the long or medium term	not available, planned over the long or medium term
Expertise	own expertise available, expertise only available within the company, expertise to be built up. Should not be handed over to others.	available from the supplier, only available from the supplier and protected by patents/licenses.
Quality	supplier cannot guarantee required level of quality	in-house production cannot guarantee required level of quality
Independence	would be compromised by external procurement	would not be compromised by external procurement
Reliability of supplier	not guaranteed	guaranteed
Company secrets (technical and commercial)	would be compromised by awarding external contract	would not be compromised by awarding external contract
Countertrade	not aspired to/not possible	desirable
Deadline	in-house production can achieve required delivery date – supplier	supplier can achieve required delivery date – in-house pro-

	cannot	duction cannot
Formulation of technical changes	quick reaction to complaints necessary, strong influence required	unproblematic part, changes not necessary
Sales volumes (inventory / capacity risk)	constant	fluctuating, quick adaptation necessary
Manufacturing technology	development of a technology	quick adaptation to latest manufacturing technologies, use of cost and functional advantages.
Supply market	no / few suitable suppliers	multitude of suitable suppliers with excess capacity

A subsequent reworking of production programs takes place in the case of

- restricted or undifferentiated programs for obtaining sufficient representative initial data for the dimensioning of the plants as well as in the case of
- definitive programs with a large variety of parts/products to be processed by the production plants being planned for efficient processing (dimensioning calculations) of the large data quantities.

The compilation of parts families has proved useful in reducing the amount of data. They represent a number of products that fall into a category or family as a result of having certain common features. The parts families provide the basis for the planning of structural areas in which different operating resources can be combined in a production group or island. Table 3.10 shows how parts families are formed.

Table 3.10 Formation of Parts Families

Requirement	Resource	Effect
Consistency of form (A)	Repeat part use	Reduction in the time and effort required for design and production scheduling
Similarity of form, same or similar operating processes (B)	Additional series Mechanical flow production line	Reduction in throughput time
Same operating processes with similar workpieces (C)	Machines with higher levels of automation	Reduction in the lot sizes

b) Type representative method

A type representative is an actual or ideal product (complex part) that to a large extent combines the requirements of the entire product group. It unites features such as geometry, surface area, mass, volume, structure, effort, costs and composition. So-called type representatives are formulated using the type representative method for the calculated reduction of product variety to product groups. This involves the formation of groups of technologically-similar parts/products, the determination of one type representative per group and the conversion of the parts/product quantities of each group on the basis of the type representative. The sum of all type representative-related individual batches stands for a "representative" quantity of all parts/products in the group and constitutes the input information for the production system project planning (type representative project planning).

The following are reasons for applying type representative project planning:

- the range of products in the production program is too numerous,
- the production program contains products that are still in the development phase,
- the client's planning specifications are based on type representatives.

Selection conditions for a type representative (TR):

- the number of products for which a type representative is to be defined should be as large as necessary and as small as possible.
- This means that there is a trade-off between the formation of large groups in order to keep the number of TRs and thus the project design effort as low as possible, and a high level of similarity of the products in a group in order to keep mistakes to a minimum.
- the manufactured object for which the result of effort and quantity is greatest will be defined as the type representative.
- This restriction should result in the project design parameters being determined accurately for the product that makes the greatest contribution to outlay in the group, and thus to the error for the product group being lessened.
- products that are in development are not suitable type representatives.

Ranking of type representative selection criteria:

1. Working time requirement (WTR), as primarily the numbers of production areas (N_{PA}) and workers (workforce) (N_{WF}) need be determined
2. In-house activity
3. Price/Costs
4. Geometry (length, width, height, etc.)
5. Mass
6. Volume

Example: Procedure for determining a type representative

Determination of the type representative	Effort • Quantity → Maximum	
Determination of the conversion factor f_{TR}	f_{TRi} = Effort $_{Prd.i}$ / Effort $_{TR}$	(3.1)
Determination of the reference lot size n_{TRi}	$n_{TRi} = n_i * f_{TRi}$	(3.2)
Determination of the reference lot size for the product group n_{TR}	$n_{TR} = \Sigma\, n_{TRi}$	(3.3)
Determination of the effort required for the entire product group	Effort $_{PG}$ = Effort $_{TR}$ • n_{TR}	(3.4)
Determination of additional parameters for the type representative or the individual products	There are various possibilities for this: The type representative's parameters are used (relatively imprecise, if the TR has extreme values). The TR's parameters are the arithmetic mean values of the corresponding parameters of all products in the corresponding product group. The TR's parameters are the weighted arithmetic mean values of the corresponding parameters of all products in the product group. The weighted parameter W should be selected in accordance with the project planning aim.	
	$W = \Sigma\, W_i$	(3.5)
	$P_{TRk} = \{\, \Sigma(P_i \bullet W_i) \,\} \,/\, \Sigma\, W_i$	(3.6)
	P_{TRk} – parameter to be determined for the type representative	

c) Variable production program

If the production range is stochastic in character then variable production programs must be created. Here the special feature is that a certain tolerance range will be set for each product for a probable production quantity. The production program that results in the greatest time load must be determined for each manufacturing workstation group.

Note: the sales market, which is subject to fluctuations, has an influence on the layout of the production facilities. For this reason a degree of flexibility must be introduced into the production and performance programs.

3.2.2 Determination of functions and processes

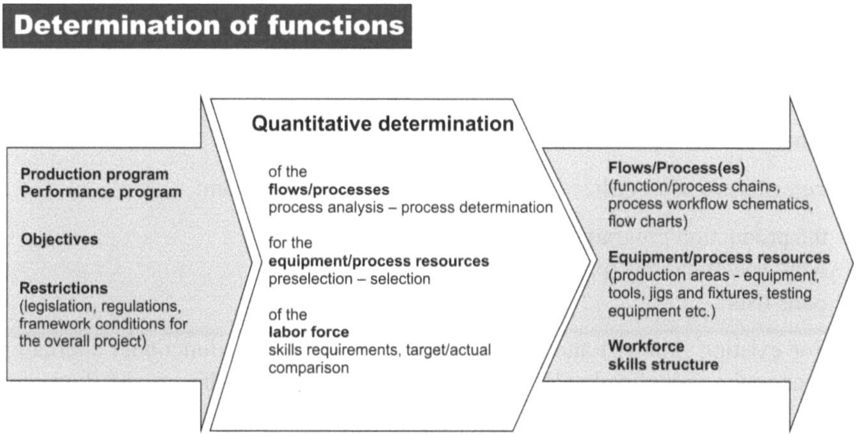

Fig. 3.15 The determination of functions planning step

Determination of functions is the functional determination of technological sequences and their processes for product creation together with the associated equipment/plants.

Its basis is the technological planning of the processes and includes:

- how the goods should be produced, with what, under which conditions, using which technological processes and equipment and in what order,
- how materials, workpieces, parts, assemblies and products should be modified to achieve the desired final state.

The determination of functions is subject to technological change and is decisive for the adaptability of processes, equipment and plants. Each basic technology must be mirrored by innovative technologies. These are defined by new materials (e.g. fiber composites instead of steel), power and environmentally-efficient production, low-cost utilization of raw materials and substances, changed work contents in connection with mechanization and automation, plus integration of production and logistics processes (and technology).

When determining functions and processes the following questions must be answered:

- which functions need to be fulfilled by the production facilities being planned?
- which processes take place in the production plants being planned?
- which installations fulfill these functions?
- what level of adaptability must processes and their functional implementation possess?

The following form the basis for the determination of functions:

- a specific product
- the design and design documents
- the technology (procedures, process workflows, times and equipment/operating resources)
- production quantities, times, quality and costs
- other parameters

Documents are to be utilized or drawn up for this process from:

- the production program and product structure
- work specifications (process steps and objectives)
- parts lists

For existing (known) and new technologies (production functions) alternative options to the predefined solutions should be considered in the case of short-term changes.

The determination of functions can be broken down into five sub-steps.

(1) Analysis of the production program and the product structure

The product structure (unfinished part/purchase part, individual part, assembly, main assembly, product) and the determination of the production stages (prefabrication/manufacture of parts) and assembly (component assemblies, main component assemblies, final assembly) are presented in the form of a schematic diagram in fig. 3.16.

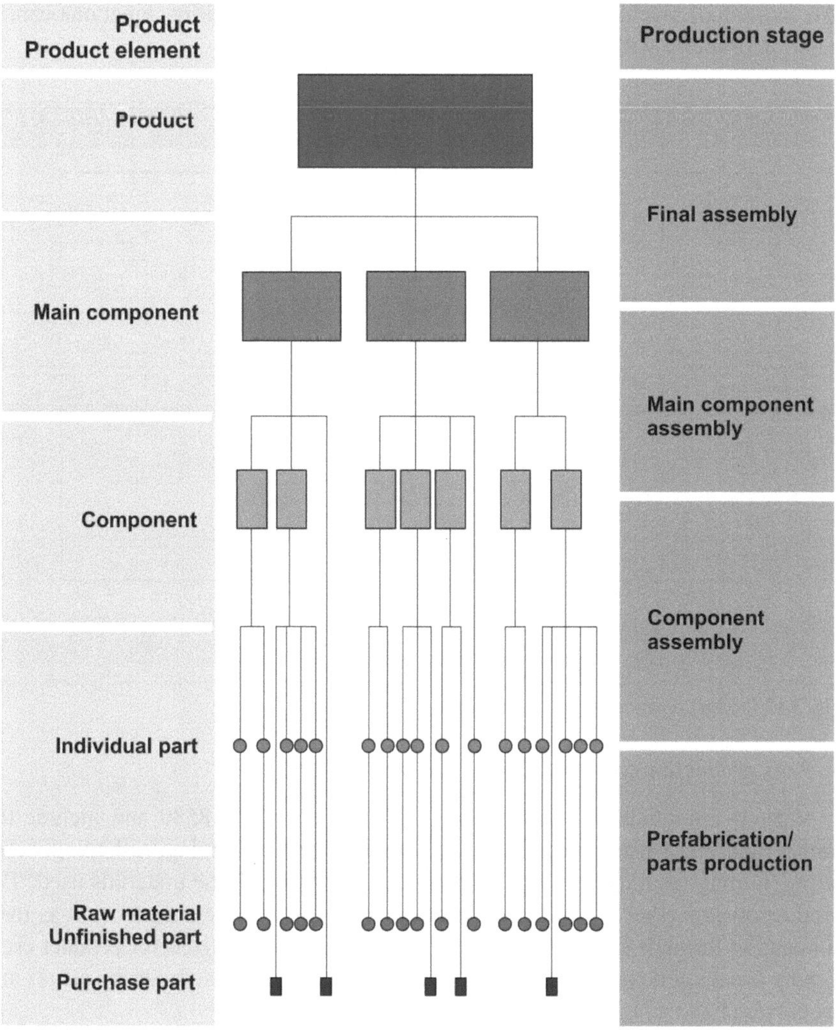

Fig. 3.16 Product structure and manufacturing stages (schematic diagram, Grundig 2000, p. 67)

(2) Determination of the transformation, transportation and storage production functions

The manufacturing concept specifies the transformative manufacturing processes to be used for manufacturing the product, i.e. the production functions (transformation, transportation, storage) (fig. 3.17).

The production functions should always be considered as a whole in the technological process flows.

An individual machine without transport, storage and handling functions cannot function properly within the process.

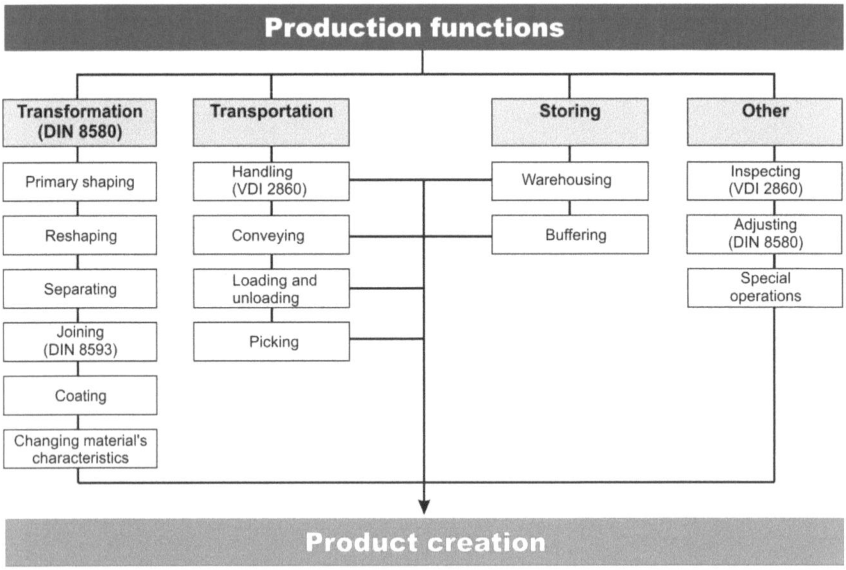

Fig. 3.17 Product creation production functions

- Parts production (manufacturing process)

Various manufacturing processes are specified in DIN 8580 and include the transformation of workpieces/materials through primary shaping, reshaping, separating, joining, coating and changing the characteristics of the materials used. The production processes for the manufacture of parts and aggregates as well as their connection through transportation and storage can be visualized for product creation by breaking down the parts lists, product structures, manufacturing stages and processes (fig. 3.17).

- Assembly/aggregate production (assembly processes)

Assembly is the collection of all procedures that are used to combine geometrically-determined elements. Assembly is joining objects together in conjunction with workpiece handling in accordance with VDI guideline 2860, inspecting (VDI 2860), adjusting (DIN 8580) and special operations in accordance with fig. 3.18.

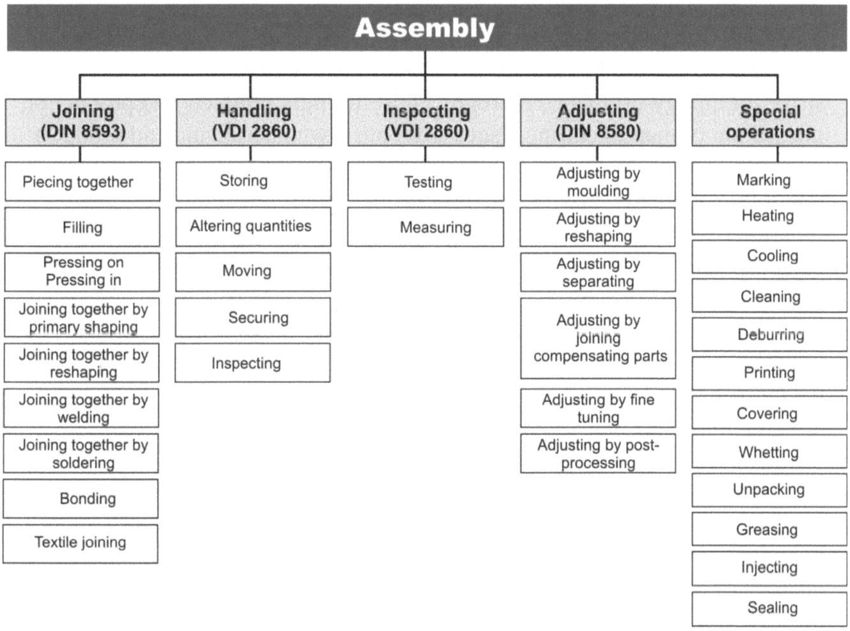

Fig. 3.18 Assembly functions (Lotter & Wiendahl, 2006)

Production and assembly alter the products. Logistics technology (transport, storage and warehousing) is part of the overall production process.

- Transport, handling and storage – THS (logistics process)

This is the whole of the THS chain between the equipment/plants and storage. The process of transporting the goods/workpieces between source and destination points by conveying and handling (transferring, loading and unloading), and the storing of the goods/workpieces by holding them for short term supply in the production area or stores within or between facilities at different levels in the system is known as the transport, handling and storage process (THS process).

(3) Determination of the production process including the assembly process and the workflows

Processes are determined in order to describe the (production) functions and the number of processes (process quantity) in the future production facilities in factual and commercial terms in such a way that the dimensioning and structuring processes can build upon them. The object of production facility and production plant planning is the design of the production processes.
The following principle applies: "Process determination only in as much detail as necessary."

NB: Optimal process determination

- total process design throughout the product life cycle
- strive for energy-efficient, environmentally-friendly products and production
- integration of manufacturing stages (for both raw material and end product)
- integration of several processes and functions in one piece of equipment (e.g. multiple-technology machinery, processing and production cells with multi-function parts)
- assembly-friendly product design, e.g. according to the modular or building block principle
- integration of final production stage for individual parts in the assembly
- integration of moving workflows (transport, handling and storage facilities) in the machinery
- integration of core and auxiliary processes in one facility
- integration of plants and production buildings
- standardization of interfaces within and between the flow facilities
- miniaturization of products, processes and equipment (microtechnology) in conjunction with high quality standards
- change of the materials used for the purposes of lightweight construction through composite and plastic components in place of steel and iron materials (change of materials)
- weighing of mechanization and automation solutions from the perspectives of job creation and the use of hybrid automation solutions

The manufacturing process can be divided into manufacturing stages, operations, operational stages and operational elements. Manufacturing processes, stages and operations are the object of production facility project planning according to their level of detail (cf. table 3.11).

Table 3.11 Process determination stages

Manufacturing processes	Object of production facility project planning
Production stages	
Production operations	
Production operation stages	Detailed planning of workstations
Production operation elements	(Ergonomics)

In the context of planning, process determination comprises two basic tasks:

- process workflow planning (determination of the order of operations and processes)
- operation description (process workflow and selection)

The processes are checked for effectiveness in accordance with the principle of improvement. This is absolutely essential. The results of the determination of processes are summarized in tabular form, and in planning practice are mostly designated as work, manufacturing, assembly and transport plans etc.

In process workflow planning, the value-added chain is defined as a consequence of the operations (workflows, assembly steps, transport processes, handling operations etc.). Workflows can be represented as flow diagrams, work process charts, flow charts and in matrix form. An example of workflow planning is shown in the workflow diagram in fig. 3.19.

	In-house produced part / Equipment	Auger foot housing	Housing cover	Drive housing	Drive shaft	Drive pinion	Worm gear	Drive wheel	Output shaft
1	(Abrasive blasting)	10	10	10		10		10	
2	(Corrosion protection)	20	20	20		20		20	
3	Sawing				10		10		10
4	Turning			30	20	30	20	30	20
5	Broaching					40		40	
6	NC hobbing					50	30	50	
7	Milling	30	30	40					
8	Milling/drilling/ countersinking	40	40						
9	Deburring area	50	50			60	40	60	
10	(Carburizing)					70	50	70	
11	Circular grinding				30	80	60	80	30
12	CNC hobbing					90		90	
13	(Hobbing)						70		
14	Testing	60	60	50	40	100	80	100	40

Fig. 3.19 Workflow diagram (sectoral) – procedures (equipment) for product groups (O = process number)

There are three options for determining processes:

- takeover of existing processes
- modification of existing processes
- development of new processes

(4) Derivation of the functional schematic

The spatial arrangement of the individual sectors with their selected manufacturing processes is summarized in a functional schematic based on the workflow diagram (cf. fig. 3.20).

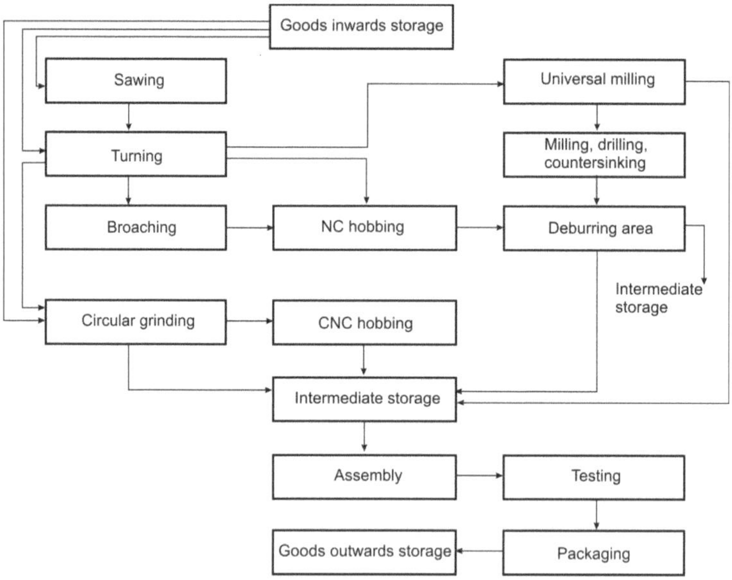

Fig. 3.20 Functional schematic (sectoral)

This shows the connection between the parts manufacturing and assembly production processes, and the flow of materials from the goods inwards store (GI) to the goods outwards store (GO).

(5) Preselection of equipment

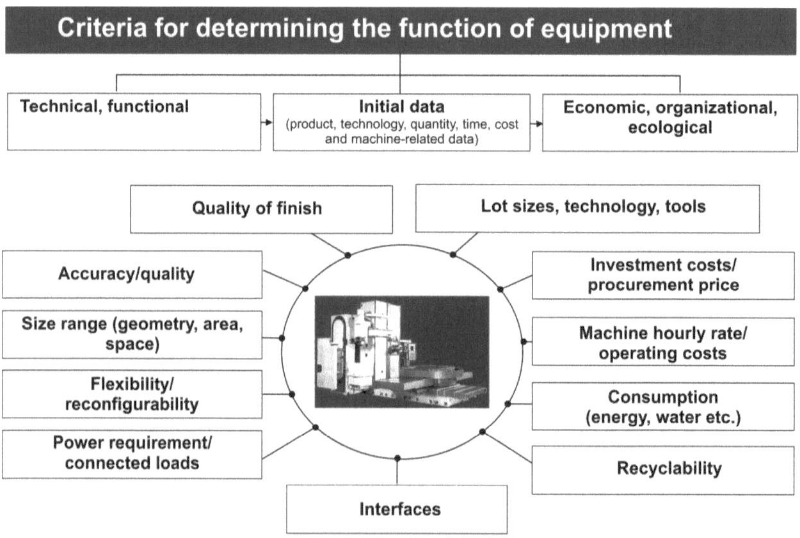

Fig. 3.21 Criteria for the selection of equipment (after Wirth)

Correctly preselecting and determining equipment is the main focus of efficient planning and implementation. It is done in accordance with technical, functional, economic and ecological criteria (fig. 3.21). The availability of machinery and plants is included in VDI guideline 3423 2002.

a) technical, functional and ergonomic aspects

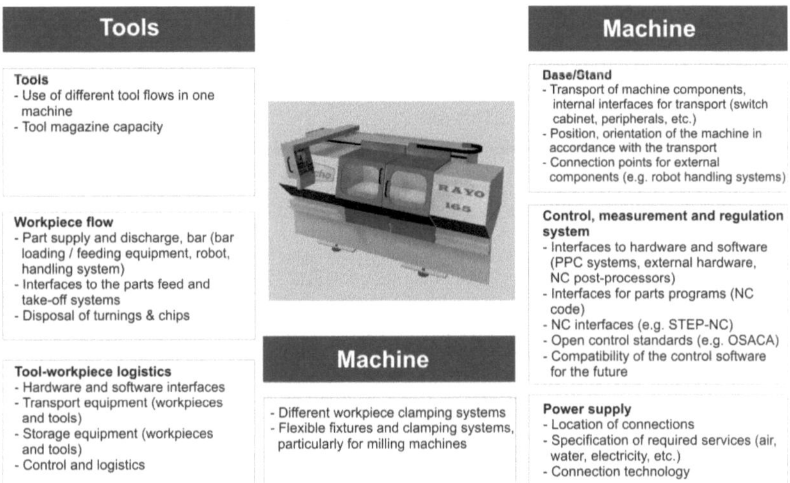

Fig. 3.22 Adaptation-relevant components of a lathe

- by means of the determined manufacturing processes,
- by technical and functional aspects (quality, dimensions, geometry, performance, load, supply, disposal and failure parameters),
- by user-friendliness and maintenance intensity
- by adaptable equipment and components selected from catalogs, offers, planning documentation. (cf. fig. 3.22)

b) Economic and ecological aspects

The final selection of equipment is then made by comparing the functionally similar equipment determined by the preselection process from an economic, ecological and organizational point of view.

In order to select the optimal equipment, the ability to fulfill technical functions, the acquisition costs and the operating costs must be calculated. If investments are to be made, forecasts regarding life cycle costs from the product idea to disposal must be examined in detail. These costs include acquisition, installation, commissioning, operating, maintenance and environmental costs (disposal costs.) Significant elements of the life cycle costs can be seen in fig. 3.23.

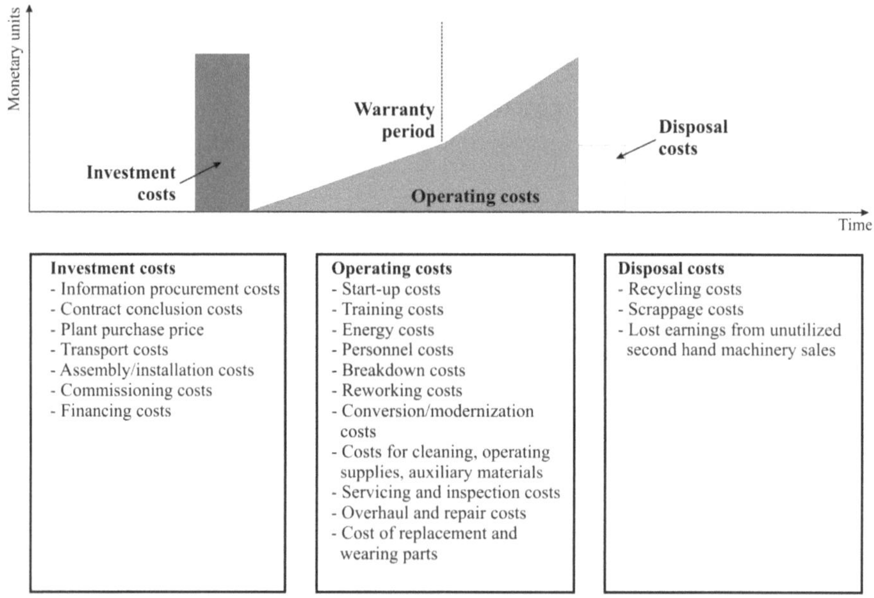

Fig. 3.23 Life cycle costs (Lay 2005, p. 87)

A machine is adaptable if new functional units/components can be integrated, if it can be modified, extended and reprogrammed and has standardized interfaces and connection points. This requires modular construction of equipment and plants.

Examples of interfaces/connection points with connecting elements are contained in Figs. 3.24 to 3.27. (Hildebrand 2005, p. 86 ff)

Example: Mechanical interfaces with stabilization functions (positive fit, friction fit and bonded connections - fig. 3.24)

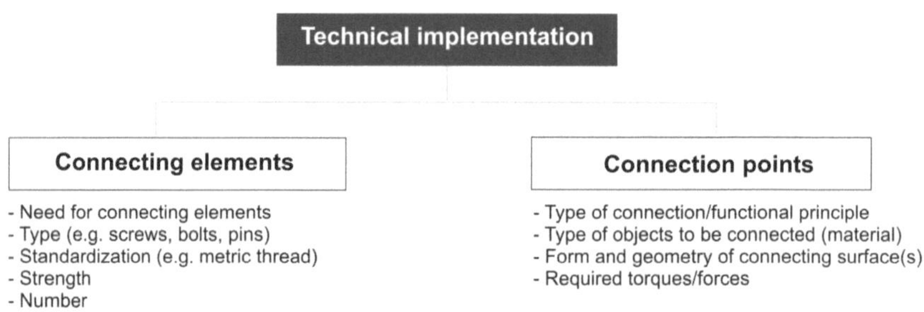

Fig. 3.24 Mechanical interface design factors

Example: Mechanical transfer interfaces (fig. 3.25)

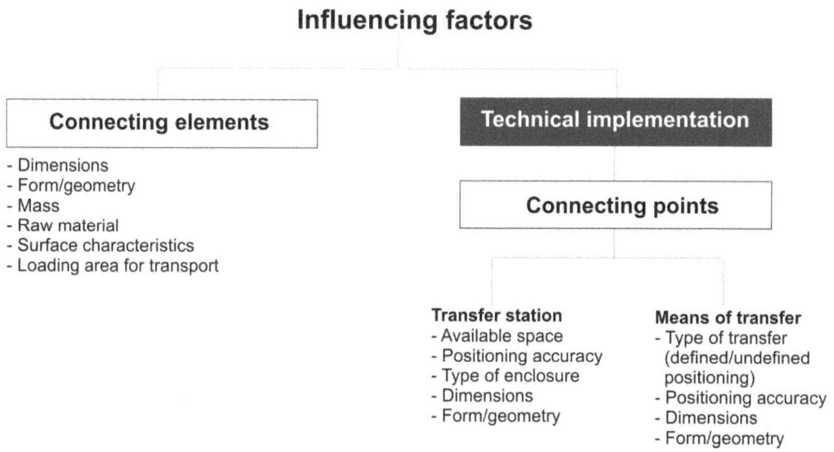

Fig. 3.25 Mechanical transfer interfaces

Example: Hydraulic and pneumatic interfaces (fig. 3.26)

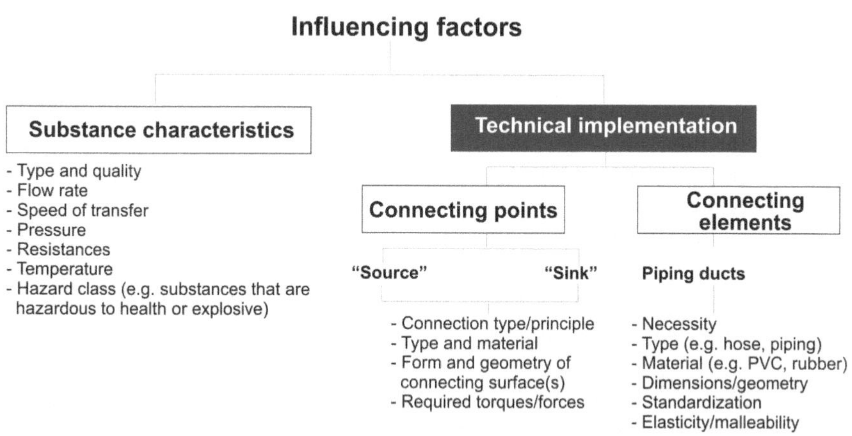

Fig. 3.26 Hydraulic and pneumatic interfaces

Example: Electrical interfaces (fig. 3.27)

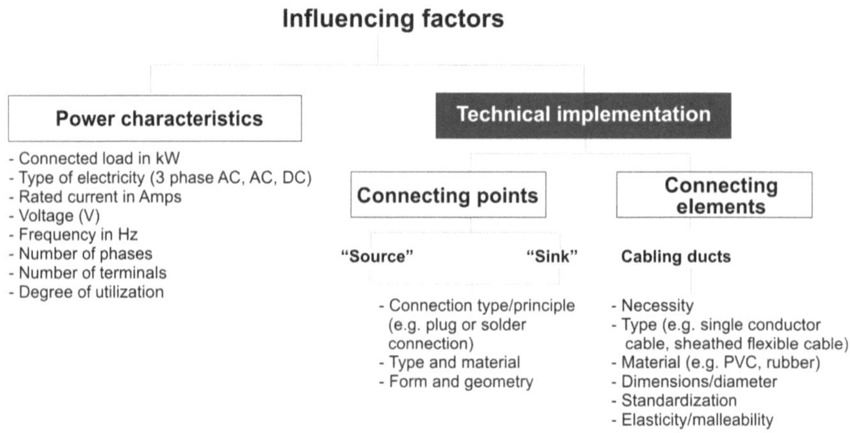

Fig. 3.27 Electrical interfaces

Information-related interfaces provide data and signal exchange for the control of equipment, process control and monitoring (own standards and wireless WLAN technology).

a) *Specific equipment selection for parts manufacture* by product and equipment-related components (Conrad 2002).

Product and workpiece-related components

- workpieces: dimensions (size, shape, geometry, sensitivity); complexity/parts (unfinished part, individual part, complex part); material, weight, handling
- quantities (lot sizes), variety of workpieces, lot, series size (make-to-order, series production)
- quality/accuracy (dimensional, form and position tolerances) for contours, form elements, surface
- envelope volume, processing space (L x W x H)
- processing (type, number, sequence): individual, multiple, complete (standard, special); manufacturing workflow and technologies (processing and finishing, THS technologies and methods)

Equipment-related components

- production type: make-to-order, multi-unit (group), complete, complex and special purpose production
- production principle: workshop, island, group, serial and continuous flow production (manual, mechanized, automated)

- main equipment: manual tools (equipment), individual machines (single, multipurpose, special purpose), processing center, manufacturing cell, production systems/transfer lines (flexible, fixed); weight, floor space and room dimensions, manufacturer, type, model, performance parameters
- size: workspace dimensions, basic and main unit with clamping, assembly, tool unit, adjustment ranges, use of jigs and fixtures, manufacturer.
  Basic unit: stand, main drive (construction type, performance data, revolutions, number of spindles, gear ratio).
  Main unit: manufacturing processes (primary shaping, reshaping, separating, joining, coating, changing of material's characteristics).
  Auxiliary and additional facilities: tools/sets of tools, jigs and fixtures, testing equipment, retaining equipment (changing, storage, handling equipment), chip to chip time
- control/communication: type, model, upgradeability, hardware and software, manufacturer;
  Control type: CNC, SPC, number of axes, network service/Internet
- type of installation: fixed/flexible with/without foundations
- installation location: floor, foundations
- connections/connection points/consumption: power, gas, water, air, data/Internet
- performance capabilities/parameters: productivity, availability (set-up, utilization, maintenance, waiting times)
- occupational safety/usability (ergonomics), flexibility/adaptability, mobility, reconfigurability/modularity, environmental compatibility (emissions, operating materials)
- profitability: acquisition price, operating costs, costs per workpiece, cost/benefit ratio, amortization, life-cycle costs
- delivery times (contracts), training, service

Equipment selection examples (parts manufacture)

Example 1: modules and features of a piece of equipment (machine tool) as the basis for the design variants for adaptable production areas

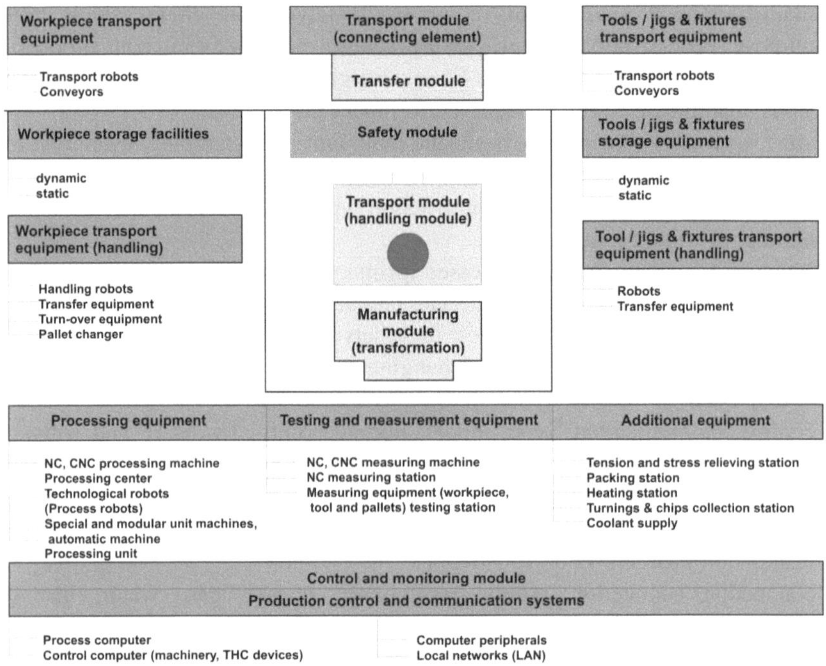

Fig. 3.28 Modules and features of a production area – selection (Wirth 1989, p. 34)

In the center is the production module (the processing facility), that can only fulfill its transformative function if the transport and storage facilities for the workpiece, tool, jigs and fixtures and testing equipment are in place. An operating facility can be assembled in modular form according to the production task using the equipment mentioned.

Example 2: components of a processing center

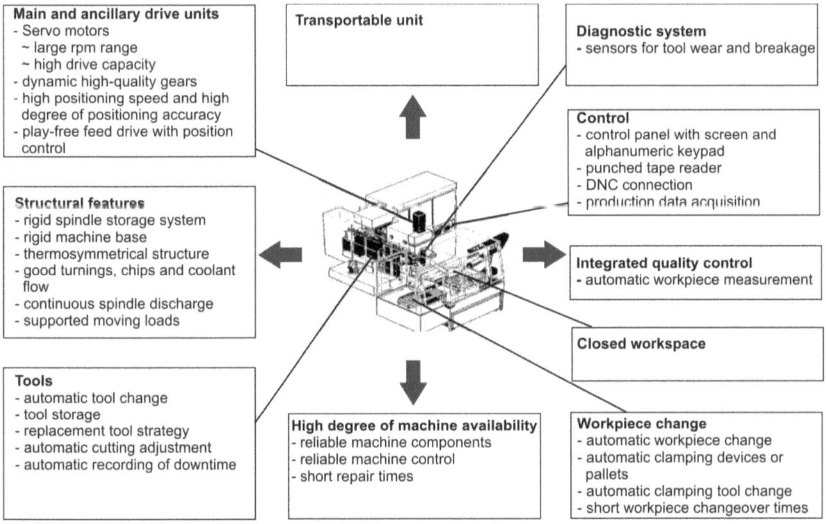

Fig. 3.29 Components and characteristics of a processing center (Eversheim & Schuh 1999, p. 11 ff)

Example 3: automation and layer model of the means of production

(1) Selection of operating resources according to the degree of automation

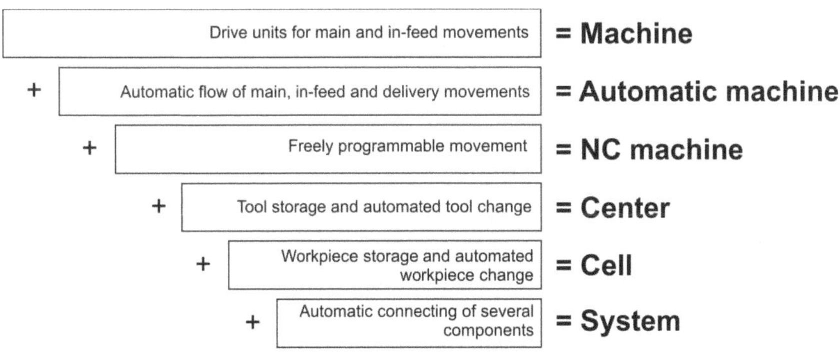

Fig. 3.30 Operating resource selection according to degree of automation (Hirsch 2000)

(2) Selection of equipment using the layer model

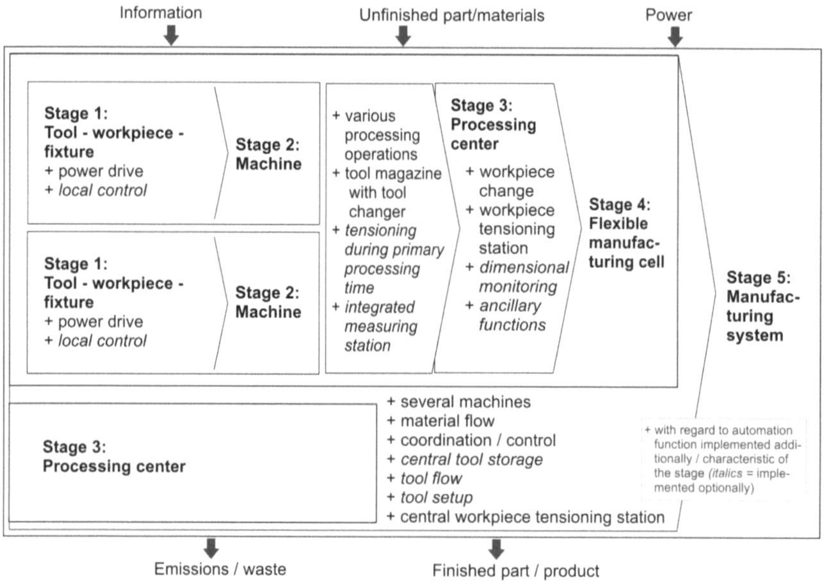

Fig. 3.31 Layer model of the means of production (Eversheim & Schuh 1999, p. 10)

b) *Specific operating resource and equipment selection for assembly (component assembly production)* by product and equipment-related components

Product/assembly unit-related components

- assembly unit dimensions (size, shape, geometry, sensitivity); complexity/type of component assembly (component assembly, system component assembly, (end) product), materials, weight, handling
- quantities (lot sizes), variety of assembly units, series size (make-to-order, series production)
- quality/accuracy (parts, component assemblies)
- product/assembly unit area:
  > 1,500 cm^2 (e.g. machinery, plant, domestic appliances, cars)
  < 1,500 - 250 cm^2 (e.g. domestic appliances, component assemblies for machinery, plant/automobile construction, electrical appliances, hydraulic components)
  < 250 cm^2 (e.g. precision tools, electrical/electronic component assemblies, pneumatic components)
- product/assembly unit envelope volume, workspace (L x W x H)
- assembly (type, number, sequence): individual, multiple, complete (standard), assembly workflow and technologies (joining, handling, transport, storage, testing and packaging technologies)

Equipment-related components

- assembly type: piecemeal, routine, combined, special
- assembly principle: installation site, single location, group, island and continuous flow assembly in the form of manual, mechanical (automated) and hybrid assembly
- main equipment: manual workstations with auxiliary equipment, single-station machine (assembly by means of robots, special purpose machinery), multi-station machine (continuous, intermittent cycle time), automatic assembly machines, assembly centers, cells and systems
- size: workstation/workspace dimensions for basic unit (stands, workpieces/assembly unit holders, drives, control, peripherals)
  Main/joining units (press, screw, rivet, welding/laser, soldering, measuring unit)
  Auxiliary and additional units (supply, storage/magazine, fixture, picking/sorting, delivery, robot, conveyor/interlink, organizational, adjusting equipment)
  Facilities for special operations (cleaning, cooling, numbering, deburring, marking) and operating facilities/equipment (operator console, standing and seated workstations)
- control/communication: type, model, upgradeability, hardware and software
- control type: CNC, SPC, network service and Internet
- type of installation: fixed/flexible, circular, linear, sequential, series and parallel interlinkage
- installation location: floor, assembly bench/workbench, conveyor, machines
- connections: power, gas, water, air, data/information
- performance capabilities: performance parameters, productivity/cycle time, availability, maintenance, servicing times)
- occupational safety: usability, flexibility, mobility, modularity, environmental compatibility (emissions, operating materials)
- profitability: acquisition price, operating costs, costs per assembly unit/component assembly, cost/benefit ratio, amortization
- delivery times (contracts), training, service

Below are four examples showing the equipment selection for the assembly workstation (Fig. 3.32), the assembly cell (Fig. 3.33), the components of an assembly system (Fig. 3.34) and mechanical assembly facilities (Fig. 3.35).

Example 1: components of a manual assembly station

Manual assembly stations consist of modules and equipment as shown in fig. 3.32.

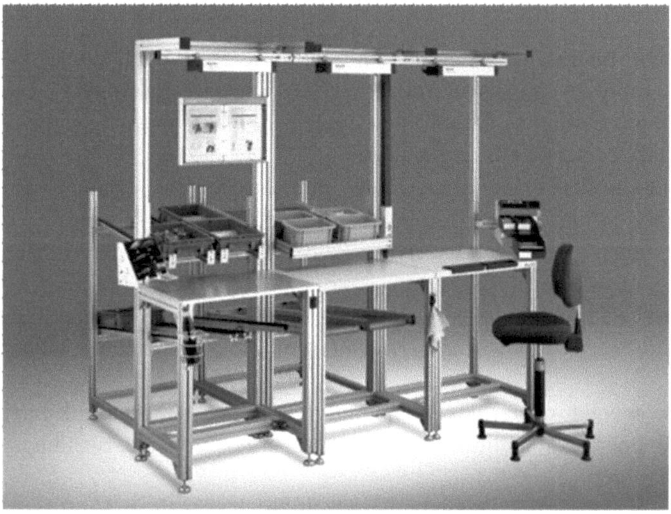

Fig. 3.32 Example of an assembly station (Lotter & Wiendahl 2006, p. 137)

Example 2: components of an assembly cell

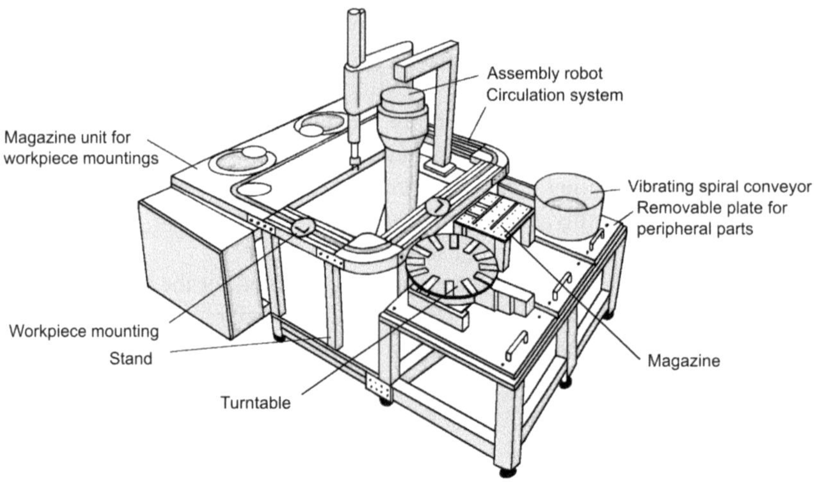

Fig. 3.33 Example of an assembly cell (Lotter & Wiendahl 2006, p. 224)

Example 3: components of an assembly system

Equipment components \ Assembly technique	Assembly cell	Assembly line			Automatic assembly machine	
	Cell with assembly robots	Manual assembly line	Assembly line with robots	Hybrid assembly line with robots	Transfer machine without robots	Transfer machine with robots
SCARA robots	required	not required	required	required	not required	required to some extent
Vertical articulated arm	required	not required	required	required	not required	required
Linear gantry robots	required to some extent	not required	required to some extent	not required	not required	required to some extent
Area gantry robots	required to some extent	not required	required to some extent	required to some extent	not required	not required
Motion modules	required to some extent	required to some extent	required	required	required	required
Interlinking means	not required	required	required	required	required to some extent	required to some extent
Joining module	required to some extent	not required	required	required	required	required
Joining stations	required	required to some extent	required to some extent	required	not required	not required
Automatic joining machines	required	required to some extent	required to some extent	required	not required	not required
Workpiece mounting	not required	required to some extent	required	required	required to some extent	required to some extent
Handling modules	required	required to some extent	required	required	required	required
Single purpose gripper	not required	not required	required	required	required to some extent	required
Multi-purpose gripper	required to some extent	not required	required	required	not required	required to some extent
Gripper changing system	required	not required	required to some extent	required	not required	not required

required — required to some extent — not required

Fig. 3.34 Typical equipment components for assembly systems (Lotter, Wiendahl, 2006, p. 223)

Example 4: mechanical assembly facilities

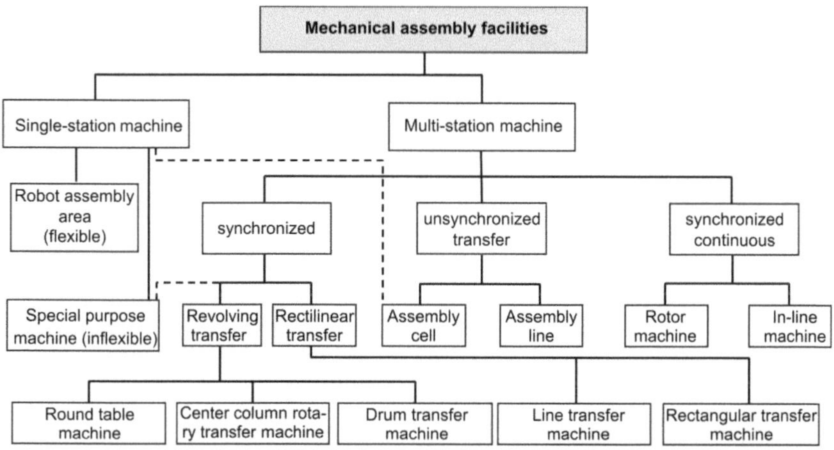

Fig. 3.35 Classification of mechanical assembly facilities (Lotter & Wiendahl 2006, p. 225)

c) *Specific operating resource and equipment selection for transport, handling and storage (logistics)* according to product and equipment-related components

Product-related components

- object type (goods for transport/storage; transport/storage securing device; empties/empty unit)
- quantities (lot sizes), time and quantity-based occurrence, variety
- quality
  Shape characteristics (design (geometry, dimensions, surface configuration)
  Aggregate state (the state in which goods exist in the logistics process) – gaseous, liquid, plastic/highly viscous, solid (bulk goods), solid (packaged goods); shape behavior (in fixed aggregate state) – form stability, flow behavior, angle of repose
  Mass characteristics (single mass, density, center of gravity)
  Material characteristics
  Damage sensitivity
  Damage behavior (mechanical load (resistance to breakage, shocks, pressure, surface damage)
  Climatological stress resistance (resistance to temperature, moisture, internal deterioration, dust, dirt)
  Deterioration behavior (explosiveness, flammability, combustibility, toxicity, corrosiveness, nauseating, infectious, afflicted by pests)
- envelope volume, density, dimensions (l, w, h), bearing surface, center of gravity, particle size, bulk density, angle of repose
- operations (transport, handling, storage, sorting, identification, packaging etc.)

Equipment-related components

- mode of operation (continuous/synchronized, intermittent/synchronized, intermittent)
- main equipment (movement and suitability of the goods for handling (rollable - with what, liftable – with what), up-take (active, passive), movement (horizontal, vertical), storage (static, moving), transfer (active, passive), accessibility to source and destination (from above, from above and the side, drivable)
- size: working area, lifting capacity (kg or t), support
- control/communication (type, model, upgradeability)
- mobility (mobile, stationary, transferrable, guided, portable)
- connections/connection points/consumption: power, compressed air, data etc.
- performance capabilities/parameters: productivity, availability (set-up, utilization, maintenance, waiting times) capacity, throughput
- occupational safety/usability (ergonomics), flexibility/adaptability, mobility, reconfigurability/modularity, environmental compatibility (energy-efficiency)
- profitability, acquisition price, operating costs, asset liquidation proceeds, costs per part, cost/benefit ratio, amortization
- delivery times (contracts, training, service)

Examples for logistics equipment including THS technology can be found in chapter 4.

Evaluation of equipment: there are various methods of variant and profitability calculation for the economic evaluation of equipment. A preferred method of choosing between variants is calculation on the basis of the cost breakdown according to the machine's hourly rate (example).

Example: selection of a piece of equipment by calculating the machine's hourly rate

Given: initial data for 2 machines
Sought: machines with low operating costs on the basis of hourly rate

As part of the determination of functions, two possible machines have been selected for executing a manufacturing process. The aim is to select the most cost-efficient machine using the machine hourly rate calculation. The following data on the individual machines was determined taking the N_S (number of shifts) into account:

Table 3.12 Initial data for selected machines

Machine		Universal lathe UDM 180	Universal lathe UDM 200
Procurement price [T€]		17	19
Procurement price incl. installation [T€]		18	20
Resource utilization time T_{SM} [h/a]		5000	3500
Machine service life [a]	N_S=1	10	10 - 8
	N_S=2	8	
Target operating time of the machine [h/a]	N_S=1	1900	1900
	N_S=2	3600	3600
Target operating time of the operator [h/a]	N_S=1	1850	1850
	N_S=2	1780	1780
Interest rate [%]		9	9
Floor space requirement [m^2]		6.8	7.2
Imputed rental price [€/m^2 • a]		30	30
Electrical power [kW]		6	8
Price for electricity [€/kWh]		0.10	0.10
Cost rate for maintenance [%/a]	N_S=1	8	8
	N_S=2	14	14
Hourly wage rate (€/h)	N_S=1	15	15
	N_S=2	20	20

Table 3.13 Cost breakdown for the machine hourly rate according to REFA

Imputed depreciation costs [€/h]	$\frac{Procurement\ price[€]}{Service\ life[a]} \bullet \frac{1}{Operating\ time[h/a]}$	(3.7)

Imputed interest costs [€/h]	$\frac{Procurement\ price[€]}{2} \bullet \frac{Interest\ rate[\%/a]}{100} \bullet \frac{1}{Operating\ time[h/a]}$	(3.8)
Floor space costs [€/h]	$Floorspace\ requirement[m^2] \bullet imp.Rental\ price\left[\frac{€}{m^2 \bullet a}\right] \bullet \frac{1}{Operating\ time[h/a]}$	(3.9)
Power costs [€/h]	$Power\ requirement\ [kW] \bullet Power\ costs\ [€/kWh]$	(3.10)
Maintenance costs [€/h]	$Procurement\ price\ [€] \bullet \frac{Maint.Costs\ [\%/a]}{100} \bullet \frac{1}{Operating\ time\ [h/a]}$	(3.11)

Machine hourly rate [€/h] = Total of individual costs [€/h]

Calculation for UDM 180 (N_s=1):	
1. Imputed depreciation costs = 17 T€/10 a • 1/1900 h/a	= 0.89 €/h
2. Imputed interest costs = 17 T€/2 • 9 %/a/100 • 1/1900 h/a	= 0.40 €/h
3. Floor space costs = 6.8 m^2 • 30 €/m^{2}a • 1/1900 h/a	= 0.11 €/h
4. Power costs = 6 kW • 0.10 €/kWh	= 0.60 €/h
5. Maintenance costs = 17 T€ • 8 %/a/100 • 1/1900 h/a	= 0.72 €/h
Machine hourly rate UDM 180	*= 2.72 €/h*

Calculation for UDM 200 (N_s=1):	
1. Imputed depreciation costs = 19 T€/10 a • 1/1900 h/a	= 1.00 €/h
2. Imputed interest costs = 19 T€/2 • 9 %/a/100 • 1/1900 h/a	= 0.45 €/h
3. Floor space costs = 7.2 m^2 • 30 €/m^{2}a • 1/1900 h/a	= 0.11 €/h
4. Power costs = 8 kW • 0.10 €/kWh	= 0.80 €/h
5. Maintenance costs = 19 T€ • 8 %/a/100 • 1/1900 h/a	= 0.80 €/h
Machine hourly rate UDM 200	*= 3.16 €/h*

According to the calculation the more cost-effective machine is the UDM 180 universal lathe, which will be selected for use (variant comparison), if there is no expectation that variability or larger workpieces will also be needed.

Note: when determining the production function with equipment (machinery, plants and facilities) and personnel, adaptability should be considered and planned for in advance. Change enablers are:
universality with reference to product, process, (technology), structural change and human resource development
mobility with respect to spatial relocation of people and goods (objects)
reconfigurability and scalability with respect to technical and structural changeability in accordance with the "plug and produce" principle
Modularity - efficient interchangeability of module and building block units, including interfaces
compatibility with regard to minimal installation and adaptation effort for materials, information, power and workflows, including interfaces

life cycle costs from the product idea, the acquisition costs (purchase price, installation and commissioning costs), operating, maintenance and environmental costs, taking resource and energy flow efficiency into account.

3.2.3 Dimensioning

Dimensioning is the quantitative determination of capacities/resources such as the number/measurements of all of the equipment, workforce, floor space, rooms/buildings and the calculation of costs (cf. fig. 3.36).

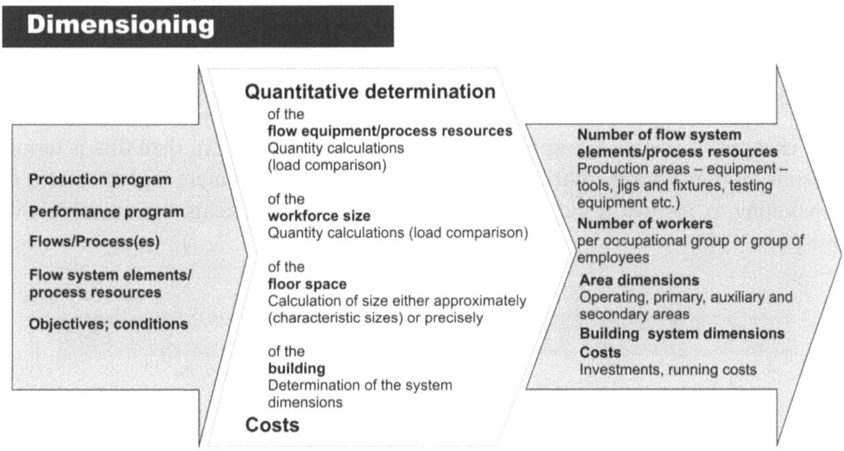

Fig. 3.36 The dimensioning planning step

Dimensioning seeks to answer the question regarding

how many functional installations (pieces of equipment) and workers with the required level of skills are needed?

The objective of dimensioning is to specify the future production facilities' installations/equipment in factual and commercial terms in such a way that

- decisions about investments and production costs can be made
- the necessary requests for tender, solicitation of quotations, meetings with suppliers, orders etc. can be arranged.

In the planning context, dimensioning comprises the quantitative determination

- of the required equipment (objects),
- the necessary personnel,
- the floor space needed and

- the costs

for the future production facilities / plants.

The results of dimensioning are summarized in tabular form, and in planning practice are mostly known as “requirement lists” for machinery, plant, equipment, floor space, storage facilities, conveyors, power, utilities and personnel etc. This means that any investment and cost decisions can be substantiated and the necessary requests for tender, solicitation of quotations, meetings, orders etc. can be justified and initiated.

The basic calculation method for dimensioning when planning production facilities is the balance approach for a specified timeframe (e.g. shift, day, week, month or year). It assumes that the workload capability (capacity) to be installed should be the same or greater than the anticipated load:

load capacity ≥ load
capacity of an item of equipment • quantity ≥ load

If changes to the load over time are taken into consideration, then this is termed dynamic dimensioning, otherwise, it is known as static dimensioning. Static dimensioning is always a simplification, as ultimately all loads are variable over time (fig. 3.37).

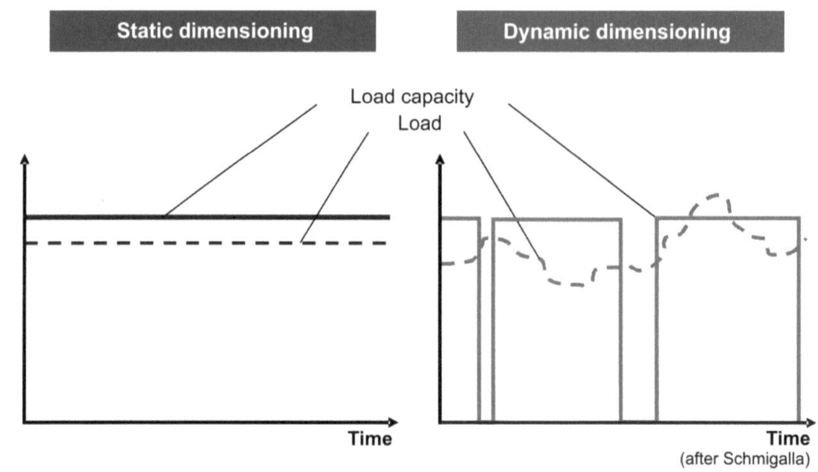

Fig. 3.37 Static and dynamic dimensioning

3.2.3.1 Static dimensioning of equipment and workforce requirements

a) Equipment

In this way, in general terms, the calculated number of items of equipment N^*_E is determined as a quotient of the required performance (capacity, load) P_E and the available effective, installed performance (capacity, load capability) of the equipment or production area P_{Ev}. The required performance can also be calculated in terms of equipment utilization time. This is derived from the product processing time per item of equipment.

$$N^*_E = \frac{P_E}{P_{Ev}} \tag{3.12}$$

Here N^*_E must be rounded to a whole number N_E, whereby the following applies:

$$N_E \geq N^*_E \; (/1.1 \; for \; N_S = 1;2) \tag{3.13}$$

In the case of single and double shift working time patterns, a maximum 10% overloading of the equipment is permitted (operational increase in the number of shifts during peak periods).

The degree of utilization of the equipment in terms of time η_E is used as a measurement of the quality of the dimensioning, and is calculated as follows:

$$\eta_E = \frac{N^*_E}{N_E} \tag{3.14}$$

P_E and $P_{E\,v}$ can be calculated on the basis of various reference values:

- time (h/RefP) → use in production engineering,
- mass (t/h),
- quantities (units/h).

A common load is the amount of effort in terms of time that must be managed during a particular time segment. The amount of machinery required can be calculated from this.

$$Amount \; of \; equipment \geq \frac{Total \; production \; time \; outlay \, / \, time \; segment}{Capacity \; in \; terms \; of \; time \; of \; a \; machine \, / \, time \; segment}$$

The basis for calculating the required capacity in terms of time is the required effort in terms of time; that is, the operating resource requirement that is needed to carry out the tasks. It is made up of the standard amount of time required for the task and the amount of additional time required.

Figures 3.38 and 3.39 show the breakdown of time according to REFA (1991).

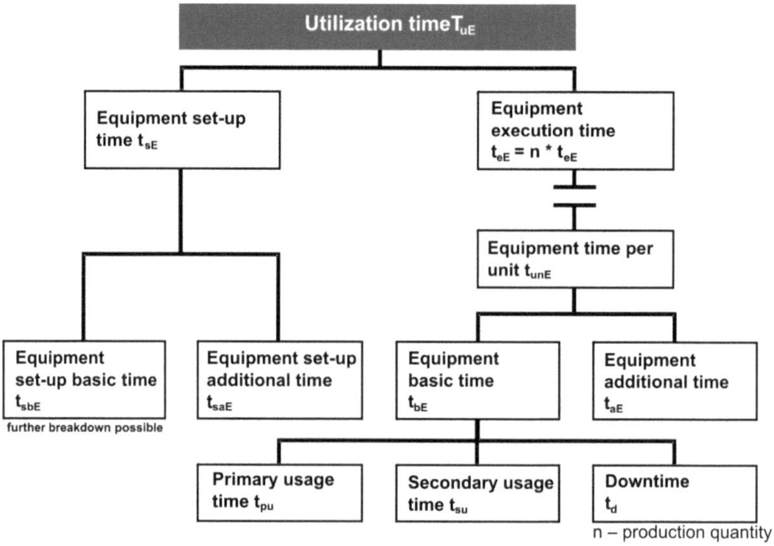

Fig. 3.38 Utilization time breakdown

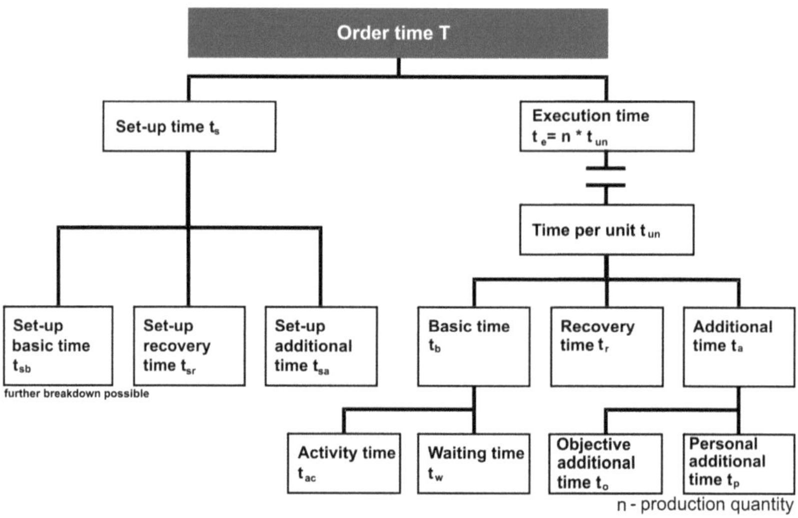

Fig. 3.39 Order time breakdown

The calculation formulae for determining static equipment (see table 3.14) and workforce requirements (see table 3.15) are derived according to this breakdown.

Table 3.14 Calculation formulae for the static dimensioning of equipment

No.	Variable	Formula		Notes
1	Utilization time for product group i with equipment group j where $N_{MMO} = 1$ T_{uEij} (h/RefP)	$T_{uEij} = n_j * t_{unEij} + N_{Li} * t_{sEij}$ $N_{Li} \geq N^*_{Li} = n_i/n_{Li}$	(3.15) (3.16)	n_j = No. of units per product group (units/a) n_{Li} = Product group lot size (units/lot) N^*_{Li} = Calculated number of lots per product group N_{Li} = Number of lots per product group t_{unEij} = Equipment time per quantity unit (unit) of the product group (min/unit) where $N_{MMO} = 1$ t_{sEij} = Equipment set-up time (min/lot)
2	Required equipment utilization time per product group i with equipment group j where $N_{MMO} > 1$ T_{uEij} (h/RefP)	$T_{uEij} = \frac{f_{lj}}{f_{pj}} \left(n_i \bullet t_{unEij} + N_{Li} \bullet t_{sEij} \right)$	(3.17)	Equipment utilization time taking following factors into account: f_l loss factor includes e.g. organizational downtimes for MMO, e.g. $f_l = 1.14$ where $N_{MMO} = 2;3$ f_p Productivity factor
3	Required equipment utilization time per equipment group T_{uEij} (h/RefP)	$T_{uEj} = \sum_i T_{uEij}$	(3.18)	
4	Calculated amount (number) of equipment per equipment group N^*_{Ej}	$N_{Ej} = T_{uEj}/T_{Evj}$	(3.19)	T_{Evj} target operating time of a piece of equipment in this equipment group (available working time fund) (h/RefP)
5	Amount (number) of equipment to be employed per equipment group N_{Ej}	$N_{Ej} \geq N^*_{Ej}$	(3.20)	N_{Ej} is derived from N^*_{Ej} rounded up to a whole number
6	Degree of utilization in terms of time of the equipment n_{Ej}	$n_{Ej} = N^*_{Ej}/ n_{Ej}$ $n_{Ej} = \frac{T_{uEj}}{Z_{Ej} \bullet T_{Evj}}$	(3.21) (3.22)	

b) Workforce

The number of workers required (N_{WF}) is determined analogously to the amount of equipment. The following applies:

$$N_{WF}^* = \frac{\text{Required performance } (P_{WF})}{\text{Available performance } (P_{WFV})} \tag{3.23}$$

Here the available performance is determined by the nominal and actual operating time.

The *nominal* operating time of the workforce is calculated from the workdays (D_w) and hours per working day (h_{WD}) based on the year.

$$D_{WF}\ nom = D_W \bullet h_{WD}\ (D_{W\_}\ workdays\ per\ year) \tag{3.24}$$

$$D_W = D_C\ [D_{Sat} + D_{Sun} + D_H + D_V - D_{STOP}] \tag{3.25}$$

Stoppage days D_{STOP} include extra vacation, reduced working hours, company holidays and pregnancy.

The available operating time of a worker ($D_{WF,V}$) takes the production stoppage days $D_{STOP,P}$ into account

$$D_{WF,V} = D_W\ h_{WD} - D_{STOP} \tag{3.26}$$

Stoppage days = D_{STOP} in hours:

- times for paid breaks
- vacation overrun
- average sick time
- time for civic obligations (e.g. military service, political representation, sports, fire department)
- deployment for operational reasons (e.g. natural disaster etc.)
- special responsibilities
- reduced working time (e.g. by the hour)
- shut-down (e.g. Christmas, New Year's), etc.

The number of workers required per equipment group j and occupational group K (skills) is calculated using the following formulae.

Table 3.15 Calculation formulae for the static dimensioning of workers

No.	Variable	Formula		Notes
7	Calculated number of workers per equipment group j and occupational group k N^*_{WFkj}	$N^*_{WFkj} = T_{uEj}/(T_{WFvkj} \cdot N_{MMOkj})$	(3.27)	D_{WFvj} Target operating time of a worker [h/RefP] N_{MMOkj} Amount (number) of equipment, that is operated by a worker from an occupational group
8	Calculated number of workers per occupational group k N^*_{WFkj}	$N^*_{WFk} \sum_{i} N^*_{WFkj}$	(3.28)	For all equipment groups that are operated by the same occupational group
9	Number of workers to be utilized per occupational group k N_{WFkj}	$N_{WFk} \geq N^*_{WFk}$	(3.29)	N_{WFk} is derived from N^*_{WFk} rounded up to a whole number
10	Degree of utilization in terms of time of the workers per occupational group k n_{WFk}	$n_{WFk} = N^*_{WFk} / N_{WFk}$ $n_{WFk} = \dfrac{T_{uEj}}{N_{WFk} \bullet D_{WFvk} \bullet N_{MMOkj}}$	(3.30)	

When carrying out a detailed determination of the number of workers N_{WF}, a distinction should be made between:

- *multiple work* – several persons operate a single piece of equipment (e.g. press). The multiple station coefficient c_{MST} is calculated using the following equation:

$$c_{MST} = \frac{\sum N_E}{1 \cdot N_{E1} + 2 \cdot N_{E2} + \ldots + n \cdot N_{En}} \tag{3.31}$$

- *multiple machine operation* – one person operates several items of equipment (individual multiple machine operation) or several people operate several items of equipment together (group/team multiple machine operation). The multiple machine operation coefficient c_{MMO} is calculated using the following equation:

$$c_{MMO} = \frac{\sum N_E}{1 \cdot N_{E1} + 2 \cdot N_{E,MMO2} + \ldots + n \cdot N_{E,MMOn}} \tag{3.32}$$

Reference values exist for c_{MST} and c_{MMO}.

In the case of mechanized and automated production, technological, quality assurance and logistical functions are assigned to items of equipment. The worker carries out pick-and-place, operating, monitoring and technology preparation functions (servicing and maintenance). In the case of *individual work* the individual employee carries out his or her tasks largely independently of others.

In the case of *group work*, a group in decentralized organizational structures carries out tasks jointly and autonomously.

These are planning, control, executive/operating, monitoring and production support activities. The responsibility for times, costs and qualities lies with the group. In relation to group work, the continuous improvement process (CIP) has been introduced to the work processes.

The difference between classical and adaptive team work is shown in table 3.16.

Table 3.16 Specifics of classical and adaptive team structures (Selection) – (Hildebrand 2005, p. 167)

classic team structure	Team existence	long-term	medium-term	short-term
	Team size	fixed	variable: downsizing	variable: expansion
	Team skills	fixed	variable: upskilling (continuous / spasmodic)	variable: simplification of skills
	Task contents (qualitative)	fixed		variable
	Task contents (quantitative)	fixed	variable: slightly divergent	variable: strongly divergent
	Possible degree of variation (work tasks)	low		high
	Organizational connection of a team to a team task	fixed		variable
adaptable team structure	Team existence	long-term	medium-term	short-term
	Team size	fixed	variable: downsizing	variable: expansion
	Team skills	fixed	variable: upskilling (continuous / spasmodic)	variable: simplification of skills
	Task contents (qualitative)	fixed		variable
	Task contents (quantitative)	fixed	variable: slightly divergent	variable: strongly divergent
	Possible degree of variation (work tasks)	low		high
	Organizational connection of a team to a team task	fixed		variable

Note: simply determining the number of workers is not enough. The required work task must be compared with the workforce's skills structure. The specific requirements for individual activities and group/team capabilities (e.g. multiple-machine operation) must be identified.

Technical/technological adaptation challenges are primarily determined by the skills, proficiency and readiness of the workers. Adaptability requires continuous up-skilling.

NB: principles for the dimensioning of equipment and workforce (personnel) requirements

Dimensioning must be carried out in such a way that

Costs → minimum
N_E → minimum
N_{WF} → minimum

physical load on the workforce → minimum, i.e. reduction of the number of shifts. If there is a conflict of objectives, the cost criterion is key.

(1) N_E is to be calculated on the basis of the equipment group or production area group (P_{AG}) and also of the type of production area (P_{AT}).

A production area group is a group of technically and economically equivalent machines (production areas). For each production area group there is a characteristic available fund of time T_{Evj}, which is dependent on the operational conditions (e.g. maintenance, servicing). T_{Ev} is dependent on the working days D_W, the number of shifts N_{Sh} and the down times D_{STOP}. Down times may be technical, organizational, work-related and physical in nature.

$$T_{Ev} = D_W \cdot N_{Sh} - D_{STOP} \qquad (3.33)$$

h/a • E

The available equipment time fund T_{Ev}, should be calculated using the following formula:

$$T_{Ev,target} = \left[D_c - \left(D_{Sat} + D_{Sun} + D_H + D_{STOP}\right)\right] \cdot N_{Sh} \cdot h_{sh} \qquad (3.34)$$

h/a • E

D_c = calendar days/a
D_{Sat} = Saturdays/a
D_{Sun} = Sundays/a
D_H = public holidays/a
D_{STOP} = stoppage days/a
N_{Sh} = number of shifts

(2) N_{WF} should be determined separately, but nevertheless in conjunction with N_E.

($T_{Ev} \neq T_{WFv}$ → workers can be deployed more flexibly than equipment, equipment stoppage days do not correspond with operating personnel stoppage days in terms of scope and timing; in terms of time, the workforce can be utilized to greater capacity - $\eta_{WF} \geq 90$ %)

(3) N_{WF} should be calculated on the basis of the occupational group

(the personnel in an occupational group may operate different equipment, i.e. it is possible to plan in the simultaneous or consecutive operation of different pieces of equipment during one shift → multiple-machine operation). In this case N_{WF} is calculated as follows:

Occupational group k is able to operate the following production area groups

Example:

PAG 1 $N^*_{WF1} = 0.28$ → $N_{WF1} = 1$

PAG 2 $N^*_{WF2} = 1.14$ → $N_{WF2} = 2$

.

.

PAG n $N^*_{WFn} = 0.75$ → $N_{WFn} = 1$

$N_{WF} = 4$ (without MMO → low utilization)

$\Sigma N^*_{WF} = 2.17$ → $N_{WF} = 3$ (with MMO, 3 people are to be employed for OG k and to be allocated accordingly to the PAG → 1 for PAG 1; 2 for PAG 2

By changing T_{uE} a reduction in N_E and N_{WF} can be sought.

Example: changing of the equipment utilization times

Given: processing machines of different sizes
Sought: optimal machine utilization and minimum number of machines

Table 3.17 Adjusting equipment utilization times to improve capacity

PAT	N^*_{Ej}			N_{Ej}		
DRT 36	1.2	-0.2		2	1	1
DRT 50	1.3	+0.2	-0.5	2	2	1
DRT 80	1.3		+0.5	2	2	2
Σ	3.8			6	5	4

Arithmetically, 3.8 lathes are required → however practically, 6 must be employed. Poor capacity utilization due to unfavorable allocation of utilization times

→ Adjusting T_{uE} can lead to machinery savings; in the example shown, instead of 6 only 4 machines are required

NB if adjusting T_{uE}:

- only from machines with a smaller technical operating area to machines with a larger technical operating area
- from machines with low levels of precision to machines with higher precision
- savings in the number of workers needed by
  - changing from low to high degree of automation
  - changing from a smaller to larger number of MMO
  - reduction in the number of shifts
- a high level of workforce capacity utilization should be aimed for ($\eta_{WF} \geq 0.9$).

NB: the number of workers per equipment group in occupational groups N^*_{WFkj} (calculation variable 7, table 3.15) takes a worker's target operating time T_{WFtarg} and actual available working time (T_{WFv}) into account.

3.2.3.2 Dynamic dimensioning of equipment and workforce requirements

The load on a production plant varies over time. These time-related changes are neglected by static dimensioning, which assumes an even distribution of the required and available capacity and a synchronization between the two. Furthermore, planning should not neglect the complex product structure and temporal dependency of the production processes.

With dynamic dimensioning, the temporal relationships between facilities are taken into account. Using dynamic methods, it is possible to depict the complex processes in operation in the production facilities, and the temporal influences can be taken into account in terms of their dynamic interactions.

This means that the load on the equipment and workforce over the operating life of the production plants can be determined. The outcomes of dimensioning are then derived from this.

The example indicates the manufacturing utilization times in accordance with table 3.18.

Example: 4 production areas and utilization times

Table 3.18 Production area utilization times and indicators

Order	PA1	PA2	PA3	PA4	PT	TPT	ICP	Feed-in time	Feed-out time
PO1	1 h	2 h	2 h	-	5 h	5 h	1	0	5
PO2	2 h	-	2 h	-	4 h	6 h	0.67	1	7
PO3	1 h	2 h	-	1 h	4 h	4 h	1	3	7
PO4	1 h	-	-	1 h	2 h	4 h	0.5	4	8

In order to represent the temporal load, diagrams known as Gantt diagrams (or timeline charts and also bar charts) are used (see Fig. 3.39).
All the other variables can be derived from the Gantt diagram (Fig. 3.40).

- feed-in times of the manufacturing orders
- feed-out times of the manufacturing orders
- throughput time (TPT) or cycle time of the manufacturing orders
- the throughput time is the time from the in-feed of a product into a production plant to the time when it leaves the production plant.
- in-cyclical parallelism (ICP). *In-cyclical parallelism* indicates the ratio for an order of the sum of its equipment utilization times to its throughput time This indicator is suitable for use comparing orders with uniform processing times with respect to their throughput characteristics. For orders without wait time, the ICP = 1; in engineering companies which predominantly produce to order, the ICP = 0.1 – 0.05. (i = manufacturing order)

$$ICP_i = \frac{\sum T_{PAi}}{TPT_i} \tag{3.35}$$

- working times of the production areas (PA)
- capacity utilization of the production areas within the throughput time or working time
- number of required storage areas
- determination of the number of transports

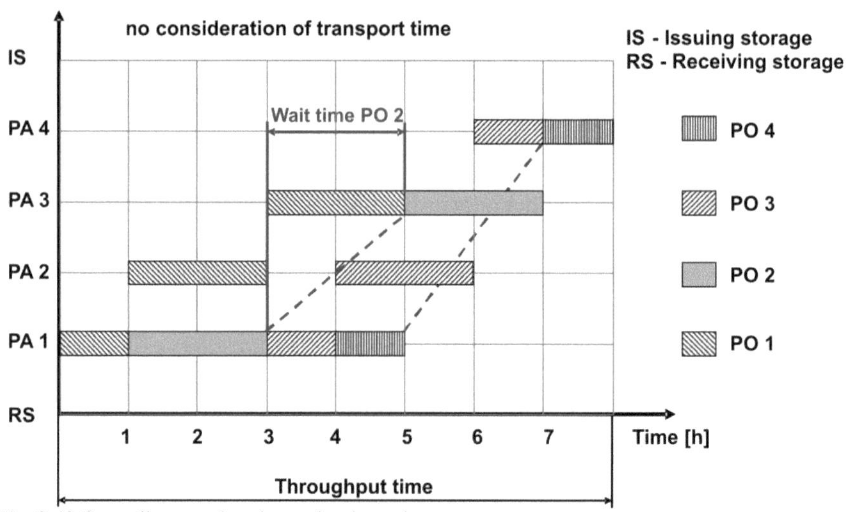

Fig. 3.40 Gantt diagram showing order throughput

Equipment and personnel, which are subject to fluctuating capacity requirements, can be organized in different ways:

- installation of available capacity over the peaks of demand,
- installation of "average" capacity,
- installation of capacity corresponding to demand.

These three ways are illustrated in fig. 3.41 and are compared with respect to investment costs, flexibility, throughput times and planning and control effort. The first and second ways can be implemented using static methods; for the third way, dynamic dimensioning methods are required. In the field of production plant planning, simulations are used to investigate dynamic interdependencies.

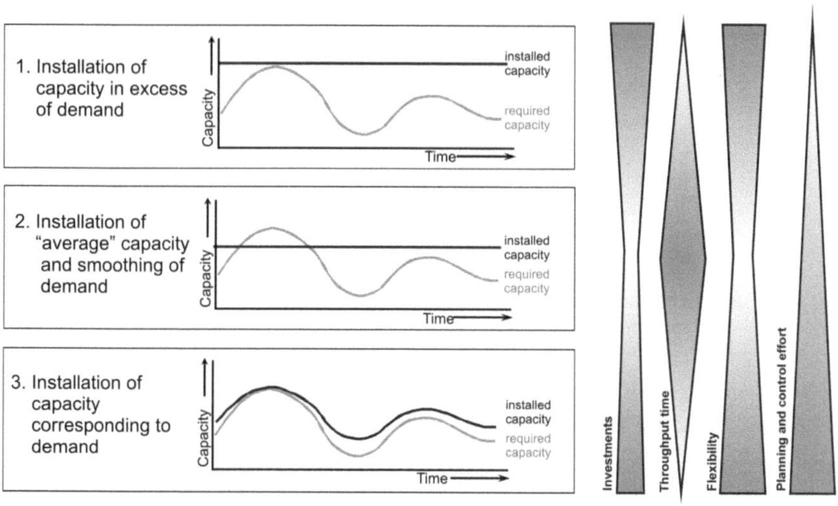

Fig. 3.41 Methods for the capacity-related organization of resources (Kobylka 2000)

Personnel and equipment requirements can be planned on this basis. Fig. 3.42 summarizes the preferred solutions for human resource development measures for different requirements and objectives.

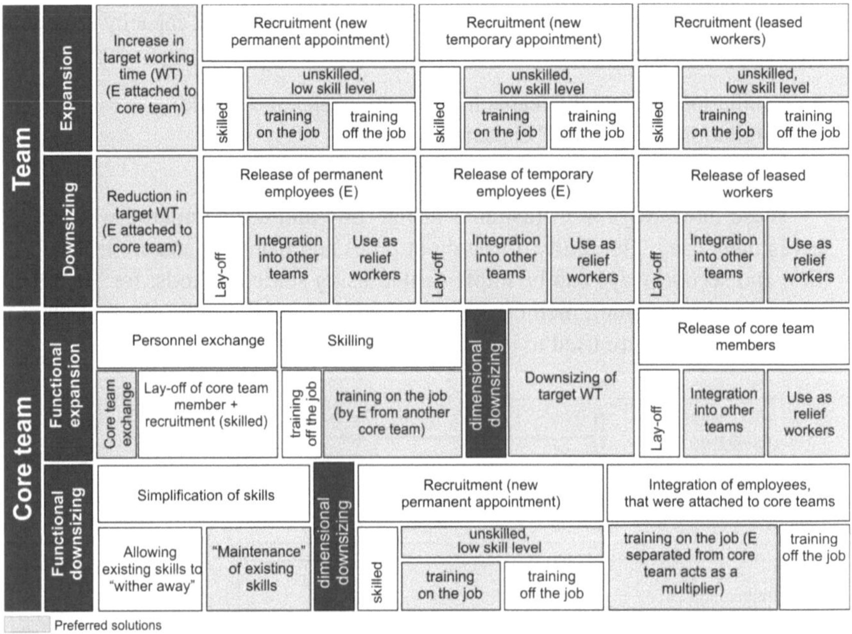

Fig. 3.42 Possible human resource development measures; identification of preferred solutions (Hildebrand 2005, p. 165)

Note: strategies for the technical and work organization-related design of change processes require human resource development concepts that correspond to the work of teams and core teams. (See fig. 3.42)

3.2.3.3 Area dimensioning

Area categorization

There are official schemes for categorizing areas, and many companies also have in-house area categories of their own The designations and their definitions are set out in standards and guidelines. A generalized breakdown of operating areas, in continuation of DIN 277 and REFA 91 can be seen in fig. 3.43.

They include the plot, building, divisional and departmental areas. For production facility planning, the divisional, departmental (primary and secondary workshop areas) and workstation areas (parts manufacture and assembly) are significant elements of the building floor area. Their determination follows the hierarchical breakdown with different levels of detail in effective production/primary areas, additional production/effective areas (with transport and handling areas), administrative, storage and other auxiliary and secondary production areas. Divisional, departmental, sectoral and workstation areas all consist of standardized primary/effective and additional/secondary areas, which are broken down into standard sub-areas.

For example, for the divisional area, a distinction is made between production, storage, administrative and secondary areas. The production area (primary, effective and additional/secondary areas) incorporates all required floor space for the accommodation of the workforce, supplies and workpieces as well as fixtures, tools and testing equipment for production and assembly. Furthermore, logistics areas for floor-based transport, parking and interim storage areas in the production area must all be taken into account as well.

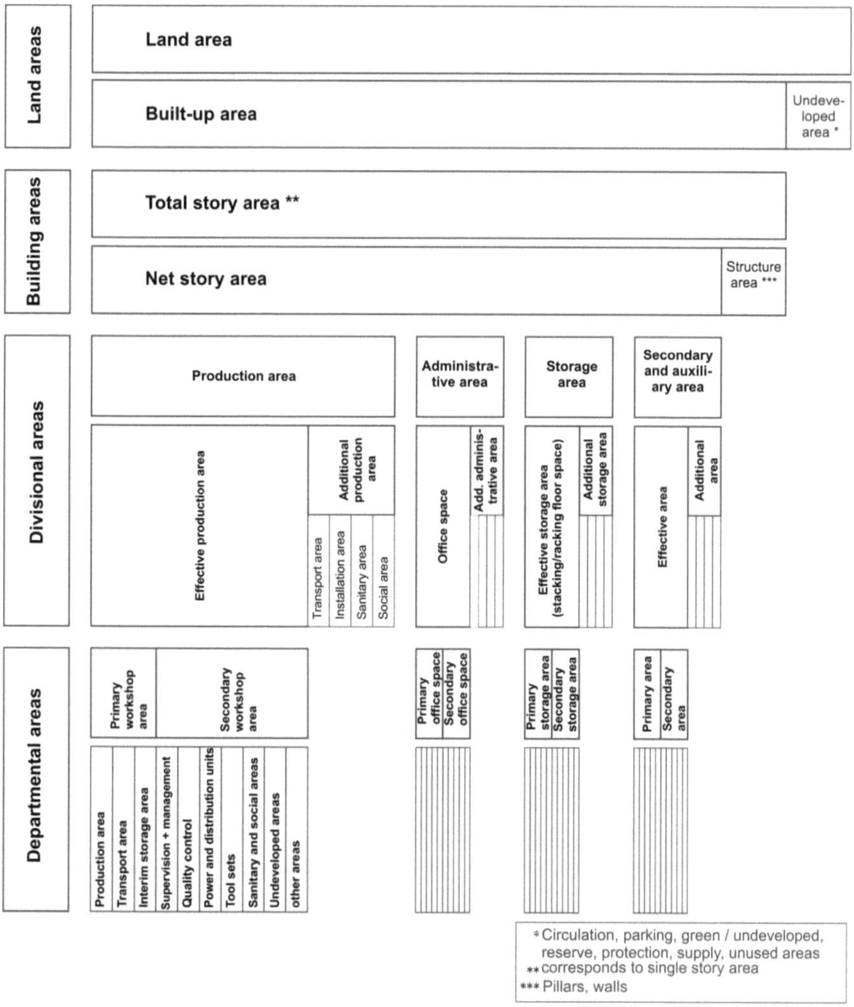

Fig. 3.43 Area categorization based on DIN 277 and REFA 1991

Storage areas are required for storage per se, for transport routes, picking and staging as well as for incoming and outgoing goods areas.

Administrative areas are used to accommodate accounts, human resources, purchasing, sales and marketing and management, for example. Some auxiliary functions, such as production scheduling and manufacturing control can also be allocated here.

Auxiliary and secondary areas fulfill the space requirements for production-related secondary areas, central building services, supply and disposal, central conduits and social and sanitary facilities.

On the one hand subdividing these types of area constitutes classification, and on the other hand very different structural conditions result (e.g. effective height, floor load) from the different types of utilization of floor space. Floor space can be determined globally (approximately) or in detail depending on the point in the planning schedule and the required level of accuracy of the data.

Note: Areas can be categorized in standard terms according to different perspectives from the workstation to the building area.

3.2.3.3.1 Global area dimensioning

Areas are dimensioned globally using key indicators in accordance with the following calculation method:

$$Area = base\ variable \bullet floor\ space\ indicator \tag{3.36}$$

Important indicators include:

Area/employee	[m^2/empl]
Area/machine	[m^2/mach]
Area/turnover	[m^2/€]
Area/production volume	[m^2/unit]; [m^2/t] etc.

Area indicators

- are obtained by means of floor space analyses combined with an analysis of the base variables in a certain company at a certain point in time,
- are strictly speaking only valid for a certain company at a certain point in time,
- can only be transferred for use in other companies (in the same sector) and at other time periods by using correction factors that take such dynamic determining factors as productivity, vertical integration and degree of automation etc into account.

The global (approximate) area can be determined using general floor space indicators, sector indicators and top-down / bottom-up calculations.

a) Global floor space indicators for selected examples (cf. table 3.19)

Table 3.19 General floor space indicators (Kettner 1984, p. 65)

Area	Indicators
General development planning	
Land area/employee	60 - 90 [m^2/empl.] [m^2/Besch.]
Undeveloped area/land area	30 - 40%
Transport area/land area	20%
Parking space/car incl. entrance and exit	25 [m^2]
Parking space/LGV	35 [m^2]
Reserve area/land area	20 - 30%
Building area/employee	44 [m^2/empl.] [m^2/Besch.]
Story area/employee	32 [m^2/empl.] [m^2/Besch.]
Gross floor space in production buildings - Low-rise building - Multi-story building	 95% of the story area 75% of the story area
Gross floor space in storage buildings - Hall structure - Multi-story building	 89% of the story area 84% of the story area
Effective area	75% of the gross floor space
Secondary area	15% of the gross floor space
Functional area	5% of the gross floor space
Social area	5% of the gross floor space
Production	
Production floor space	50% of the story area
Production floor space/employee	35 [m^2/empl.]
Machine work stations - Turret lathe - Center lathe, light - Center lathe, medium - Center lathe, heavy	 6 [m^2/machine] [m^2/Maschine] 6 [m^2/machine] [m^2/Maschine] 12 [m^2/machine] [m^2/Maschine] 15 [m^2/machine] [m^2/Maschine]
Transport area	30% of the machine working floor space
Storage	
Area for goods acceptance/shipping	3% of the effective area

Storage area	22% of the effective area
Auxiliary area	
Area for auxiliary systems	6 - 15 % of the story area
Area for test bay	3% of the effective area
Testing area/employee	5 -8 [m²/Besch.]
Area for heating, ventilation, air conditioning systems	6 - 7 % of the story area
Area/workstation	1.3 – 1.5 [m²/Besch.]
Area for training workshop/apprentice	15 [m²/Besch.]
Administration	
Administrative area	15% of the effective area
Office space/employee - in open plan offices - in individual offices	 11.5 [m²/empl.] 12.5 [m²/empl.] [m²/Besch.]
Social area	
Social floor space	5% of the gross floor space
Kitchen floor space/diner	0.8 [m²/empl.]
Dining room floor space/diner	1 [m²/empl.]
Total canteen floor space/diner	1.5 [m²/empl.]
Sanitary room floor space/workstation	0.5 [m²/empl.]
Toilet space - Men - Women	 0.15 - 0.25 [m²/empl.] 0.20 - 0.33 [m²/empl.]

b) Sector-specific floor space indicators (table 3.20)

Table 3.20 Sector-specific floor space indicators for the overall floor space requirement (IREGIA 2004)

Sector	Indicator range [m²/empl]	Avg. value	Sector	Indicator range [m²/empl]	Avg. value
Heavy engineering/foundry			Electrical engineering		
Assembly of heavy machinery	40-80	60	Electrical machinery and transformers	20-30	25
Energy machinery	50-80	70			
Repair of heavy machinery	50-80	70	Assembly of electrical products	20-40	30
Transport equipment	50-100	70	Electrical products for	20-40	30

Other equipment and machinery	50-100	70	the consumer market		
			Cables and electrical materials	20-50	30
Forged, pressed and die cut parts	60-100	70	Equipment and devices for communications engineering	20-50	30
Steel construction	60-100	80	Other electrical products	20-50	30
Iron and steel forging	68-120	90	Other electrical items for production	20-50	30
Non-ferrous metal casting	68-120	90			
Machine tools, forging and pressing equipment	68-120	100	Switchgear	20-50	30
			Repair of consumer electrical products	30-70	50
General engineering			Precision engineering/optics		
Assembly of general machinery	20-50	30	Office machinery	7-15	10
Food industry	30-50	40	Measuring equipment and clocks	7-15	10
Repair of general machinery	25-60	40	Measuring equipment and clocks for the consumer market	7-15	10
Other machinery/appliances	30-70	50			
Machine parts, tools	30-70	50	Other precision engineering products	7-15	10
Fittings	30-70	50			
Light industry	50-80	60	Optical instruments	7-15	10
Metal working industry	40-100	70	Optical instruments for the consumer market	7-15	10
Machines and appliances for primary industry	50-110	80	Diamond tools	7-15	10
Construction industry	60-110	80	Measuring equipment and clocks for production	10-20	15
Metal goods			Metal goods		
Technical ironware	60-100	80	Equipment for industry, agriculture/trades	70-110	80
Blacksmithery, metalworking	60-100	80			
Other workshops	60-100	80	Wire and wire products	80-150	120
Repair workshops for the consumer market	70-110	80	Metal commodities for the consumer market	80-150	120

c) Top-down and bottom-up calculation using minus factors. Top down calculation: an area category of a higher order is converted to an area category of a lower order through multiplication by a minus factor (see fig. 3.44).

Gross story area	**100**					
Structure area	**8**					
Net story area	**92**	**100**				
Circulation area	12	**13**				
Effective area	80	**87**	**100**			
Secondary area	9	10	**11**			
Primary area	71	77	**89**	**100**		
Area for R&D	1	2	2	**2**		
Office space & administration area	8	8	10	**11**		
Storage area	9	10	12	**13**		
Production area	53	67	66	**74**	**100**	
Auxiliary production area	19	20	23	26	**35**	
Primary production area	34	37	43	48	**65**	**100**
Interim storage area	1	1	2	2	2	4
Area for prefabrication	18	20	22	25	34	53
Area for intermediate fabrication	3	3	4	4	6	9
Area for assembly	12	13	15	17	22	34

Fig. 3.44 Area reference values for top-down calculation (example from precision engineering/optics)

Bottom-up calculation: higher order area categories are calculated from lower order area categories through multiplication by a plus factor (plus calculation). For more on this see fig. 3.45.

Gross story area	294	191	141	125	**109**
Structure area	24	16	12	10	**9**
Net story area	270	175	129	**115**	**100**
Circulation area	35	23	17	**15**	
Effective area	235	152	**112**	**100**	
Secondary area	25	16	**12**		
Primary area	210	**136**	**100**		
Area for R&D	5	**3**			
Office space & administration area	23	**15**			
Storage area	28	**18**			
Production area	**154**	**100**			
Auxiliary production area	**54**				
Primary production area	**100**				
Interim storage area	**4**				
Area for prefabrication	**53**				
Area for intermediate fabrication	**9**				
Area for assembly	**34**				

Fig. 3.45 Area reference values for bottom-up calculation (example from precision engineering/optics)

3.2.3.3.2 Detailed area dimensioning

The main objects of this process are departmental/workshop and workstation areas (A_{WS}) for the production of parts (manufacturing area A_{PA}) and assembly (assembly area A_{AA}), including the primary and secondary areas.

a) Parts production

Detailed area dimensioning is done on the basis of the smallest area category that cannot be further subdivided (footprint of a piece of equipment) using the following calculation method:

$$Area = base\ variable \bullet plus\ factor \tag{3.37}$$

The (projected) footprint of a piece of equipment is the area of the projected perimeter in all of its final working positions. It should not be equated with the standing base.

Other areas can only be obtained through bottom-up calculation, i.e. the workstation area (machine/assembly workstation floor space) is calculated from the machine footprint, and from this further area categories of higher orders can be calculated through multiplication by plus factors.

Plus factors

- are obtained through area analyses in certain enterprises,
- are largely independent of a certain enterprise and are therefore transferrable,
- are largely independent of dynamic influences and only need slight corrections,
- are dependent on the size of the machine footprint: small footprints require large plus factors and vice versa,
- are dependent on the type of production and the spatial structure: workshop structures require larger plus factors than storage structures,

There are both footprint-related (f_F) and workstation area-related (f_W) plus factors.

The floor space of a production area (A_{PA}) (fig. 3.46) is comprised of primary and secondary areas:

Primary area:

- footprint of the production equipment (machine footprint A_{FP})

Secondary area:

- area for repairs and assembly A_{RA}
- area for servicing and cleaning A_{SC}
- operating area (supervision and management) A_O
- staging area for jigs and fixtures, tools, testing equipment A_{FTT}
- staging area for materials and workpieces A_{WP}
- area for transport, handling, storage A_{THS}

- area for turnings, chips and waste (supply and disposal) A_{SD}

The sub-areas can be transferred to the sectoral and divisional levels.

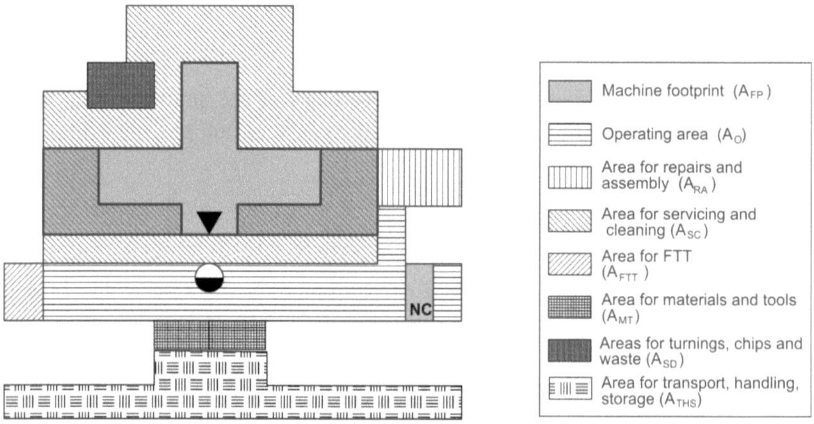

Fig. 3.46 Production area floor space and its constituent parts

Calculation of the production area floor space A_{PA}

The production area floor space A_{PA} can be calculated, depending on the required level of accuracy and subject to the available input information, using different methods, generally on the basis of the (projected) object footprint A_{FP}:

1. Floor space or plus factors

Starting from the machine or equipment footprint A_{FP}, the next floor space complex, the production area floor space A_{PA}, can be calculated through multiplication by a plus factor f_F. The prerequisite for this is that dimensioning of the machine requirements (specification of type and size as well as the calculation of quantity) has been completed.

$$A_{PA} = A_{FP} \cdot f_F \tag{3.38}$$

The floor space factor f_F includes the additions for staging, operation, servicing, plus supply and disposal in the production area. (Woithe 1977, p. 122; Rockstroh 1982, p. 54)

A more detailed calculation of the production area floor space can be achieved using plus factors for the individual sections of floor space.

The advantage of this method lies in its ease of use. However it is relatively imprecise and therefore should only be used for approximate calculations in rough planning.

2. Substitution areas

The starting point of the substitution area method is the smallest bounding box of the projected footprint of the object together with width and depth measurements. By adding to this base a rectangular substitution area is formed for the production area (Wirth 2000, p. 52) (cf. fig. 3.47).

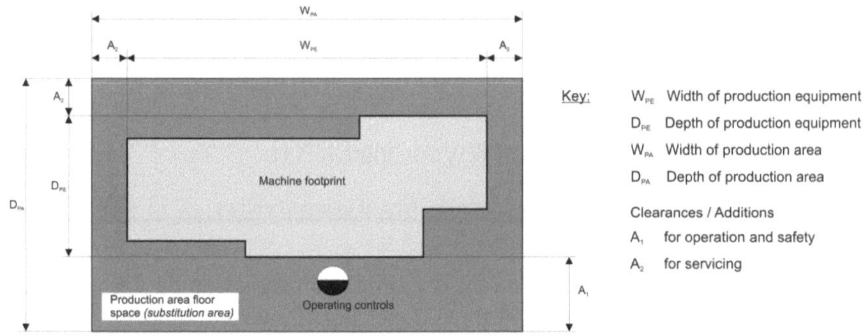

Fig. 3.47 Substitution area method (based on Kettner et al, 1984)

On the operating sides A_1 the additions = 1.0 m and on all other sides for servicing A_2 = 0.4 m. Thus the production area floor space A_{PA} can be calculated as follows:

$$A_{PA} = W_{PA} \cdot D_{PA} = (W_{PE} + 0.8) \cdot (D_{PE} + 1.4) \quad [m^2] \tag{3.39}$$

The floor space areas that are not captured by this can be covered by the staging area A_{SA}. In the process $N_{TU} = 4$ is assumed for the number of transport units.

$$A_{SA} = A_{TU} \cdot N_{TU} \tag{3.40}$$

$$A_{WS} = A_{PA} + A_{SA} \tag{3.41}$$

In general, the following applies for the evaluation of the workstation (A_{WS}) (parts manufacture and assembly) taking four transport containers into account (based on Rockstroh 1982, p. 53):

$$A_{WS} = (l_{Ob} + 2 \cdot A_s) \cdot (w_{Ob} + A_o + A_s) + 4\, l_{TUT} \cdot w_{TUT} \tag{3.42}$$

A_{AP} - Substitution area
w_{Ob} – Width of the object (of the machine)
A_s - Addition for servicing
w_{TUT} – Width of transport utilities
l_{Ob} - Length of the object (of the machine)
A_o - Addition for operation
l_{TUT} - Length of transport utilities

3. Determination of functional area

With this method, a mathematical connection is made between the individual areas, meaning that the higher order areas can often be determined quickly and with adequate precision. It is used particularly for assembly.

Taking into account functional dependencies that exist between the object footprints and workstation areas, the floor space requirement is determined by means of regression functions. (Woithe 1977, p. 125). The process is useful for quickly determining the workstation area, this being determined as a function of the:

- projected footprint of the object,
- the type of workstation and
- the type of transport utility (TUT) (Wirth 2000, p. 53)

$$A_{WS} = f(\text{WG, type of workstation, type of transport utility}) \quad (3.43)$$

4. Trial layout

The workstation floor space and all remaining sub-areas can be visualized using a workstation layout. Two and three dimensional machine models, templates (model projection) and virtual, computer-assisted, digital 2D, 3D and VR models and animations are used. By taking the measurements, the scale floor space required can be determined relatively precisely.

Evaluation: Subject to the data quality (initial data) and abstractions of the floor space determination, the use of the substitution area method for parts manufacture is optional for determining workstation areas. Fig. 3.48 shows the large field of application in the floor space determination process for the substitution area method. From this it can be derived that this method is applicable for both the approximate and detailed determination of floor space requirements. For assembly, the functional area determination method is preferable.

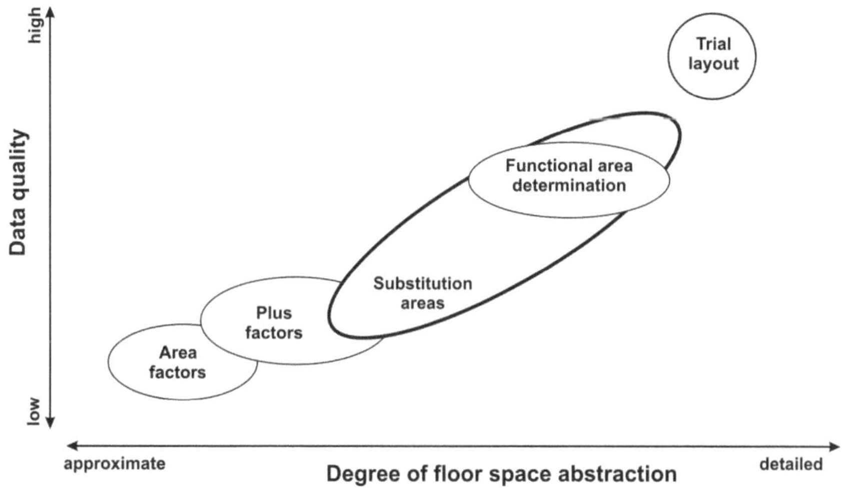

Fig. 3.48 Ranking of procedures for the detailed determination of floor space requirements (Börner 2008)

Example: calculation of the production area AP (rough calculation)

Starting from the smallest complex area building block – the production area floor space A_{PA} – the production area (main and effective areas) A_P of a workshop together with its auxiliary and secondary areas can be calculated through multiplication by a plus factor f_W. The prerequisite for this is that the production area floor space has been calculated in advance or determined in another way (e.g. via floor plans from quotations).

$$A_P = \Sigma A_{PA} \bullet f_W \tag{3.44}$$

The multiplication factor fW includes the additions for secondary areas for storage, transport, quality assurance, supply and disposal as well as control and management in the workshop (table 3.21 contains the floor space factors, Fig. 3.49 the elements of the production area.)

Table 3.21 Production area floor space-related factors f_W

Area category	min. – max.	According to Kettner	According to Rockstroh	According to TGL
Area for staging and intermediate storage	1.20 - 1.40	1.40	1.20 - 1.30	1.20
Area for transport	1.25 -1.45	1.40	1.25 - 1.30	1.45
Area for quality assurance	1.13 -1.20	-	1.10 - 1.20	1.13
Area for technical supply equipment	1.10 - 1.20	-	1.10 - 1.20	1.10
Area for production control and management	1.10 - 1.20	-	1.10 - 1.20	-
Unused residual area or additional areas	1.10 - 1.20	1.20	-	1.10
Factor f_W	1.88 - 2.38	2.00	1.55 - 1.80	1.98

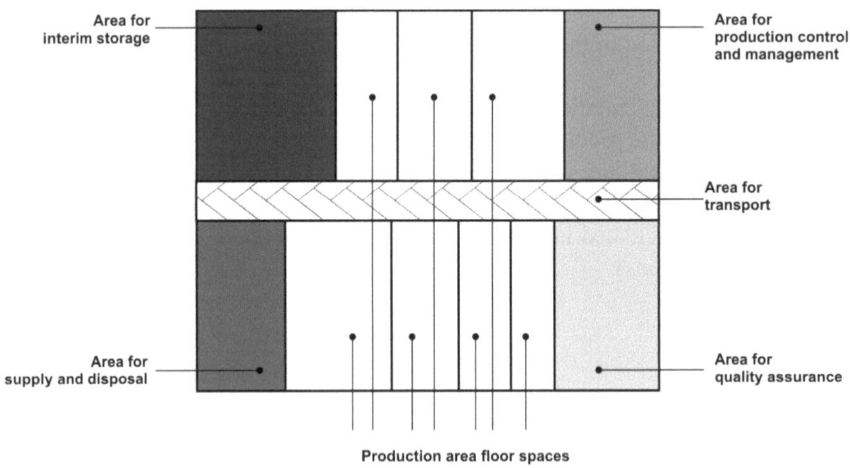

Fig. 3.49 Production area and its constituent parts

A more detailed calculation of the production area can be achieved by adding together its constituent areas (cf. fig. 3.50).

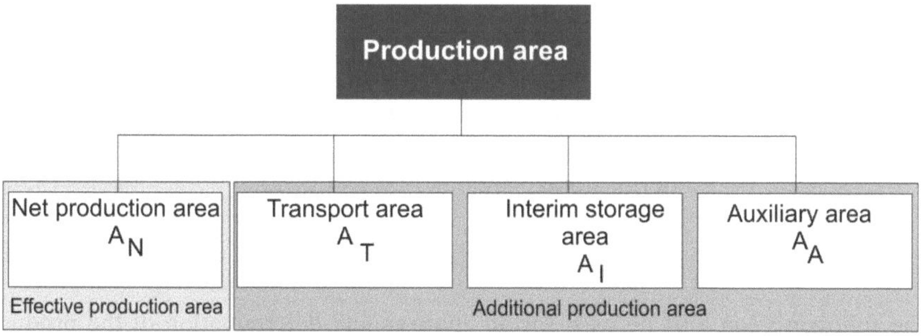

Fig. 3.50 Production area breakdown

The *production area* in parts manufacture ($A_{P,PM}$) is thus calculated from the sum of its sub-areas.

$$A_P = A_N + A_T + A_I + A_A \tag{3.45}$$

which are, in part, determined using plus factors (after Woithe):

- *Net production area* A_N is the sum of the areas that exclusively serve the execution of operations that progress production (e.g. for primary, auxiliary and secondary production) including the areas for maintenance on site.

$$A_N = \Sigma A_{PA} \bullet (1-\eta_{Oe}) + A_{un} \tag{3.46}$$

η_{Oe} *external overlap factor* (approx. 0.1- 0.2) – measurement of the overlap of the production areas without compromising function (e.g. partial overlapping of floor space sections of neighboring production areas for successive servicing of machinery)

A_{un} *undeveloped floor space* – for additional floor space requirements that cannot be planned (approx. 0.1- 0.2 ΣA_{PA}) among other things due to unfavorable layout of the structural units, structural restrictions (pillars, separating walls) and required flexibility with a view to layout or extension

- Transport, interim storage, auxiliary areas
  Transport area A_T is the sum of the labeled but not firmly delimited (in structural terms) neighboring areas for the circulation of personnel and transportation of materials.
  Interim storage area A_I is the sum of the neighboring incomplete production storage areas (staging within the production workshop).
  Auxiliary area A_A is the sum of the areas that are needed for carrying out management and control functions as well as for tool distribution and supply/disposal.
  These area sections can be roughly determined using floor space factors. The

production area floor space is used as the basis. With the plus factors (empirical values) for the machine construction division:

$$\text{transportation area: } A_T = 0.4 \bullet \Sigma A_{PA} \tag{3.47}$$

$$\text{interim storage area: } A_I = 0.4 \bullet \Sigma A_{PA} \tag{3.48}$$

$$\text{auxiliary area: } A_A = 0.2 \bullet \Sigma A_{PA} \tag{3.49}$$

The following can thus be used for an approximate determination of the production area:

$$A_P \approx 2 \bullet A_N \tag{3.50}$$

The precise floor space requirement is ultimately determined using the real layouts for the production areas (in the Design planning step).

- Overlapping of area sections

When determining the areas the possible overlaps should be taken into account using the overlap factor. The overlap factor indicates the degree to which the functional floor spaces of a production area (internal overlap factor η_{Oi}) or the floor space of several production areas (external overlap factor η_{Oe}) may overlap, without this impairing their function.

$$\text{Internal overlap factor } \eta_{Oi} = \left(1 - \frac{A_{\text{PA actual}}}{A_{\text{PA calculated}}}\right) * 100\% \tag{3.51}$$

$$\text{External overlap factor } \eta_{Oe} = \left(1 - \frac{A_{\text{P actual}}}{A_{\text{P calculated}}}\right) * 100\% \tag{3.51}$$

Guideline values for the external overlap factor η_{Oe} subject to the level of multiple-machine operation (N_{MMO}).

N_{MMO}	1	2	3
η_{Oe}	0	0.1	0.2

When determining floor space the aim should always be to achieve an overlap of area elements both within and between neighboring production areas (above all in the case of MMO) (fig. 3.51).

Approaches for predetermining floor space (parts manufacture)

1. Determination of the method for calculating the floor space
2. Calculation of the individual production areas A_{PA} on the basis of the substitution area determination method
3. Taking into account of area overlap
4. Determination of the indirectly attributable areas (A_T, A_I, A_A) with associated overlap
5. Determination of the production floor space A_P for parts manufacture

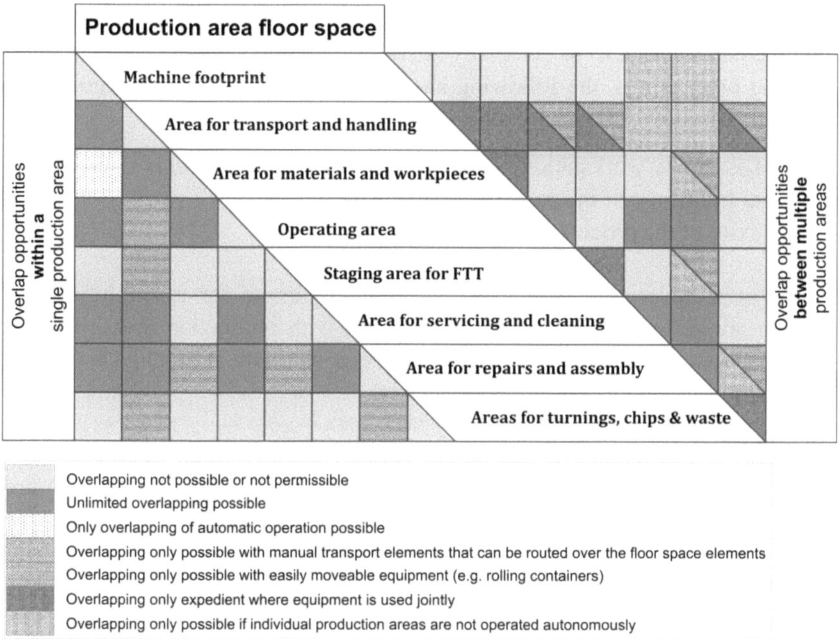

Fig. 3.51 Possible functional area overlaps within a production area and between multiple production areas.

As the result of the pre-determination of floor space, an area layout plan must be created.

b) Assembly (component assemblies, finished products)

The assembly production area ($A_{P,As}$) is made up of the sum of the assembly workstation areas (A_{AA}) that constitute the net area (A_N) plus the transport area (A_T), the interim storage area (A_I) and the auxiliary area (A_A).

$$A_{P,As} = A_N + A_T + A_Z + A_H \tag{3.53}$$

A_N	=	Net area (equals the sum of all of the assembly workstations)
A_T	=	Transport area (~ $0.15 \cdot A_N$)
A_I	=	Interim storage area (~ $0.1 \cdot A_N$)
A_A	=	Auxiliary area (control, tool distribution)

The floor space for assembly workstations is determined in accordance with the "functional" area determination. In the process the assembly workstation floor space is determined both by the assembly unit (assembly product) and the assembly utilities (e.g. floor, tables). An assembly unit can grow internally (envelope volume remains constant) or externally (envelope volume increases in size).

Among other things, the following have a particularly strong influence on the dimensions of assembly workstations:

- type of assembly workstation, manual workstation, machine workstation
- size and frequency of the parts being processed
- complexity of the process.

Note: in accordance with fig. 3.52, constituent parts of the assembly workstation are defined differently in comparison with production areas. There are commonalities in the working area (operating area), servicing/cleaning, staging/storage and transport/handling areas.

Due to the individuality of the assembly workstations, when calculating floor space it is possible to distinguish between six different types of assembly area $A_{AA(1\text{-}6)}$ (fig. 3.52) (Rockstroh 1982, p. 56, TGL 13384, BG 1 and 2):

- type 1: assembly on the ground (floor space of the assembly unit), $A_{AA(1)}$
- type 2: work at assembly benches (area = assembly bench), $A_{AA(2)}$
- type 3: work in testing areas (area of assembly unit and/or testing facility), $A_{AA(3)}$
- type 4: work on conveying equipment (area of the conveying equipment), $A_{AA(4)}$
- type 5: work at workbenches (area of the workbenches), $A_{AA(5)}$
- type 6: workstations on machines and plant (area of the machinery). Machine assembly workstations are determined according to the substitution area method as is the case with the manufacture of parts, $A_{AA(6)}$

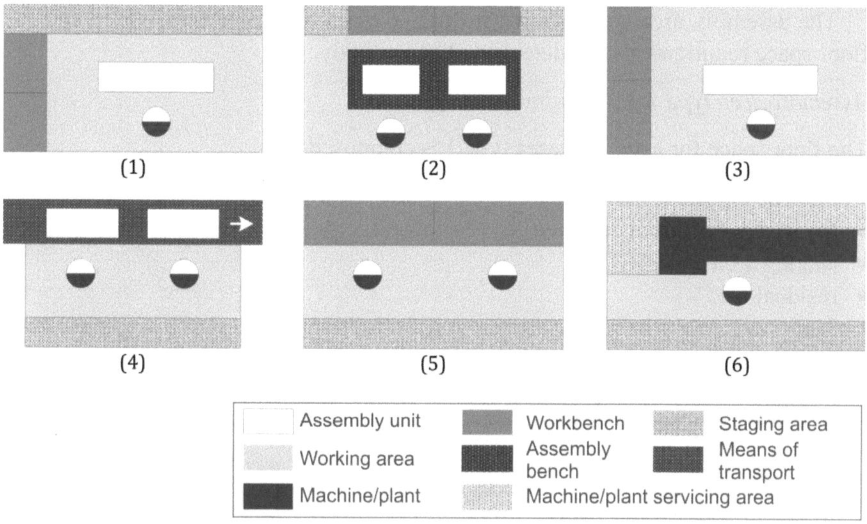

Fig. 3.52 Types of assembly area ($A_{AA(1\text{-}6)}$)

The individual assembly area types are distinguished according to the features detailed in table 3.22.

Table 3.22 Classification features of assembly area types

Type	Location of the assembly unit	Work-benches	Organizational form
1	Directly on the ground, in assembly pits, on assembly stands or on conveyor equipment with access to the assembly unit from all sides.	Yes*	Optional
2	On assembly benches	Yes*	Optional
3	Directly on the floor, on test benches or on run-in stands	No**	Assembly units and workers stationary or assembly units mobile and workers stationary
4	On conveyor equipment with single and double-sided access	No	Assembly unit mobile and workers stationary
5	On work benches	Yes	Assembly units and workers stationary
6	-	No	-

* for fitting and joining work, ** only for storage of testing equipment

The assembly area types shown in fig. 3.52 can be further subdivided and the floor space requirements are determined accordingly.

Assembly area type 1 – Assembly on the ground

The floor space for assembly area type 1 is composed of

- the assembly unit floor space
- the workbench floor space
- staging area
- residual area

The floor space requirement is thus determined using the following formula:

$$A_{AA(1)} = A_{AU} + A_{WB} + A_{SA} + A_R \tag{3.54}$$

$A_{AA(1)}$ = floor space for assembly area type (1)
A_{AU} = assembly unit floor space
A_{WB} = workbench floor space
A_{SA} = staging area floor space
A_R = residual floor space

The dimensions of the working area are dependent on its design. In addition to the primary working side, up to three secondary operating sides are possible. Depending on the number and layout, corresponding length and width additions are required. Consequently, assembly area type (1) can be subdivided into six further assembly area types (fig. 3.53).

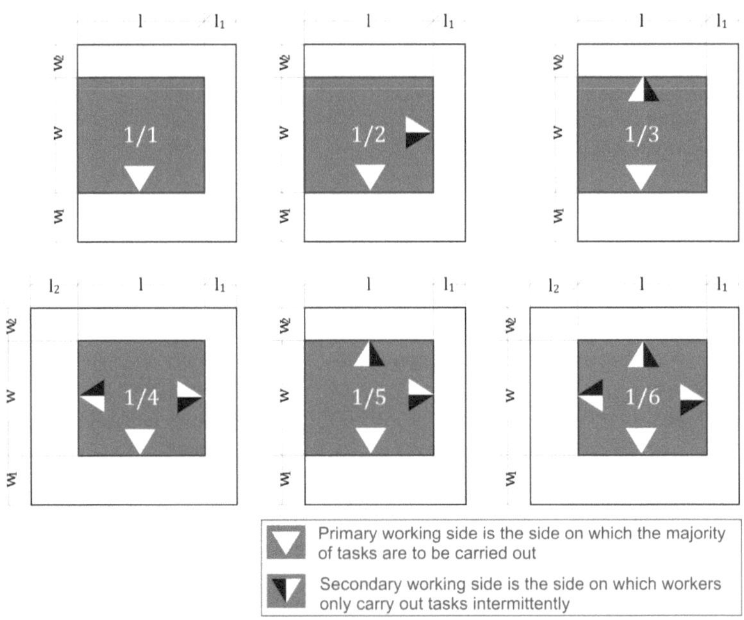

Fig. 3.53 Assembly area 1 sub-types

When determining the floor space requirement for assembly area type 1, the layout of several similar assembly areas in one group (single row, two rows or three to n rows) is taken into account. A corresponding plus factor f_l is determined together with the number of workers, each of whom works on one assembly unit. The plus factor value can be obtained from tables; it increases with the number of workers, but decreases with an increase in the number of rows. The usual range of values falls between 0.5 and 1.0. The floor space of the assembly unit A_{AU} is calculated as follows:

$$A_{AU} = (l_{AU} + l_1 + l_2) \bullet (w_{AU} + w_1 + w_2) \bullet f_l + N_{AU} \qquad (3.55)$$

l_{AU} = length of the assembly unit
$l_1; l_2; b_1; b_2$ = working area according to workstation type and working height
N_{AU} = number of assembly units
w_{AU} = width of the assembly unit
f_l = plus factor according to number of workers and rows

If the assembly units need to be assembled on assembly stands or in assembly pits, and their length and width exceed the assembly unit dimensions, then for l_{AU} and w_{AU}, the length and width of the assembly stand or assembly pit should be used. The required additions for length and width can be taken from table 3.23.

Table 3.23 Assembly area additions

Type	Secondary operating sides		Working height [m]	Assembly area additions [m]			
	opposite	at the side		l_1	l_2	w_1	w_2
1/1			≤ 1.3	0.6		1.2	0.6
			> 1.3 < 3	0.6		1.6	0.6
1/2		1	≤ 1.3	0.85		1.2	0.6
			> 1.3 < 3	1.6		1.6	0.6
1/3	1		≤ 1.3	0.6		1.2	0.85
			> 1.3 < 3	0.6		1.6	1.6
1/4		2	≤ 1.3	0.85	0.85	1.2	0.6
			> 1.3 < 3	1.6	1.6	1.6	0.6
1/5	1	1	≤ 1.3	0.85		1.2	0.85
			> 1.3 < 3	1.6		1.6	1.6
1/6	1	2	≤ 1.3	0.85	0.85	1.2	0.85
			> 1.3 < 3	1.6	1.6	1.6	1.6

The workbench floor space A_{WB} in m^2 is calculated from the number of assembly units and the number of workers per assembly unit.

$$A_{WB} = 1.2 \bullet N_{AU} \bullet N_{WAU} \qquad (3.56)$$

N_{AU} = number of assembly units N_{WAU} = number of workers per assembly unit

The staging area A_{SA} is determined as follows:

$$A_{SA} = (A_{LP} + A_{MSP}) \bullet N_{AU} \tag{3.57}$$

A_{LP} = area of the large parts A_{MSP} = area of the medium and small parts

The residual area A_R in m² can be determined as follows:

$$A_R = \left(w_{AU} + 1.4 + \frac{1}{N_R} \right) \bullet 0.6 \bullet N_R \tag{3.58}$$

Assembly area type 2 – Assembly at assembly benches

Assembly area type 2 is similar to assembly area type 1, but differs in the fact that assembly takes place on an assembly bench. This assembly area type gives rise to the following sub-areas:

- area for assembly bench
- area for work benches
- staging area
- residual area

The formula for calculating the floor space is as follows:

$$A_{AA(2)} = A_{AB} + A_{WB} + A_{SA} + A_R \tag{3.59}$$

$A_{AA(2)}$ = floor space for assembly area type 2 A_{AB} = floor space for assembly benches

The floor space for the assembly benches A_{AB} is calculated as follows:

$$A_{AB} = a_{AB} \bullet \frac{N_{AU}}{N_{AUAB}} \tag{3.60}$$

The workbench floor space A_{WB} can be calculated in the same way as for assembly area type 1

$$A_{WB} = 1.8 \bullet N_{AU} \bullet N_{WAU} \tag{3.61}$$

The staging area A_{SA} is determined using the following formula

$$A_{SA} = A_{MSP} \bullet \frac{N_{AU}}{N_{AUAB}} \tag{3.62}$$

Here the floor space for medium and small parts A_{MSP} should be determined depending on the assembly bench length, by multiplying the bench length by 1m and adding 0.5 m^2. The residual floor space A_R is between 1 m^2 and 4 m^2 subject to the layout, i.e. one, two or three rows.

Assembly area type 3 – Work in testing areas

The floor space for assembly area type 3 can be determined from the sum of the testing areas:

$$A_{AA(3)} = \sum A_{TA}. \quad (3.63)$$

$A_{AA(3)}$ = floor space for assembly area type 3 A_{TA} = floor space for testing areas

When determining the floor space for the testing areas, assembly area type 3 is subdivided into two types. If the assembly unit is set on the ground for testing or run-in, the calculation used is as follows:

$$A_{TA}=(l_{AU}+1.7)\cdot(w_{AU}+2.05)\cdot n_{AU}\cdot f_O+1.2\cdot N_{AU} \quad (3.64)$$

The factor f_O for the overlap is 1 for one assembly unit and 0.75 for multiple assembly units. If the assembly unit is set up on a testing or run-in stand for testing or running in, then the testing area floor space is calculated using the following formula:

$$A_{TA} = (l_t + 1.7) \bullet (w_t + 2.05) \bullet \frac{N_{AU}}{N_{AUT}} \bullet f_o + 1.2 \bullet \frac{N_{AU}}{N_{AUT}} \quad (3.65)$$

l_t = length of the testing or run-in stand
w_t = width of the testing or run-in stand
n_{AUT} = number of assembly units per testing or run-in stand

The factor for the overlapping of testing or run-in stands is analogous to that for the assembly units.

Assembly type 4 – Work on conveying equipment

The floor space for assembly area type 4 is calculated as follows:

$$A_{AA(4)} = \sum A_S \bullet N_{SC} + \sum A_{SA} \quad (3.66)$$

$A_{AA(4)}$ = floor space for assembly area type 4 A_S = floor space for one station
N_{SC} = number of stations per conveyor unit A_{SA} = central staging area

Subject to the layout and set-up of the assembly workstations on conveyor equipment, for which structuring knowledge is required, five different types arise within assembly area type 4. Examples of these assembly area types are shown in fig. 3.54.

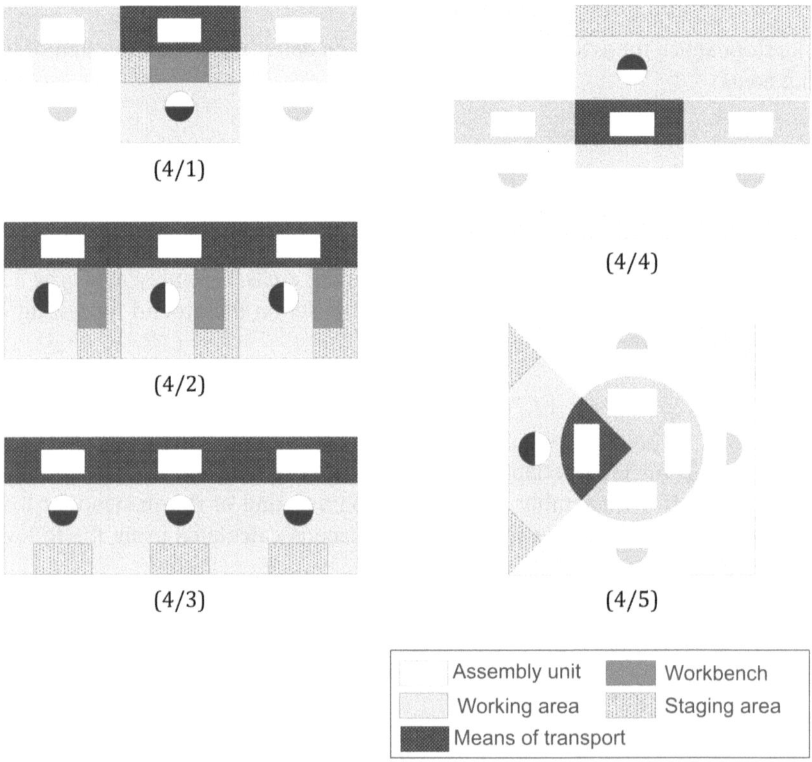

Fig. 3.54 Assembly area 4 sub-types

The values for A_S and A_{SA} can be obtained from table 3.24 in accordance with the assembly area type. If the assembly area type cannot be clearly defined, A_S is to be specified in accordance with the "double-sided" column in table 3.24.

Table 3.24 Determination of the station staging area

Type	Comment	Dimensions of the assembly unit			A_s* [m²]			A_{SA} [m²]
		projected footprint [m²]	$\frac{l}{b}$	Mass [kg]	single-sided	double-sided	Altern-ate	
4/1	Predominantly seated workstations, set out single or double-sided	≤ 0.06	≤ 1.5	≤ 8	2	3	-	4
		≤ 0.24	≤ 1.5	> 8	3	4	-	6
4/2	Seated or standing workstations, set out single or double-sided	≤ 0.2	≤ 1.25	≤ 5	2.5	4	-	4
		≤ 0.24	≤ 1.5	> 5	3.5	6	-	6
4/3	Standing workstations, set out single or double-sided	≤ 0.24	≤ 1.25	≤ 60	4.5	6	-	6
		≤ 1.1	≤ 3	> 60 ≤ 150	7.5	9.5	-	12
4/4	Standing workstations, set out alternately	≤ 0.4	≤ 1	≤ 80	-	-	8.5	8
		≤ 0.8	≤ 1.6	> 80 ≤ 100	-	-	10.5	10
4/5	Maximum of four seated or standing workstations	≤ 0.16	≤ 2	≤ 40	5	-	-	4

*dependent on the type of layout of the workstations on the conveyor equipment

Assembly type 5 – Work on workbenches

The assembly area floor space for assembly area type 5 is calculated as follows and can be subdivided into the types depicted in fig. 3.55 as examples.

$$A_{AA(5)} = \sum A_{AA} \bullet N_{WS} \tag{3.67}$$

$A_{AA(5)}$ = floor space for assembly area type 5

A_{AA} = floor space for one assembly area

N_{WS} = number of workers per shift

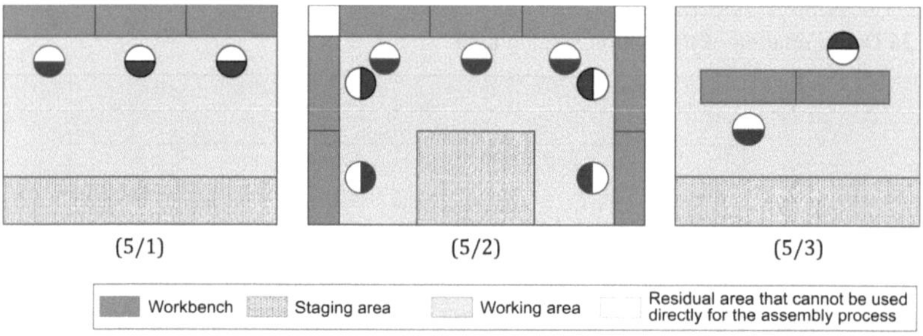

Fig. 3.55 Assembly area 5 sub-types

The assembly area floor space A_{AA}, consisting of the floor spaces for workbenches, working area and the staging area for items being worked upon is determined subject to the workbench group and staging type for assembly units that are listed in table 3.25. Here the workbench groups are composed as follows:

- workbench group I with lengths of 1.25 m to 1.5 m and widths of 0.7 m or 0.8 m,
- workbench group II with a length of 2.0 m and widths of 0.7 m or 0.8 m and
- workbench group III with lengths of 2.4 m or 2.5 m and widths of 0.8 m or 0.7 m.

Table 3.25 Determination of workstation floor space for assembly area type 5

Type	Assembly unit staging type	A_{AA} by workbench group [m²]		
		I	II	III
5/1	on transport pallets	4.65	6.6	7.75
	on racks	4.4	5.98	7.05
5/2	on transport pallets or individually	-	5.8	8
5/3	on transport pallets or individually	9	11.1	13
	on racks	8.6	10.7	12.6

If the workbench group cannot be determined unequivocally, then workbench group II should be selected.

Assembly type 6 – Workstations on machines and plant

The floor space for assembly workstation type 6 is determined through application of the procedure that has already been described in detail for the determination of floor space in parts manufacture.

Procedure for pre-determination of floor space (assembly)
(1) Determination of the assembly area type and model
(2) Calculation of the assembly area floor space A_{AA} on the basis of the functional area calculation
(3) Taking into account of overlap (cf. parts manufacture)
(4) Determination of the indirectly attributable areas (A_T, A_I, A_A)

(5) Determination of the production floor space A_P for assembly

NB: differing workstation types

Lay out *manual workstations* according to ergonomic, user-friendly considerations.

Hybrid assembly facilities (add the floor space of the automation equipment and manual workstations together)

Automatic assembly facilities (floor space of the assembly robots - as for parts manufacture equipment)

Manual assembly workstations with assembly equipment - assembly processes are carried out by workers using suitable workstation equipment (e.g. auxiliary mechanical equipment) within a certain gripping range. Individual assembly (piecemeal, routine assembly e.g. for electrical equipment, electronic component assemblies). Individual assembly according to the one-piece-flow principle is an individual part flow assembly.

Hybrid assembly equipment with automatic stations – for assembly of components in automatic stations manual workstations are combined (e.g. changeover times, transfer layout). Note floor space overlap.

Automatic assembly machines – assembly operations (e.g. welding, riveting, bolting, bonding) run automatically in connection with the feed-in and feed-out units (e.g. robots). These are predominantly interlinked flows. (cf. parts manufacture).

Note: production areas (parts manufacture and assembly) are calculated both from the approximate to the detailed, as well as inversely through calculation and dimensional design of the layout. Geometry, load, interference, supply and disposal parameters all have an influence of the floor space and room (building) (cf. Appendix 6). In addition to the production area, floor space for logistics (THS), administration, secondary areas and auxiliary areas must also be taken into account. Flexibilization and adaptability must be taken into account by means of reserve areas (10-20%), foundation-free flexible machinery installation, floor space overlap, floor space integration (of, for example, parts manufacture and assembly as well as transport and storage areas). Floor space expansion options must be retained in at least two directions.

3.2.3.4 Variant and cost evaluation

Costs are the value of the goods (materials consumption, depreciation etc.) and services (salaries, social costs etc.) consumed for the creation and sale of the company's products as well as for maintaining operational readiness.

In cost accounting, the classification of effort and costs plays a significant role, as costs, unlike effort, only represent consumption of resources in the context of the company's production of goods and services (operational orientation). For efficient operation, costs must be lower than revenues. To this end, a solid cost accounting basis and cost management are essential (Götze 2000).

Effort is the entire recorded consumption of the work, capital and material production factors (consumption of resources) during an accounting period. When calculating the profit and loss, effort is deducted from earnings (turnover revenues, inventory increase) in order to determine success (profit or loss). For efficient operation, effort must be less than revenue.

Conversely, expenses in the form of cash payments, increased debt and a reduction in accounts receivable constitute the countervalue to the purchases made by the company. For efficient operation, expenses must be less than income.

The factors detailed in table 3.26 are used to categorize costs.

Table 3.26 Cost classification factors

Cost classification features	Explanations
Behavior of costs if capacity utilization is changed	fixed and variable costs
Attributability of costs to cost objects	direct costs and overheads
Frequency of occurrence	one-off and recurrent costs
Composition of the costs	simple and composite costs
Level of resolution of coinciding costs ("type of actuality")	actual, normal and planned costs
Point where incurred (functions)	procurement, production, sales costs etc.
Type of costs (type of cause)	material, human resources and capital costs etc.

Direct costs are all costs that are caused by an individual cost object and which can be directly attributed to it (fig. 3.56).

Conversely, overheads are costs that accrue jointly for several products. They can only be attributed to specific products using allocation formulae (fig. 3.56).

Overheads that accrue when developing the company's products and services, can be split into fixed (time dependent) and variable (quantity dependent) costs in relation to the level of activity.

Fixed costs accrue to the same level independently of the rate of activity for a certain time period. As costs for operational readiness they are time dependent (e.g. building costs, calendar depreciation, etc).

Variable costs change with output. Output and costs can thus have a linear, progressive or diminishing relationship to one another (e.g. piecework rates, energy consumption, production materials).

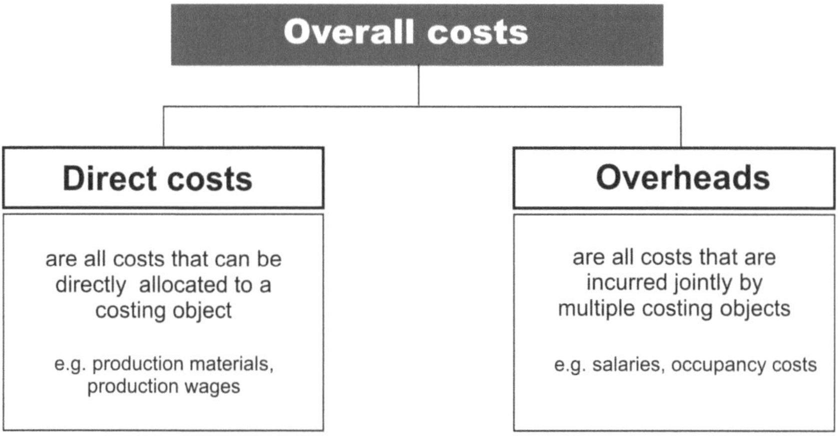

Fig. 3.56 Differentiation between direct costs and overheads

The average costs per product unit are the quotient resulting from the overall costs for an accounting period and the product units (quantities) made during this period.

$$Average\ costs = \frac{Total\ costs\ for\ the\ period}{\text{Production quantity}} \left[€ / unit\right] \qquad (3.68)$$

Over the long term a company must cover at least its average costs through the price of its products if the company's assets are to be retained and its existence is not to be jeopardized.

Cost accounting incorporates all costs that accrue for the procurement, storage, production and sales of the operating output during an accounting period in an enterprise. Cost type accounting is the apportionment of overall costs to individual cost types. As a result it is possible to carry out a largely cause-based allocation of the individual costs to cost objects and cost types. According to their origin, costs can be apportioned to the following different cost type groups (cost type accounting):

Material costs – are incurred through procurement, storage and consumption of materials in the context of production of goods and services. They can be subdivided into:

- production material costs (raw materials) → direct product costs
- auxiliary material costs (e.g. coating materials, cleaning materials) → "pseudo" production overheads
- operating material costs (e.g. electricity, gas, lubricants, oil) → "real" production overheads (via machine hourly rates)

Labor costs – are both the gross wages and salaries to be paid directly to the employee as well as the social costs to be borne by the employer. These are subdivided into:

- direct labor costs (production wages) – the work performed for this is directly attributable to the product (e.g. worker on a drilling machine)
- overhead wages (auxiliary wages) – this work serves to further the production process (e.g. warehouse worker, in-house transport worker, cleaning personnel)
- salaries – the work performed is for the administration and management of the production process (e.g. stock controller, bookkeeper, production manager)
- additional allowances – these are awarded for particular achievements (e.g. long service award)

Capital costs – are caused by the use of capital assets (invested assets such as buildings, machinery, patents and bank loans). They can be subdivided into:

- depreciations, which take into account the reduction in value of capital assets through wear and tear and increasing age.
- interest rates (imputed), which are calculated on the basis of the overall capital employed.
- risks (imputed), which take into account production, storage, transport, trading and financial risks using risk premiums (empirical values).

Costs for external services – are incurred through the use of services provided by third parties (e.g. for rent, lease, transports, licenses, association fees, repairs, etc.).

Social costs – through taxes, charges and contributions that are strictly interdependent on the operational production of goods and services (without this they would not be incurred). These include property tax for the company's assets, tax on operating capital, land tax, property transfer tax and vehicle taxes.

The price for a product is built up on the basis of its costs (fig. 3.57).

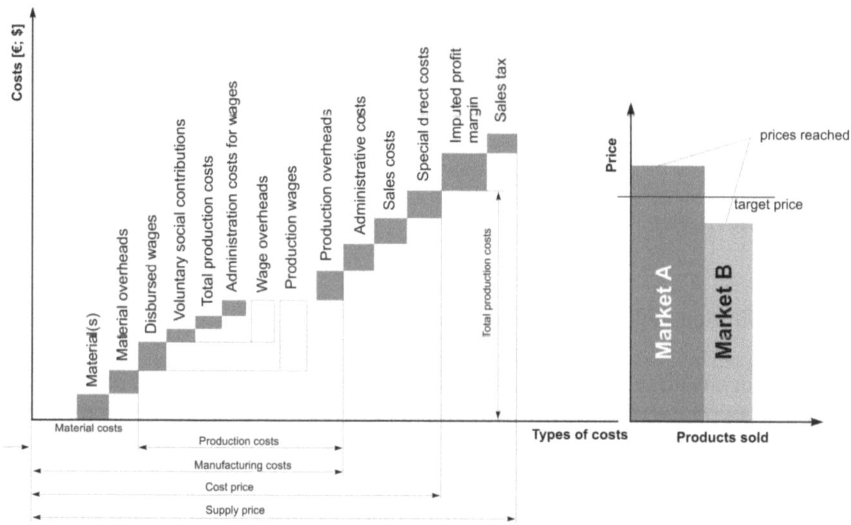

Fig. 3.57 Pricing on full cost basis (after Warnecke 1993)

In addition to cost type accounting, cost center accounting, cost object accounting, marginal costing, planning costing, process and target and life cycle cost accounting are also used (Götze 2000).

Ways of influencing costs:

Production program

- few influencing options (constructive, technological changes)

Determination of functions

- selection of equipment with favorable hourly rate (see calculation examples)

Dimensioning

- number of production areas (N_{PA}), number of workers (N_{WF}), number of shifts (N_S), minimizing floor space
- changing of equipment utilization times (T_{uE})
- striving for overlap areas (η_{Oi}, η_{Oe})
- striving for multiple machine operation (MMO)

Structuring

- striving for optimal layout of equipment and storage to minimize transport costs

Design

- striving for maximum correspondence between ideal and real layout

Note: all of the options for influencing a product's costs must be exploited by the planner. A planning objective may, for example, be that the price of the product to be made (see production program – value basis aspect) must be lower than that of the competitors' products.

3.2.4 Structuring

3.2.4.1 Definition, objectives and outcomes

Structure means arrangement, organization and layout. The objective of the structuring step in production facility planning is to define and implement the technical and organizational structure in planning terms.

Structuring is necessary on a variety of different levels:

- structuring of supply chain networks / enterprise networks in order to organize the nodes in a network
- structuring of individual locations (general development plan with layout of buildings)
- structuring of individual divisions (manufacturing structure, assembly structure, organizational structure)
- workstation structure for organization of the elements in an individual workstation

Note: structures can be created on different levels, from the enterprise network to the structure of the individual workstation. In the context of workshop planning, the focus is on the structuring of both the workshop and the workstation.

Some important structuring objectives are outlined in fig. 3.58.

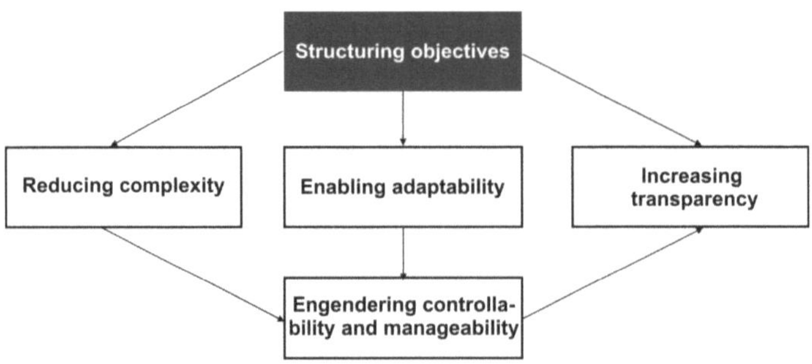

Fig. 3.58 Important structuring objectives

Further objectives include the harmonization of workflows, safeguarding and increasing performance and enabling flexibility.

Modern workshop planning combines an external overview of customers/consumers, producers and suppliers, with an internal overview of holistic and integrative considerations within the structuring step.

Generally accepted rules, principles and calculations exist to assist in structuring and are considered to be essential basic knowledge.

The "structuring" planning step includes the planning activities shown in fig. 3.59.

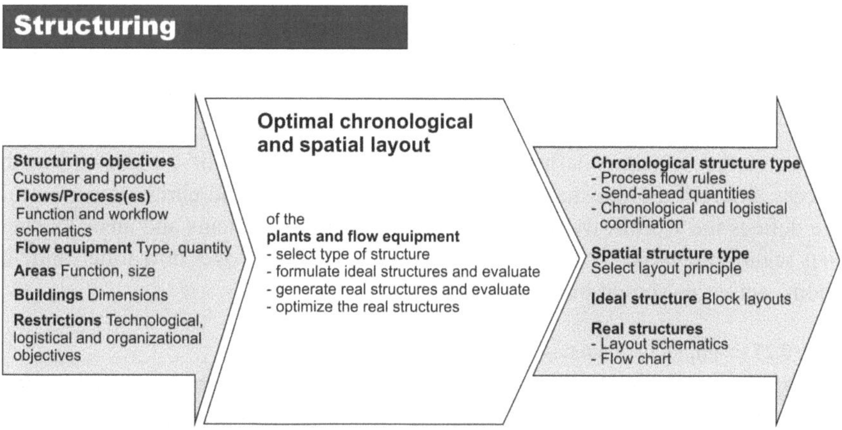

Fig. 3.59 The structuring planning step

The following is required as input information (cf. fig. 3.59):

Objectives

- structuring objectives (goals and the evaluation parameters derived from them)

Customer and product

- customer and order characteristics (number of customers, customer requirements, location of the customer order decoupling point, order structures, order quantities, performance locations)
- product characteristics (heterogeneity of the products, ABC/XYZ characteristics, materials, lot sizes, restrictive logistical characteristics)

Flows and processes

- technology characteristics (type of technologies, number of work processes and type of link to the generation of technological outcomes)

Flow equipment / objects

- machine and plant characteristics (type, number, installation conditions, conditions of use, in-feed and removal of parts)
- personnel characteristics (type, number, skills, communication matrix)

Areas and buildings

- building and area characteristics (type, areas, accesses, structural restrictions)

Additional restrictions

- statutory regulations and obligations (fire protection, safety)

3.2.4.2 Spatial structure

The spatial structure defines the configuration / outline of a plant and in the process answers the questions "who, what and where." The chronological structure defines the workflows within and between the sub-systems and answers questions regarding "what", "when", "in what order", "how", "over how long" and "by whom" activities should be undertaken.

Table 3.27 Examples of outcomes of the structuring planning step

	Material flow	Communication flow	Value flow	Energy flow	Information flow
Space (layout organization)	Manufacturing layout	Communication network	Cost center structure Center structure	Electrical power network	Computer network Data processing (DP) network
Method of illustration	Block layout, real layout	Organigram	Cost center plan	Electrical power network structural plan	DP structural plan e.g. client server architecture
Example		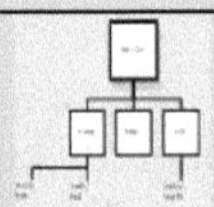	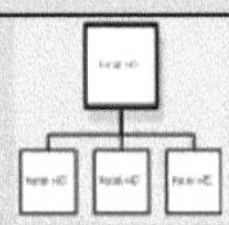	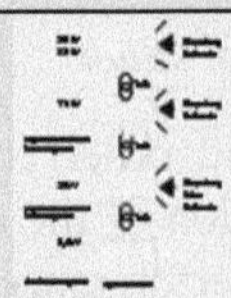	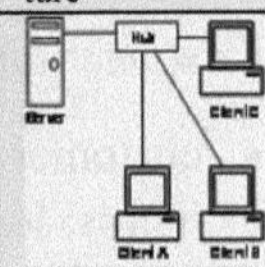
Time (workflow organization)	Type of provision and forwarding of parts	Business processes, communication processes	Cost accounting and settlement routines	Chronological resource planning	Information processing procedures
Method of illustration	Gantt diagram	EPC network plan	Cost accounting matrix	Storage and transfer plan	Program flow plan
Example	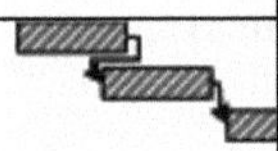	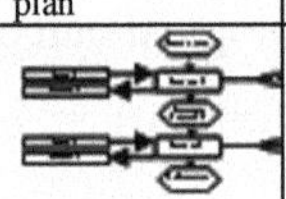	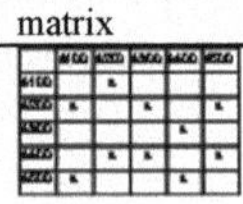		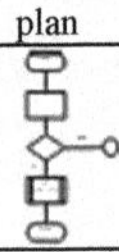

The chronological structure thus defines fundamental workflow rules. The two conceptual considerations – space and time – constitute a unit and have influence over one another.

The outcomes of structuring include various organizational layout and workflow designs, which, depending on the flow system under consideration, can be described in different ways. (cf. table. 3.27)

3.2.4.3 Chronological structure

The chronological structure combines the workflow order (sequential, parallel, overlapping) with a definition of the send-ahead quantities. In addition, a decision is made here as to how the transfer of send-ahead items will be triggered. This can take place via a push-control (material flow pushes = predecessor transfers to successor), a pull-control (material flow pulls = successor collects from predecessor) or via a control impulse (external scheduling instruction = each workstation is scheduled and supplied externally). This type of control (pull/push/hierarchy) determines the location and size of the storage/buffer and staging areas and represents possible structural boundaries.

The basic principles of the chronological structure can be seen in table 3.28.

Table 3.28 Basic principles of chronological structures (pull principles)

Basic principle	Workflow sequence	Send-ahead quantity	Illustration of principle
Sequential progression	sequential	Lots (lot size 1 to n)	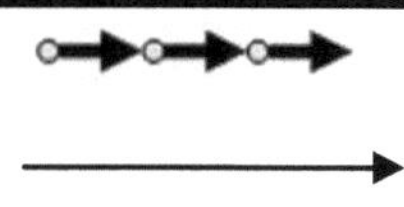
Parallel progression	parallel	Partial lots	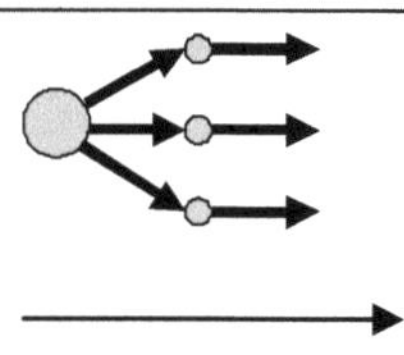
Combined progression	partially parallel (overlapping)	Partial lots e.g. transport batches	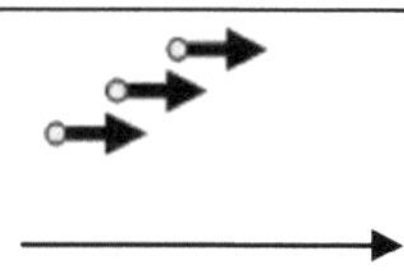

When defining the send-ahead quantity, questions arise regarding the arrival of the objects, the identification of the order and with respect to value creation:

- how will the products arrive?

- how many products will arrive at once?
- is arrival random or organized?
- what role does the identification of individual parts, batches, lots or orders play?

Staging strategy:	homogenous, picked, sequenced, hybrid
Material supply strategies:	conventional storage part delivery, direct supply, modular supply
Type of staging:	static, dynamic

The arrival and onward transfer of products and the buffers to be set up within the workflow are determined by the underlying logistics principle (selected workflow principles in chapters 4 and 6.4 as well as table 3.29 below).

Table 3.29 Some important logistics principles

Workflow principle	Explanation	Relevance for structuring
Just in time (JIT)	With just in time, materials are supplied at precisely the right time, i.e. only when actually needed.	Individual supply or small lots, prompt supply requires accessibility and floor space in the workstation
Just in sequence (JIS)	With just in sequence, materials are supplied at precisely the right time, i.e. only when actually needed and also in the correct order.	Supply of small production lots, prompt supply requires accessibility and floor space as well as ensuring parts supplied in correct order in the workstation
KANBAN	Removal of parts by the user from a buffer. Buffer filled by the KANBAN card, which in logical terms represents a fixed order for additional delivery or production	Stockpiling of fixed stock close to the consumption point requires accessibility and floor space for buffer, removal when required, buffer stores can be used as structural boundaries
Constrain work in process (CONWIP)	Larger control loops over several production stages. Constant working stock in the production facility	Control loop over several production stages Buffer stocks can be used as structural boundaries

Logical sequence relationships necessitate conditions that must be adhered to during processing and thus define both required sequences and areas where there is scope for maneuver:

- which processes or process steps are dependent upon one another?
- which processes or process steps are independent of one another?
- which restrictions should be taken into account?

Chronological sequence relationships pose the question as to "when" an activity must be carried out:

- how much time is available for the activity?
- in what chronological order should the activities be carried out?
- how are products called up?

- what requirements are there with respect to lead time and thus to time spent in the system?

Parameters for determining the chronological structure are:

- fulfillment of logistical and economic objectives
- people-friendly job contents
- high productivity through division of labor
- low transport expenses
- low storage expenses

Outcomes of chronological structuring are:

- definition of production and transport lot sizes
- determination of basic workflows through production and assembly
- determination of optimal sequence of lots
- where necessary, determination of the number of working processes (cycles) and their sequence

3.2.4.4 Connection of spatial structure and workflow principles

Through structuring, questions arise from the spatial structure and workflow principles regarding the layout (fixed or mobile) and required routes/distances. According to Schmigalla (1966) the spatial organization poses three sub-questions:

(1) which objects should have a fixed location and which objects should it be possible to move?
(2) what should be the location of the fixed objects? (See layout planning)
(3) in which ways and along which routes should the mobile objects be moved? (See material flow and logistics)

The question of mobility should be answered at this point. Products, equipment and personnel should be organized so that they are fixed or mobile under certain conditions. Table 3.30 contains several examples.

Table 3.30 Features and examples of fixed and mobile elements

	Fixed	Example	Mobile	Example
Product	Not easily managed in terms of logistics Large dimensions Fixed installation	Generator equipment Conveyors	Easily managed in terms of logistics Small dimensions Larger lot sizes	Individual parts
Equipment	Machine-determined processes	Presses Processing machinery	Equipment that is easily managed in terms of logistics	Portable drilling machines, cranes

	High set-up and installation costs High demands placed on accuracy		Equipment that is mobile by nature	
Personnel	Manual work processes, for which each movement of workers results in time lost and high costs	Manual assembly Intricate and precision work Quality control	Adjustment and monitoring tasks on machinery Multiple machine / multiple station operation	Multiple machine operation in one production cluster

The options identified in table 3.31 for the formation of structural types with respect to "fixed" and "mobile" characteristics are the result of combinatorics.

Table 3.31 Structural types in relation to objects' "fixed" or "mobile" characteristics

Type of structure	Character of elements			Comments	Terms for structural forms and examples
	Product	Equipment	Personnel		
1	fixed	fixed	fixed	• Fixed workstations • No division of labor, single-station operation, also poss. by several people where necessary	Individual station
2	mobile	fixed	fixed	• With division of labor • Single-station operation	Serial production Continuous flow production
3	fixed	mobile	fixed	• It must be possible to move equipment easily	Manual workstation with overhead crane
4	fixed	fixed	mobile	• No division of labor • Multi-station operation	Operation or supervision of semi-automatic or automatic processes or alternating deployment of personnel due to low degree of capacity utilization
5	mobile	mobile	fixed	• Uncommon	Flexible assembly cycles
6	fixed	mobile	mobile	• With division of labor • Multi-station operation	Stationary production Assembly of a plant
7	mobile	fixed	mobile	• With division of labor Multi-station operation	Group production Workshop production One piece flow (workers move with product from one piece of equipment to another)
8	mobile	mobile	mobile	• With division of labor • Multi-station operation	Line assembly Motor vehicle assembly

The structural forms highlighted in table 3.31 are explained in more detail in fig. 3.60 below.

	Form of production	Features	Layout principles (schematic)
process-oriented	**Stationary production (SP)**	- Product (WP) - fixed - Workers/working materials (W, WM) - mobile	
process-oriented	**Workshop production (WSP)**	- same / similar equipment (processes) grouped together in workshops - equipment and workshop layout independent of material flow	
process-oriented	**Group production (GP)**	- selected equipment grouped together - only a sub-phase of overall processing - material flow in cluster optional	
object-oriented	**Serial production (SP)**	- equipment layout direction-oriented in accordance with the dominant material flow - complete processing - formation of sub-groups	
object-oriented	**Continuous flow production (CFP)**	- equipment (stations) layout direction-oriented in accordance with the material flow - rigidly interlinked – synchronized material flow - loosely interlinked – buffering between the stations	

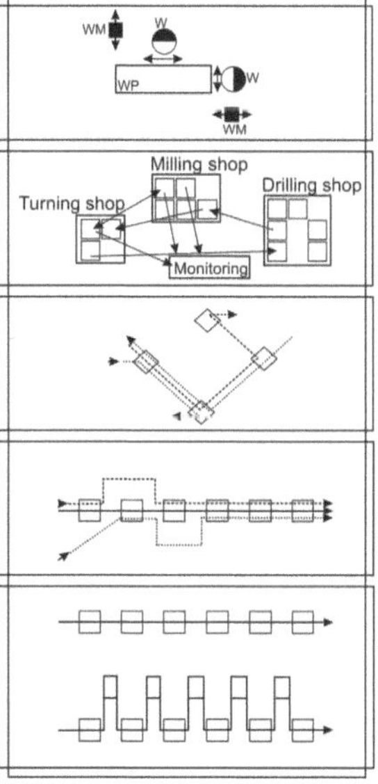

Key ☐ Machine workstation ☐ Buffer system ═ Material flow

Fig. 3.60 Typical terms for basic forms of production (Grundig 2000, p. 116)

In addition to this the following terms also exist to differentiate between forms of production according to the size of the structure, the type of production and the feasibility of complete processing. (cf. fig. 3.61)

	Form of production	Features	Layout principles (scheme)
NC processing techniques	**Flexible production cells (PC)**	- NC processing center (processing, staging, control system) - Partial or complete processing of different workpieces - Make-to-order and small batch series area	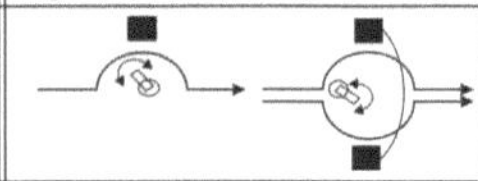
	Flexible production line (PL)	- NC processing techniques arranged in accordance with work process sequence - Complete processing of similar workpieces - Mass production area	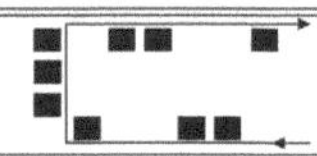
	Flexible production system (FPS)	- NC processing techniques functionally linked via integrated control, transport and storage system - Complete processing of different workpieces - Small and medium batch series area	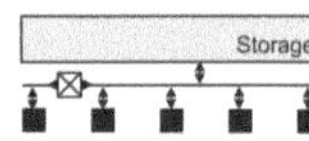
Hybrid systems	**Production island (PI)**	- Workstations (hybrid system) grouped together spatially according to main material flow - Predominantly complete processing of parts groups (production stage) - Self-direction in group work - Integration in direct production functions - Small and medium batch series area	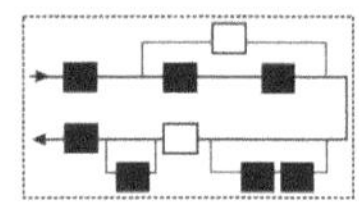
	Integrated object-specialized production sections (IOSPS) **Basic variants: - IOSPS/A** **- IOSPS/B** **- IOSPS/C**	- Workstations (hybrid system) functionally linked via integrated partially or fully automated control, transport and storage system - Complete processing of different workpieces (product lines) - Small and medium batch series area - Centrally located interim storage with separate transport system - Interim storage in the transport system - Transport in the centrally-located storage system	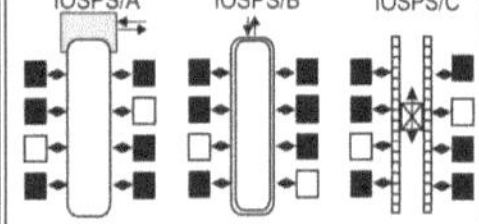

Key ■ Machine workstation □ Manual workstation

Fig. 3.61 Basic forms of production (Grundig 2000, p. 123)

These structures are basically formed via the creation of parts families according to constructive, geometric and technological criteria. This makes a reduction in set-up times possible, even with frequent order changes. The larger lot sizes enable a higher degree of automation for the equipment that is adapted to the area of activity. The integrated processes (logistics and maintenance) increase availability and autonomy. The forms of production mentioned are characterized below according to some of their features (table 3.32), whereby this characterization provides indications for selecting the form of production.

Table 3.32 Characterization of basic forms of production

	Flexible production cells/un-connected NC production	Production island	Integrated object-specialized production sections	Flexible production system	Flexible production line
Lot size per variant	low	low	low to medium	medium	high
Number of variants	high	high	medium	medium	low
Similarity of parts	low	low to medium	high	medium	high
Order change	very frequent	very frequent	medium	medium	medium to low
Example	prototype construction	tool making	small, prismatic parts	housing production	engine production

3.2.4.5 Approaches to structure formation

A mathematically-supported way of choosing the most practical type of structure is provided by the calculation of the degrees of cooperation κ (Schmigalla 1966) and integration η_{int} (Förster 1983). The structures listed by workflow type in Fig. 3.62 can all be used for parts manufacture.

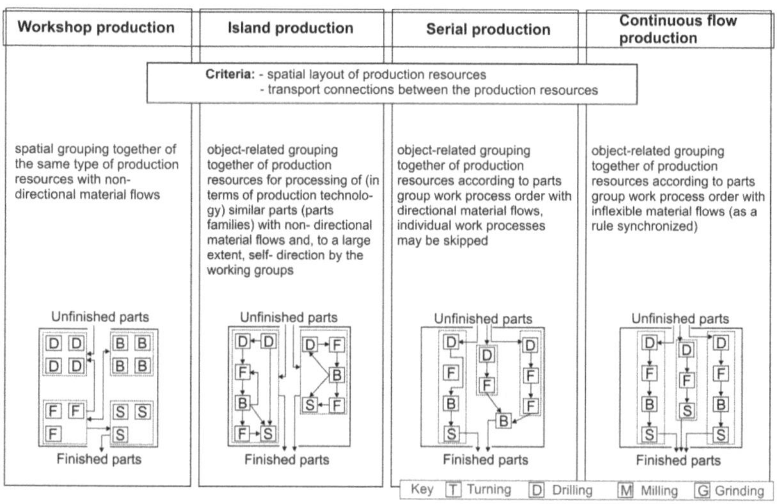

Fig. 3.62 Type of workflow in parts manufacture (Schenk & Wirth 2004, p. 60)

These classic approaches to production structuring are directed exclusively inwards, within the enterprise, and thus primarily serve improvements in efficiency. The structures formulated in this way are, as a general rule, not capable of reacting to changed requirements. The main reason for this is that influences on the company's environment (customers, competitors, suppliers etc.) are not taken into consideration in the internal organization.

Table 3.33 below contains a selection of structuring criteria. Since, for reasons of practicality, only 1 to 3 criteria can be incorporated in the structuring process, the parameters that are decisive in terms of efficiency in the specific application must be prioritized for selection.

Note: the structuring processes must be tackled from an internal and external point of view.

a) Internal view of structuring

There is often considerable potential for efficiencies in the optimization of material flows, so that the material flow determines the structuring. In the past the rule of thumb was that material flow structures are formulated correctly when the spatial layout is adapted to technological progression, i.e. the spatial layout follows the material flow.

This requires a material flow analysis as an input parameter for structuring. In turn the material flow is derived from the technological progression of operations. The technological progression of operations incorporates the work process sequence. Depending on the production type, there are also additional differing flexibility requirements.

- The technological progression of operations is strictly defined in mass production. There are only a few flexibility requirements with respect to stability and freedom from disruption.
- In series production, the technological progression of operations differs according to product type. The connections between the system elements must therefore be studied. A corresponding level of flexibility must be built in to the structures in order to fulfill even changing requirements.
- The greatest degree of flexibility with almost unrestricted connection options between the objects must be implemented in make-to-order production.

Table 3.33 can also be used as a simple checklist for the specification of relevant structuring criteria. These structuring criteria can be used individually (1 single criterion) or in any combination (2 or more criteria).

The selection of criteria is not considered to be "right" or "wrong" as a result of comparison with the structuring objectives, but only as "useful" or "not useful."

Table 3.33 Classic (internal) structuring criteria = effectiveness and efficiency-oriented

Structuring criterion	Explanation	Example of formulated structures
Product features (what)	Type of product	Product A, Product B
	Material (raw materials orientation)	Wood working, processing of plastics, aluminum working, casting work
	Mass and dimensions	Crane parts, easy to handle manually; large parts, small parts
	Geometry	Rotationally symmetrical parts; prismatic parts
	Type of construction	Assembly components, welded components
	Function	Series production part, replacement part; make-to-order production, components production
	Product structure	Pre-assembly, assembly of components, final assembly
	Special characteristics such as protection and hazard requirements	Chilled products, normal temperature products
	Parts families (e.g. link between geometry and dimensions across many products)	Large and small pressed parts
Type of production (how many, how often)	Production type	
	Mass production = low flexibility, high degree of automation	Automated lines producing product variants with high lot sizes (mass)
	Series production = medium flexibility, high degree of productivity	Flexible, interlinked equipment lines producing product variants with medium lot sizes (series)
	Make-to-order (special requirements) = highest degree of flexibility Use of ABC-XYZ analysis	"Exotic parts" group (special)
Workforce (who)	Skills, level of training, personal abilities and capabilities, employees' view of themselves	Homogenous or heterogeneous groups
	Activity-orientation	Simple tasks, highly-skilled tasks, expertise, electrical work, testing activities
	Communication	Divisions with high degree of communication to be concentrated spatially
	Teams, areas of responsibility, production organization	Structures differ according to the production organization principle
	Accessibility	Open areas, closed areas
Machines/plant (on what)	Similar types of machines (equipment-orientation)	Workshops such as turning shop, milling shop
	Degree of automation	Conventional processing and NC processing
	Conditions of installation and operation	Structures for coloring, for presses or for

	(e.g. foundations, connections, climatic conditions)	clean rooms or measuring rooms, cool areas, sterile areas
Technology (how)	Technological progression of operations = material flow-orientation	Serial production, continuous flow production
	Regularity of the technological work sequence	Sections with technological sequences that remain the same and sections with changing technological sequences
	Technology quality requirements	High quality requirements (precision work), normal quality requirements
Scope of tasks (with whom)	Integrate indirect functions	Own structures for logistics and maintenance, alternatively integrate these functions into the production structures
	Integrate planning	Own structures for planning and scheduling, alternatively integrate these functions into the production structures
Location (where)	Company has several sites	Locations are used as internal company structure and form organizational units, production building 1, building 2
	Buildings and infrastructure	Warehouse
	Structural suitability for certain tasks	

The structural forms generated according to one structuring criterion only are shown below for clarification purposes with examples. (cf. fig. 3.63)

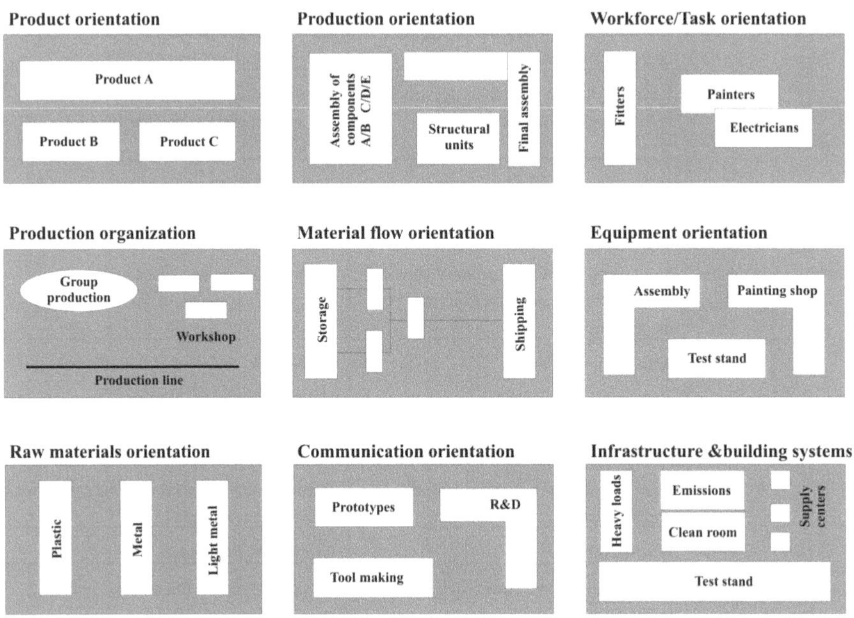

Fig. 3.63 Structuring principles (Schulte 2003, p. 97)

Structure design approaches are shown as examples for product-oriented structuring in fig. 3.64.

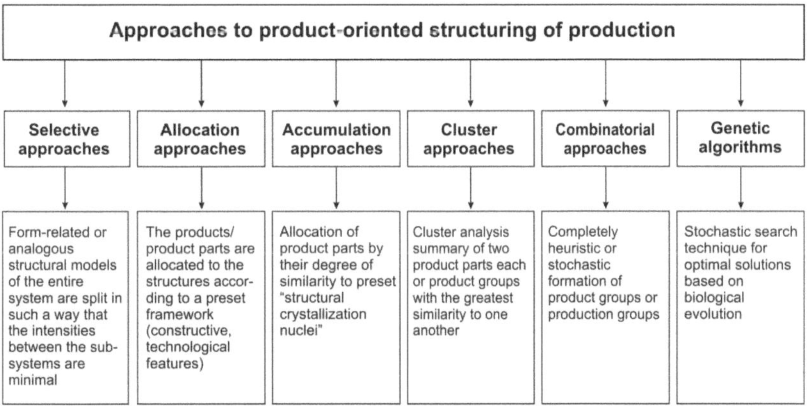

Fig. 3.64 Overview of approaches to product-oriented structuring of production

The structuring of storage by process sections in material flow facilities is shown in fig. 3.65.

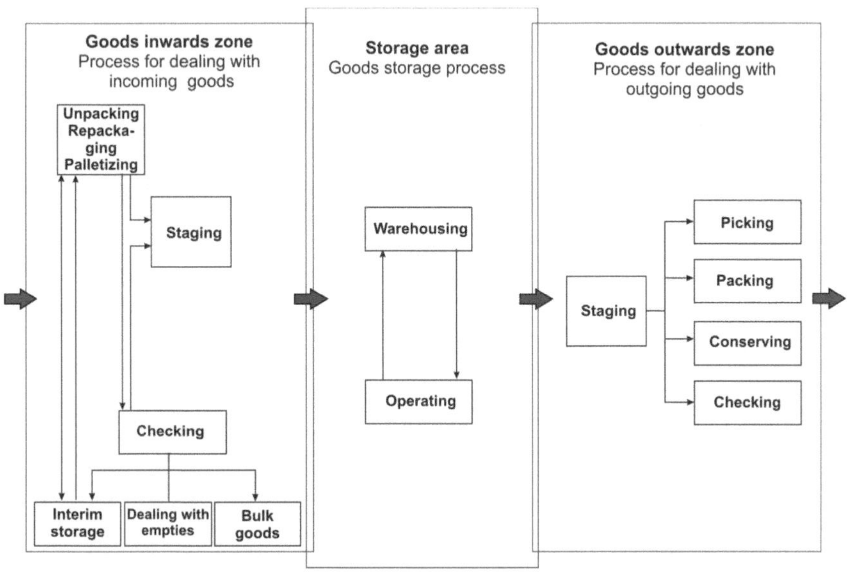

Fig. 3.65 Structuring of storage (classification according to process sections in material flow direction)

b) External view of structuring

The external view expands structuring's perspective. Every production run serves to fulfill customer needs and customer orders. For this reason the customer order features and requirements provide critical inputs for structure design.

Today, the influence of the company's environment (customers, competitors, suppliers, etc.) is considered to be an important factor in the structuring process. For the purposes of completeness, starting from the customer, the interdependencies with distribution, production (including to distributed production runs) and to procurement are also taken into account. (cf. table 3.34)

Table 3.34 Important external structuring criteria = customer and interface-oriented

Structuring criterion	Explanation	Example of formulated structures
Customer	Business enterprise	Motor vehicle suppliers broken down into "division for Daimler" and "division for VW"
	Region, country	Europe and America
	General market orientation (market, product, combinations and responsibility for costs and results)	Center organization (profit center, revenue center, cost center, service center)
Order	Number of order items	Individual and multiple order picking
	Regularity	Series production division, make-to-order production division
	Order volume	Key accounts, individual customers
	Urgency	Normal lead time, rush-lead time
Procurement	Supplier (company)	By supply characteristic. Formation of structures for permanent supply and for order-related supplies
	Type of procured goods	Large parts, small parts, open-air storage, warehouse
	Tender principle and urgency of the procurement order	Storage, cross-docking area
Competitors	Best practices	Analysis of structures in the same industrial sector and of close competitors, adoption, adaptation and development of best practices
Product	Customer order decoupling point	Customer-neutral pre-fabrication, customer-specific completion (formation of variants)
	Product group formation	Product group I and product group II
	Heterogeneity of products	Series production division, make-to-order production division
	ABC-XYZ characteristics	Hits, flops
	Accessibility	Open areas, closed areas
Distribution	Organization of distribution (time and formation of batches for distribution activities in accordance with customer requirements)	Hub/spoke structures, centralized storage with direct supply, decentralized storage in the network, direct supply ex works
	Type of transport	Rail and road transport

Regarding the customer order decoupling point:

It is important to specify the customer order decoupling point. This provides information about if and when within the company a customer-specific feature is introduced into the product, i.e. where customer-specific variants come into being. Often the customer order decoupling point is used in the following way as a structural boundary: one structure for customer-neutral production, n structures for customer-specific completion.

Regarding center organization:

Center organization targets

- an increase in overall turnover and profit
- an increase in capacity utilization of high-value equipment through acquisition of contract work
- development of entrepreneurial thinking and activity

There are four important types of center which can be characterized as follows: (cf. table 3.35)

Table 3.35 A comparison of types of center

Profit center (e.g. production operates as contract production)	Cost center (e.g. production as part of an enterprise)
– Costs as well as activity quantities of the profit centers can be influenced as the basis for entrepreneurial activity and proactivity – Profit centers are managed using cost objectives and performance standards – Accounting results are performance-cost comparisons	– Activity quantities of the cost center cannot be influenced by the responsible decision-makers – Cost centers are primarily managed by the level of costs generated (target costs, cost savings) – Accounting results are aggregated cost center results
Revenue center	**Service center**
– Costs do not constitute significant control parameters – Revenues can be influenced to a significant extent – Examples are sales and marketing structures (sales force, freelance marketing staff) – Revenue centers are primarily managed by the level of turnover, in addition certain cost parameters can be included (e.g. frequency and amount of price reductions, advertising costs)	– Special form of the cost center, service centers exclusively provide services for other divisions of the company – Examples are data processing, servicing/maintenance, financial controlling – Service centers are primarily managed by the level of costs generated (target costs, cost savings); in addition the quality of service is measured

From fig. 3.66 it is evident that in addition to the traditional link of the sales and marketing department to customers, the “Manufacturing” and “Research and Development” profit centers also build up their own relationships to customers in order to market free capacities. Figure 3.66 shows an example of center organization.

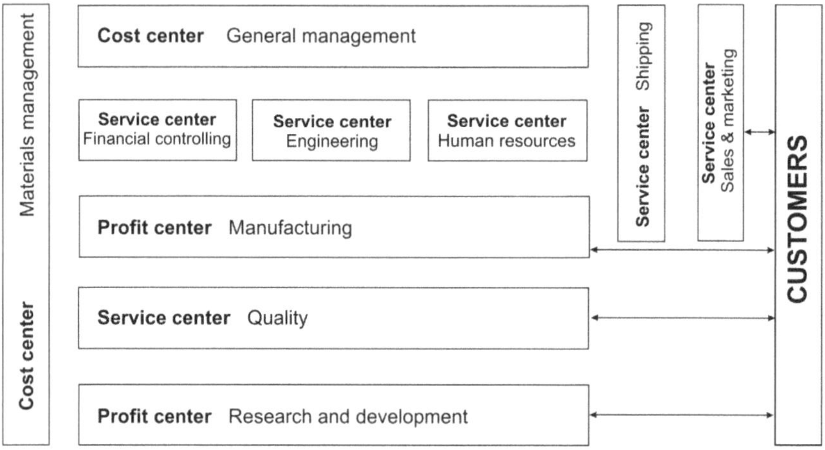

Fig. 3.66 Example of the structuring of a company in centers

In order to compile a center profit and loss statement, on the one side all goods and services must be recorded as costs. On the other side, the services performed by the structures for internal and external customers must be recorded and compared with the costs. An example for the allocation of costs (= process flow rules for the information and value flows) is shown in table 3.36.

Table 3.36 Example of the allocation of costs between centers

—	Engineering SC	Human resources SC	Manufacturing PC	Quality SC	Research/ development PC	Materials management CC	Shipping RC	Sales & Marketing PC
Financial controlling	X	X	X	X	X	X	X	X
Engineering			X	X	X		X	X
Manufacturing					X			X
Quality			X		X	X		
Research/ development			X					X
Materials management	X	X	X	X	X	X	X	X
Shipping			X		X			X

Key: PC = profit center; CC = cost center; SC = service center

Regarding the formation of segments:

Manufacturing segments are an interesting form of structure, for whose formation several structuring criteria can be drawn upon. Manufacturing segments must not necessarily only supply products to end customers, but also to customers who will process them further and who may also be part of the same company.

Structure formation criteria for manufacturing segments include:

1. Focus on markets and objectives

- creation of product – market – production combinations
- focus on competitive strategies:
  - cost leadership strategy through specialized technologies and production systems
  - differentiation strategy by providing highest quality or short lead times

2. Product

- focus on specific products in order to reduce effort required for control
  - low level of diversification
  - relatively high degree of vertical integration (as a result of aiming for complete processing)

3. Processes (several stages in a product's logistical chain)

- integration of many to all in-house value added stages for the product or production program
- difference to production cells, flexible production systems and production islands, because these concepts only comprise one stage in the logistical chain

4. Transfer of indirect functions

- reduction in interfaces and overall responsibility on site
- transfer of indirect and planning activities to the employee segments

5. Costs / profit responsibility

- segments are defined as performance centers:

  case A: segment creates marketable products that are only used in-house: due to the higher degree of integration there is extensive responsibility for costs
  - organize segment as cost or service center

  case B: segment also has market access: extensive responsibility for profits
  - organize segment as profit center

Modern computer-aided tools enable structuring and dimensioning to be linked through simulation (Schenk & Wirth, 2005).

3.4.2.6 Evaluation and optimization of structures

The following descriptive features can be used in the form of a morphological box to qualitatively describe the formulated structures. (cf. table 3.37). A goal-oriented selection must be made in order to evaluate the structures.

Table 3.37 Qualitative descriptive features of structures in factories

Descriptive features	Examples of feature characteristics
Structuring objects	Companies, locations, functional areas, machinery, personnel
	Products/services, customers
Structuring criteria	See table 3.34
Customer relationship – market relationship	Profit center, cost center, service center, revenue center
Degree of centralization	Centralized, decentralized
Durability	Long-lasting, temporary, adaptable
Structural form	Chain, converging, divergent, U-shaped, ring, star, hub/spoke, spine, network
Mobility of the structuring objects	Fixed, mobile
Degree of automation	Manual, mechanized, partially automated, fully automated
Chronological coordination	Synchronized, coordinated, not coordinated
Changeability	Static, dynamized, dynamic
Stratification with relation to logistics or value added processes	Single-stage, two-stage, multi-stage
Autonomy	Dependent, partially autonomous, autonomous
Personnel	Stand-alone, fixed teams, flexible teams
Functions	Value creation functions, control functions, maintenance functions, transport, handling and storage functions
Flow aspect considered	Material flow, communication flow, value flow, energy flow, information flow

The structures can be quantified subsequently using characteristic

- static features (e.g. number of structures formed, size of structure) and
- dynamic features in the outcome of simulations (e.g. throughput time, capacity utilization) and structure variants can be compared with one another using evaluation parameters.

Table 3.38 contains suggested evaluation parameters.

Table 3.38 Examples for quantitative evaluation parameters of a structure (cf. VDI 4400, Rietz 2001)

Evaluation parameters	Unit
Number and size of structures	
– Number of structures	[Number]
– Amount of equipment in the structure	[Number]
– Number of people in the structure	[Number]
– Number of products allocated to the structure	[Number]
Work contents within a structure	Listing of activities
Performance (e.g. throughput)	[Units/hour]
Special logistics parameters	
– Lead time	[hours or days]
– Ability to supply	[%]
– Readiness to supply	[%]
– Delivery reliability (time, quantity, quality aspects)	[%]
– Level of service	[%]
Bottlenecks	
– Number of bottlenecks	[Number]
– Location of bottlenecks	[O_x O_y O_z]
– Throughput at the bottleneck	[Units/hour]
Time	
– Average throughput time	[minutes] [hours]
– Average deviation from throughput time	[minutes] [hours]
– Average share of processing time in throughput time	[%]
– Average share of idle time in throughput time	[%]
Stock (e.g. average stock)	[hours or days] or [units]
Capacity utilization of stations and machinery (e.g. average capacity utilization)	[%]
Costs	
– Costs for restructuring	[€]
– Down-time costs	[€/h]
Degree of consistency for the evaluation of the autonomy of a structure	[%]
Number of interfaces	[Number]
Transport effort	[t*km]
Separate responsibilities for workflows	[Number]
Duplications	[Number]
Degree of control known and quantifiable	binary
Transparency?	binary

After structure variants have been evaluated, the best variants can be selected, e.g. on the basis of a cost-benefit analysis. An evaluation according to the individual criteria (cf. tables 3.37 and 3.38) serves to characterize the structure and provides starting points for a subsequent optimization.

3.2.5 Design

The Design planning stage incorporates the ready-to-implement, spatial and functional coordination of objects and their interrelationships taking particular account of the restrictions and requirements arising from the economy, ecological environment and occupational health and safety, and the detailed planning of special project areas (peripheral areas) (cf. fig. 3.67).

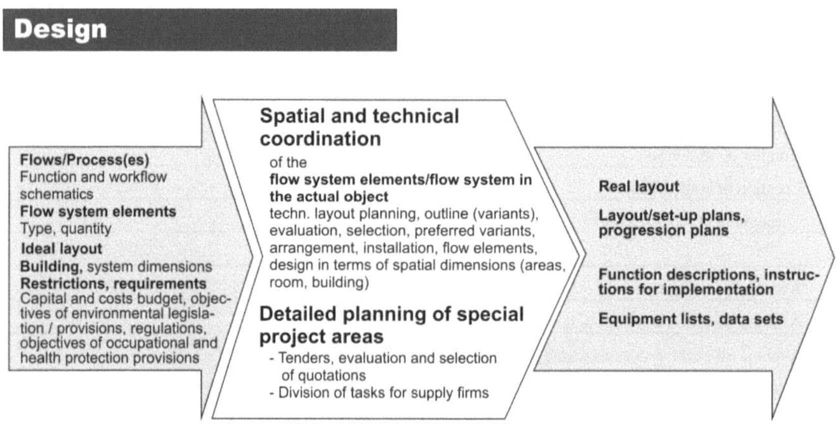

Fig. 3.67 Design planning stage

The role of the design stage is to coordinate the functional, technical and organizational specifications derived from the technological process with the tasks assigned to the workers. The outcome of the planning project is an ideal, approximate or detailed layout.

3.2.5.1 Layout planning

A layout is the graphical representation of the spatial arrangement of functional and structural units at operational level (production and assembly stations, transport, handling and storage facilities) plus the interference effects in the building.

NB: planning approaches

Processes

- reduction of the process stages and technological workflows
- striving for integrated processes
- minimizing THS processes

Area

- a reduction in the overall floor space requirement
- optimal utilization of floor space and the room (building requirements)
- provision for flexibility with a view to future expansions
- flow relationships between the objects
- the flow-oriented arrangement of facilities
- a reduction in the complexity of the transport system
- low-loss, low-cost supply and disposal system

Profitability

- investments with a favorable cost/benefit relationship, i.e. quick payback of capital
- low as possible financial expenditure when implementing the layout
- low as possible costs when operating the new structures

Robustness

- collision and hazard-free design of the objects
- provisions for / implementation of occupational and environmental protection requirements
- creation of safety, breakdown and emergency plans

Workers

- implementation of labor organization forms backed up by ergonomic science
- ergonomic design of the working environment
- taking working environment factors into account, minimizing interference effects
- creating conditions for active participation by those involved

Layout planning is executed on three levels.

Level 1: systematization of the planning principles in accordance with the planning activities and project definition.

Level 2: implementation of ideal layout planning in accordance with the project development planning activities.

Level 3: implementation of real layout planning in accordance with the project development planning activities taking real restrictions into account.

The individual levels are described in detail in Schenk & Wirth, 2004, p. 275 ff.

The *ideal layout* is the best possible solution with no allowance for restrictions. Here optimization aspects are illustrated according to the functional processes of the product flow and floor space requirement (cf. Fig. 3.68).

Example 1: ideal layout for manufacturing of equipment (fig. 3.68)

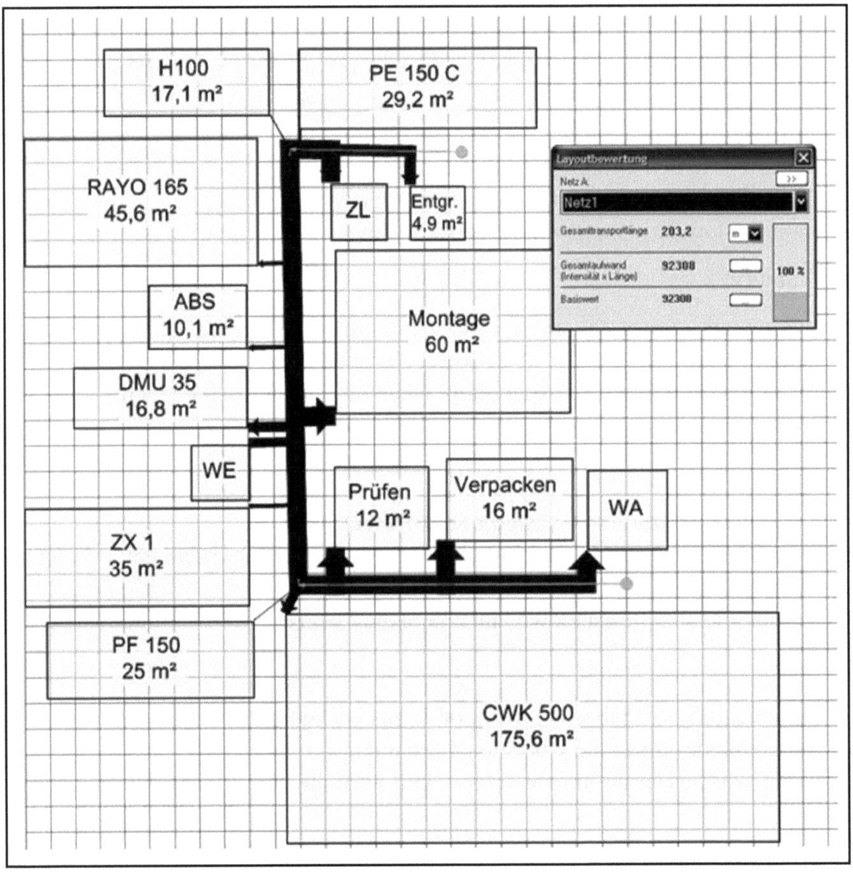

Fig. 3.68 Ideal layout (worm gear production)

The *approximate layout* is the interim step to the real layout, with particular attention being paid to the building parameters in the case of new or existing buildings. This contains the floor space arrangement of the structural and functional units, the transport routes (gates, cranes), supply and disposal flows and installation systems illustrated in block form (block layout) – cf. example, fig. 3.69.

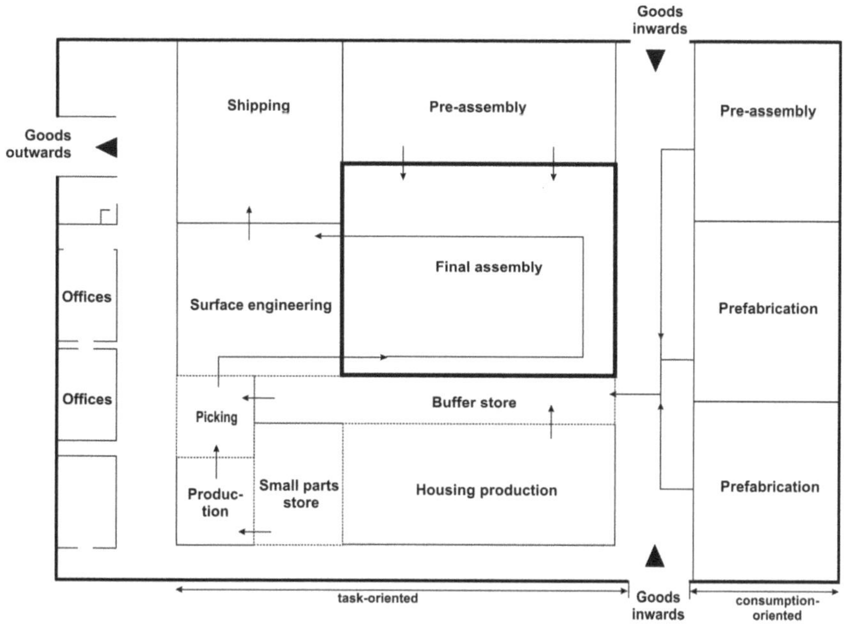

Fig. 3.69 Example of an approximate layout illustrated in block form (Schulte 2003)

An approximate layout can also be created as a machine installation plan – cf. example, fig. 3.70.

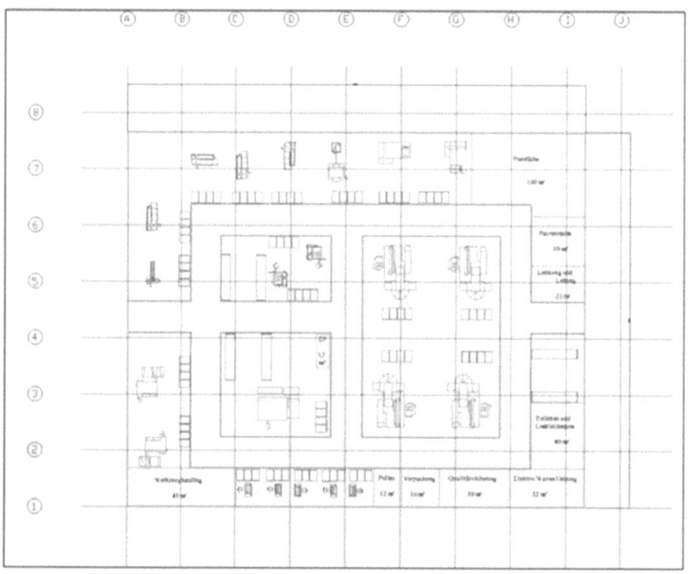

Fig. 3.70 Approximate layout (machine installation plan - worm gear production)

The *real layout* is the layout taking all restrictions into consideration. When adapting a layout from the ideal to the real, subject to the characteristics (products – production procedures/processes), a multitude of very different factors and requirements/restrictions should be taken into account as planning information for the development of the overall object. (fig. 3.71)

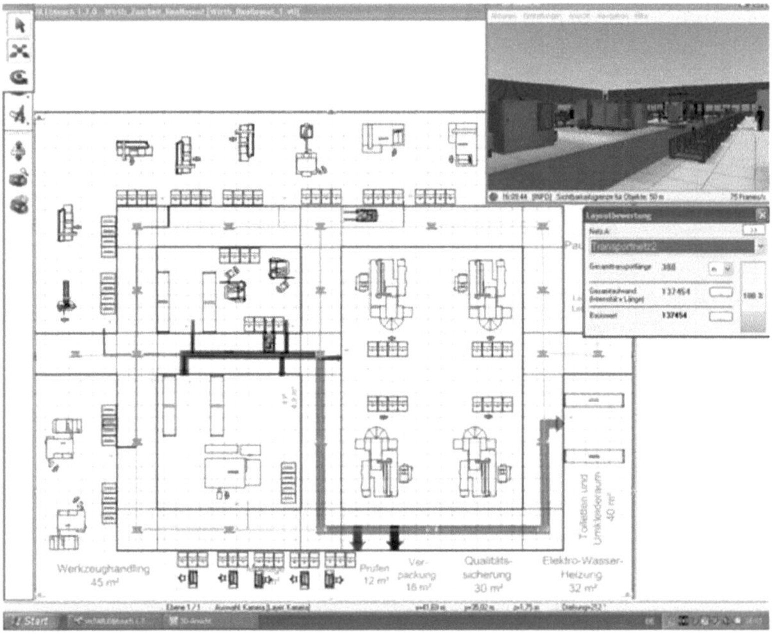

Fig. 3.71 Real layout for parts production and assembly

NB: pointers for layout design

- strive for maximum congruence between the real and ideal layouts
- keep footways for operating personnel as short as possible; keep emergency routes and safety areas clear
- ensure spatial separation of heavier and lighter, higher and lower equipment/systems to ensure better utilization of room heights and the carrying capacities of floors and ceilings in the buildings. Ensure spatial separation of production stations/equipment in the case of incompatible pressures on their environment, such as vibrations (presses, hammers), heat (drying ovens and forges), dust (grinding, polishing machines), noise (punches, presses).
- group together machines and manual workstations with similar environmental impacts or requirements (emissions, light, temperature and humidity conditions)
- minimize interference effects

- do not permit maximum workplace emission concentrations (MAK values) to be exceeded
- identify hazard areas
- ensure that equipment-related warning signals can be seen and heard from all equipment operating stations
- enable adequate flexibility (product mix, production scope) and ensure ability to expand (increasing capacity requirements) by transferring these structural units to the peripheral zones and by taking additional areas into account.

The multitude of alternative partial solutions that arise in the majority of cases will consequently lead to layout variants which should be kept in mind in the case of challenging planning projects (scope and complexity).

Evaluation – determination of preferred variants
The layout variants created are mainly evaluated using simplified methods (e.g. cost benefit analysis) using differently weighted, quantitatively and non-quantitatively ascertainable criteria, and the preferred variants will eventually be determined as the outcome of a cost / profitability comparison.

3.2.5.1.1 Technical rough planning

The specifics of real layout planning on the rough planning level are determined by the factory's structural units that are being arranged. This concerns the arrangement and flow-related interlinking of units with a view to

- material flow (main transport routes)
- energy flow (main conduits)
- information flow (main conduits, routes, coupling with material flow)
- work flow (course of movement and work processes).

Without including individual details, the approximate layout basically constitutes a fleshing-out of the general building development plan.

Over and above building floor plans and dimensions, the approximate layout provides information about the breakdown and utilization of the floor space within the building. The approximate layout should, for example, include:

- the location and size of the factory's structural units
- central facilities (e.g. sanitary and break rooms) and systems
- the location of doors, gates and main transport routes
- material flow-related interlinking facilities between the structural units
- the location of main supply and disposal routes (possibly documented in a separate plan)
- room heights, permissible floor load, differences in level
- key machinery and plants.

For certain operational functions special plans are drawn up to provide greater clarity, e.g. main trunking layouts for power supply.

a) Restrictions

The specific restrictions encountered during approximate layout planning should be recorded systematically. Tables 3.39 and 3.40 show regulatory restrictions and restrictions resulting from the location as examples.

Table 3.39 Regulatory restrictions affecting approximate layout planning (selection)

Regulatory restrictions	Sources
Transportation routes	§17 ArbStättV, ASR17
Emergency exit routes	§19 ArbStättV, ASR 19
Escape and rescue plan	§55 ArbStättV
Break rooms	§29 ArbStättV, ASR 29
Sanitary facilities	§37 ArbStättV, ASR 35, 37, 43
Doors and gates	§§10, 11 ArbStättV, ASR 10, 11

Table 3.40 Location-related restrictions affecting approximate layout planning (selection)

Location-related restrictions	Sources
Floor / ceiling carrying capacity	Building structural plans
Room heights	Building structural plans
External transport connections	General development plan
Supply and disposal connections / connection possibilities	Building mains, conduits and wiring plan
Existing structure of the building	Building structural plans
Pillars and partition walls	
Differences in level	
Existing foundations	
Existing doors and gates	

The layout of transport routes and social and sanitary rooms is also an objective and constituent part of layout planning.

b) Laying out of transport routes

The minimum width of the routes for traffic is determined by the width of the means of transport / goods being transported as well as by whether one-way or two-way traffic is to be planned for. According to the route's characteristics, additional border widths should be taken into account, as shown in fig. 3.72. These values are applicable for transport speeds less than or equal to 20 km/h. Higher transport speeds require greater additional margins. In the case of both foot and motorized traffic, additional margins of 0.75 m must be implemented. Under specific conditions, the total additions may be reduced to 1.10 m (see ASR 17).

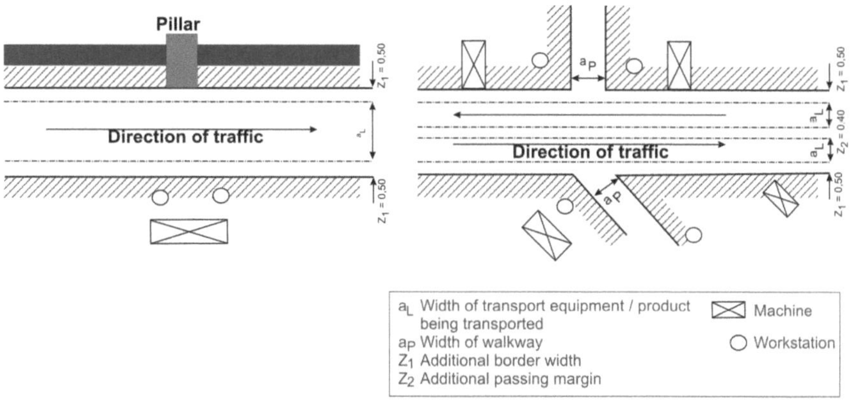

Fig. 3.72 Dimensions for transport routes according to ASR 17

Motorized traffic routes must include a minimum clearance of 1 m from doors and gateways, thoroughfares and stairways (see ASR 17). The width of footways should be specified subject to the number of people in accordance with table 3.41.

Table 3.41 Footway widths in accordance with ASR 17

Number of people (catchment area)	Path width [m]
up to 5	0.875
up to 20	1.00
up to 100	1.25
up to 250	1.75
up to 400	2.25

Transport routes must be routed in such a way that processes can take place on both sides. For this reason, transport routes should also not be routed alongside external walls, as this will result in the usability of process areas being restricted on at least one side.

c) Doors and gates

The location, number, layout and dimensions of doors and gates must be tailored to the type and use of the rooms (§ 10 ArbStättV).

Doors and gates must be arranged in such a way that from each point in the room a certain distance to the next nearest exit is not exceeded. The distance measured as the crow flies must be no more than the values shown in table 3.42.

Table 3.42 Distance from exits (ASR 10)

Room characteristics	Distance to the exit [m]
a) rooms other than as described in b to f	35
b) rooms subject to fire hazard without sprinkler systems or comparable safety measures	25
c) rooms subject to fire hazard with sprinkler systems or comparable safety measures	35
d) rooms subject to toxic hazards	20
e) rooms subject to explosive hazards, except for rooms in accordance with f	20
f) rooms subject to explosive material hazards	10

When installing the doors and gates, the building regulations of the respective country must be adhered to.

Doors and gates must be installed in such a way that when they are open they do not restrict the useable width of passing transport routes (ASR 10).

Heated buildings, which often open onto the outside, should be provided with air locks (double gates) or hot air locks. Adequate floor space must be provided.

d) Social and sanitary rooms (secondary floor space areas)

Social and sanitary facilities must be provided on the appropriate scale and with the appropriate fittings, subject to the number of employees, the genders working in the area and the type of activity. Special regulations are contained in the German workplace guidelines (ASR). Important points regarding the design of social and sanitary rooms are detailed below:

- If break rooms are to be made available, they must be arranged in such a way that wherever possible they can be reached by workers within 5 minutes. Break rooms must provide 1 m^2 of floor space per employee, and have a minimum size of 6 m^2 (ASR 29).
- Toilet facilities must be located so that they are no more than 100 m away from permanent workstations. The number of toilet facilities is dependent on the number of employees (see table 3.43). The route from permanent workstations in a building to the toilets should not travel through outdoor spaces (ASR 37). In general the floor space requirement calculated for a toilet facility including anteroom with washing facilities, access and space in which to move around is 3 to 5 m^2.

Table 3.43 Floor space values for toilet facilities

Employees	Men			Women	
	Number of toilets	Number of urinals	Floor space incl. washing facilities* in m²	Number of toilets	Floor space incl. washing facilities* in m²
Up to 5	1	-	4	1	4
Up to 10	1	1	4	1	4
Up to 25	2	2	7	3	9
Up to 50	3	3	12	4	11
Up to 75	4	4	15	6	15
Up to 100	5	5	18	7	18
Up to 160	7	7	22	10	25

* the floor space indicated corresponds to the minimum requirement if distribution is ideal; adaptation to the actual conditions necessary

- If special work clothing is required, changing rooms must be provided. The cloakroom space to be made available is 0.30 m^2 of locker space for each employee for a lockable unit and 0.50 m^2 of open floor and movement space (ASR 43).
- If required by the activity or for occupational health reasons, wash rooms must be made available. Wash rooms must be a minimum of 4 m^2 in size. The free floor space in front of a sink unit must be at least 0.7 m x 0.7 m (ASR 35). Including hand basins, space in which to move around and access, the floor space requirement per wash unit should be approx 1.00 to 1.50 m^2. If showers are also to be installed, then 2.00 to 2.50 m^2 should be provided per shower unit.

Changing, break and recreation rooms: Measurements are shown in table 3.44.

Table 3.44 Measurements of social rooms

Function	Measurements
Changing rooms	with individual lockers (hybrid system) approx. 1.4 m^2 with double lockers (separation system) approx. 2 m^2 per person in the case of dirty/clean separation approx. 2.5 m^2 per person in the case of shift operation with overlap premium (125-180 % of the larger shift)
Break rooms for recreation	measurement basis (simultaneity) approx. 10% of workforce floor space requirement: 1.5 - 3.5 m^2 per person in open plan offices: 0.3 - 0.5 m^2 per workstation

Secondary areas and areas for (central) building systems / supply and disposal / central conduits that are close to the production facilities should be taken into account at the point when the floor space requirements in the production area are determined, as these are usually closely linked in spatial terms.

Administration area – the areas for functions such as accounting, purchasing, operations scheduling and production control can be determined in detail by means of specific layout planning or via commonly-used floor space characteristic values. Table 3.45 provides useful guidelines in this respect.

Table 3.45 Characteristic floor space values for administrative areas

Type of sub-area	Dimensions and reference basis
Administrative office - Foreman - Secretary - Draftsmen, design engineers	 8-12 m^2 (14-18 m^2) per person (2 people) 12 - 15 m^2 per person 14 - 18 m^2 per person
Social room	1.8 m^2 per person
Dining area, kitchen and adjoining room	2.2 m^2 per person
Changing area	0.5 m^2 per person
Assembly and breakfast room	1.5 m^2 per person
Toilet room	1 place for every 10 people
Wash room	1 place for every 5 people
Parking area	25 m^2 per car

Secondary areas – Statutory and regulatory provisions apply when dimensioning the required secondary areas, especially for social and sanitary facilities. In particular, the German regulations governing the workplace (ArbStättV) and the associated workplace guidelines (ASR) stipulate minimum values for floor space dimensioning based on the number of male and female workers employed.

Evaluation: The layout planning must be evaluated continuously using the following criteria (based on Kettner 1984)

- production flow (material, information and energy flow)
- possibilities for expansion and change
- working conditions
- flexibility of use
- anticipated investment and operating costs
- environmental protection including energy efficiency

3.2.5.1.2 Detailed technical planning

The results of real planning (real layout) for the entire planning project must be brought to their final, fully-fledged form (detailed layout) with the aim of ensuring the

requirement-based, smooth interaction of
man ←→ machine ←→ process
at every workstation.

Detailed layout planning includes the development of the layout of a manufacturing facility, with machines, equipment, workstations and production areas being arranged according to flow system-specific, occupational safety-related and ergonomic aspects.

The detailed layout provides information about the layout and installation of machinery, equipment, workstations and production areas and contains, among other things, details of the floor space utilized by the objects in all of their working positions, their descriptions and particular characteristics. The THS facilities are also included here.

A characteristic of detailed layout planning is that the employees and subsequent operators of the production facilities are carefully factored into the planning process. As a result, not only is the acceptance of planning outcomes enhanced, but also a smoother implementation is ensured. An example of the detailed dimensioning of a machine is shown in fig. 3.73.

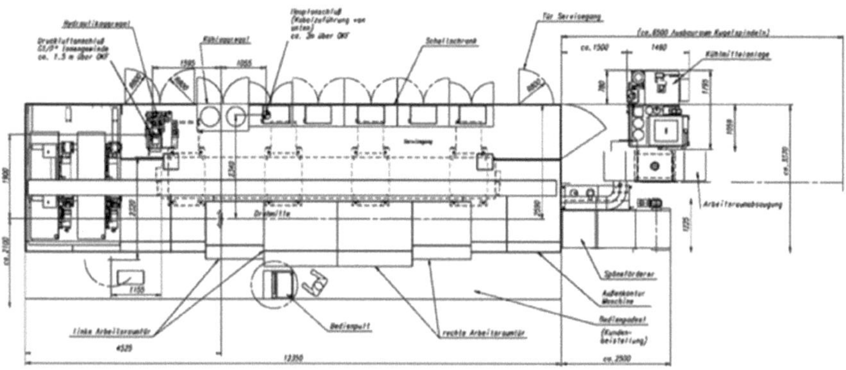

Fig. 3.73 Dimensioning of a machine (source: NILES SIMMONS)

Important objectives of detailed layout planning are:

- optimal working conditions for the workers (with respect to operation, workload etc.)
- occupational safety-compliant workstation design
- flexible, floor space-saving layout of machinery, equipment, workstations and production areas

- ensuring an effective flow of materials between the machines, equipment, workstations and production areas
- ensuring that costs are justifiable
- implementation of the production structure planning requirements.

The actual detailed layout constitutes a fleshing out of the approximate layout on the production structure level. It contains all detailed information on the objects to be coordinated and the actual environment.

Over and above the information contained in the rough layout, the detailed layout provides an insight into

- the location and size of the machines and equipment in all working positions
- connected loads for supply and disposal
- positioning of auxiliary THS equipment
- foundations required
- occupational safety facilities
- interference effects.

a) Specific restrictions
must be recorded systematically for detailed layout planning. Tables 3.46 and 3.47 show regulatory restrictions and location-related restrictions as examples.

Table 3.46 Regulatory restrictions affecting detailed layout planning (selection)

Regulatory restrictions	Source
Movement space in the workstation	§24 ArbStättV
Workstation measurements	DIN 33406
Noise	§15 ArbStättV, TA Lärm
Lighting	§ 7 ArbStättV, DIN 5034, DIN 5035
Climate	§§ 5, 6 ArbStättV, DIN 33406

Table 3.47 Location-related restrictions affecting detailed layout planning (selection)

Location-related restrictions	Source
Pillars and walls	Building plans
Light incidence	Building plans, general development plan
Transportation routes	Approximate layout
Foundations	Approximate layout, manufacturer's description
Supply and disposal connections	Approximate layout

b) Layout and installation of objects
Layout and installation are influenced, among other things, by

- the available floor space/type of floor space
- the course of transport/circulation routes
- planned multiple machine operation (MMO)

Fundamentally, all installation variants can be classified according to the following two types:

- stationary installation (commonly with foundations)
- variable installation (movable installation)

Relative to the transport route, machine installation can be planned as

- parallel installation
- installation at right angles
- diagonal installation

(see fig. 3.74).

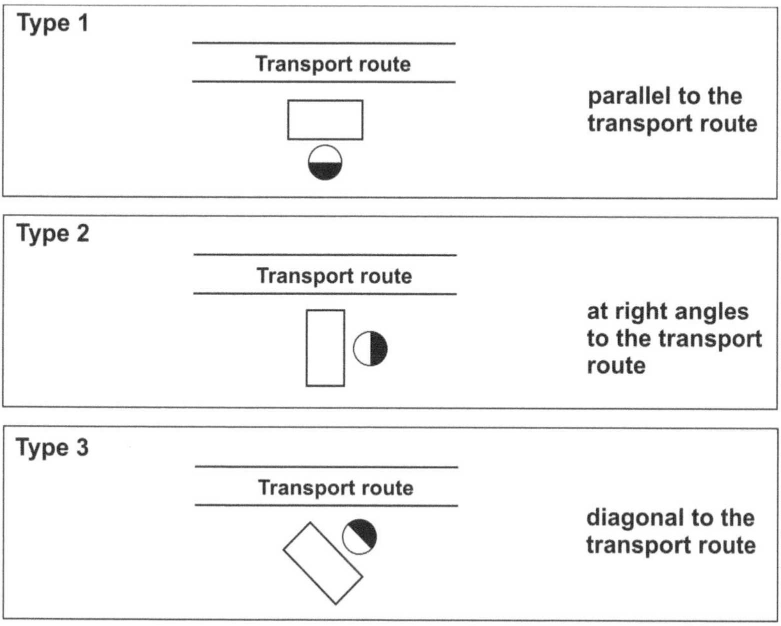

Fig. 3.74 Basic types of machine installation

The advantages and disadvantages of the installation types include:

Type 1: narrow installation width, good transfer possibilities
Type 2: broad installation width, low production area/floor space for the given transport route width
Type 3: large installation area, good material in-feed, convenient for daylight

The production areas should be laid out in basic geometric forms (line, triangle, polygon, circle) (cf. fig. 3.75).

Machine installation	Advantages	Disadvantages
Polygon 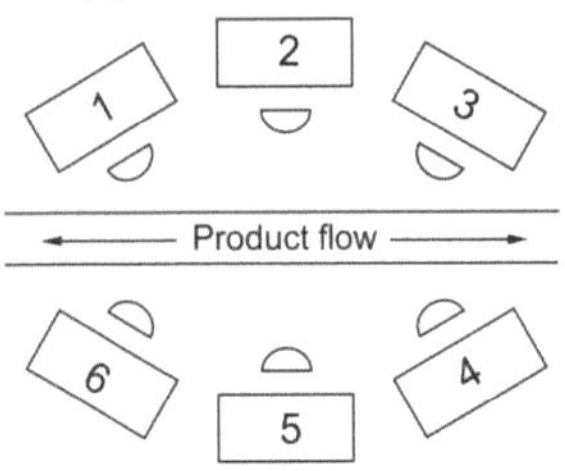 	■ shortest service aisles ■ best visibility	■ large space requirement ■ only suitable for roughly similar machine sizes ■ storage of workpieces impeded to some extent
Double row 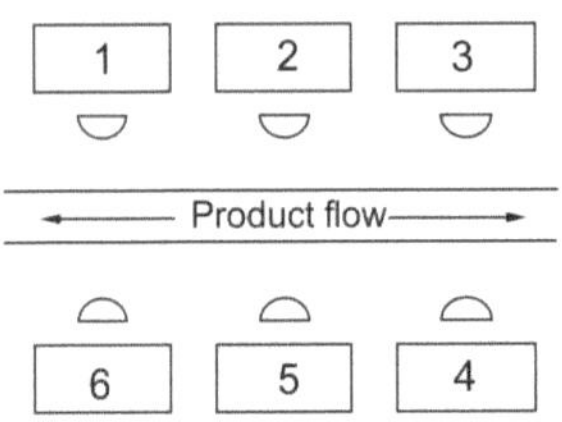 	■ convenient service aisles ■ good visibility ■ convenient workpiece storage	■ lengthening of service aisles if differing workpiece flows ■ overall visibility impeded in case of poss. requirement for diagonal installation (e.g. automatic lathes)
Row 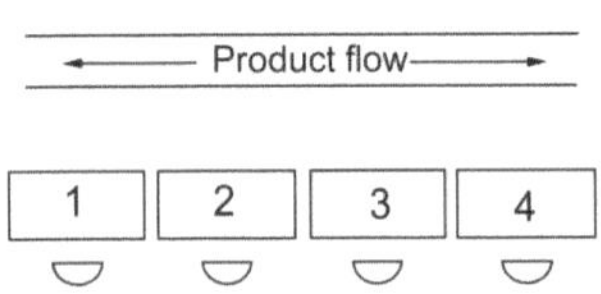 	■ lowest space requirement (transfer lines) ■ product flow rigidly interlinked	■ longest service aisles ■ overall visibility impeded

Fig. 3.75 Forms of machine installation

Decisions regarding the type of installation are governed, among other things, by

- the type and degree of permanence of the production program,
- the type, size and function of the facility being laid out

- the technical resolution of flow relationships,
- and the structural parameters.

c) Dimensional design (minimum and safety clearances)

The clearances between the equipment and objects must primarily guarantee the safety and unrestricted ability to work of the operating personnel and the unhindered functioning of the equipment (safety clearances). Simultaneously, their specification in terms of size is decisive for the optimum utilization of floor space (minimum clearances). Default measurements are available to the factory planner in the relevant tables for the minimum / safety clearances, in particular between the machining and processing equipment and other entities (e.g. walls) and transport/circulation routes (Woithe; Rockstroh; Papke et al). Overviews of recommended minimum clearances in workstations are provided for parts manufacture in fig. 3.76 and assembly in fig. 3.77 as examples (after Woithe 1965).

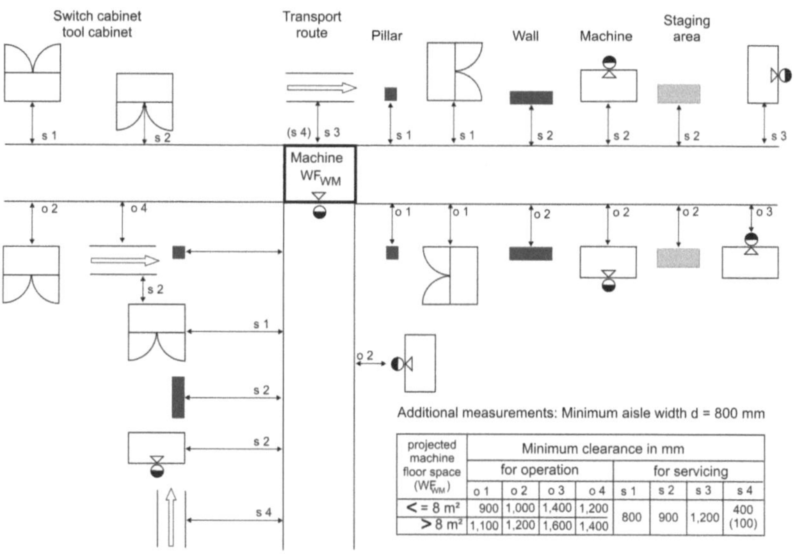

projected machine floor space (WF_{WM})	Minimum clearance in mm							
	for operation				for servicing			
	o 1	o 2	o 3	o 4	s 1	s 2	s 3	s 4
<= 8 m²	900	1,000	1,400	1,200	800	900	1,200	400 (100)
> 8 m²	1,100	1,200	1,600	1,400				

Fig. 3.76 Recommended minimum clearances for machine workstations (after Woithe 1965)

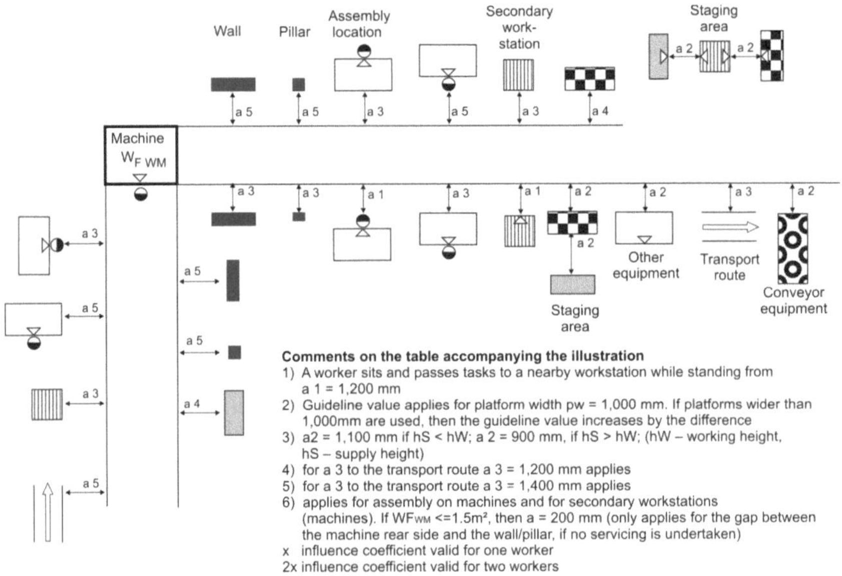

Fig. 3.77 Recommended minimum clearances for assembly workstations (after Woithe 1965)

The dimensions of the production structure are designed in close conjunction with the design of the workstation.

Adequate space in which to move around in the workstation of at least 1.50 m^2 must be provided; this must not be less than 1.00 m wide at any point. If this is not possible, then an area of the same size must be made available nearby.

Recommended minimum clearances for assembly workstations are identified in Fig. 3.78.

Determination of the clearance measurements for	Number of workers between the boundary lines	Influencing of the clearance measurements by: Accessibility only for servicing	Use of platforms	Supply height <or> working height	Seated	Standing (m < = 5 kg, l < = 30 cm)	Standing (l <= 190 cm, m < = 30 kg M; m < = 15 kg Wo)	Standing (use of hoists)	Aisle (m <= 5 kg, l < = 30 cm)	Aisle (l <= 190 cm, m < = 30 kg M; m < = 15 kg Wo)	Aisle (use of hoists)	Aisle (manually moveable floor conveyors)	Guideline values in mm
a1	1					x							1,000 1)
							x						1,000
								x					1,200
	2				2x								1,400
						2x							1,300
						x	x						1,600
							2x						1,600
							x	x					1,800
								2x					2,100
						x		x					1,800
			x			x							1,900 2)
			x				x						1,900 2)
			x					x					2,200 2)
	3					2x			x				1,900
							2x			x			2,200
							2x				x		2,400
						x	x		x				1,900
							x	x	x				2,100
							x	x		x			2,400
a2	1					x							1,000
							x						1,100
								x					1,200
									x				700
												x	1,100 3)
				x									1,100 2)
			x										1,700
	2				x				x				1,100
						x			x				1,200
						x				x			1,500
						x					x		1,700
						x						x	1,800
							x		x	x			1,200
							x						1,500
							x					x	1,800
								x	x				1,400
								x		x			1,700
a3	1				x								1,000 4)
						x							900 4)
							x						1,000 4)
								x					1,200 5)
			x										1,700 2)
	2	see a2 (2W) guideline values do not apply for a3 to the transport route											
a4	1	x											900 6)
				x									900
						x							900
							x						1,000
								x					1,200
a5	1	x											900
									x				700
										x			1,000
											x		1,200
												x	1,100
	2								2x				1,200
									x	x			1,300
									x		x		1,500

1) ... 6) see comments on the accompanying illustration
a1 ... a5 clearance measurement
m mass of workpieces and tools
l length or size of workpieces and tools

M Men
W Women
Wk Worker

Fig. 3.78 Recommended minimum clearances for assembly workstations (after Woithe 1965)

d) Foundation planning (installation, mounting, alignment and insulation)

If high weight bearing requirements arise from individual or a small number of machines and systems then uncoupled foundations can be considered as an alternative to continuous floor loading.

Uncoupled foundations are areas on the ceiling/floor that are able to absorb elevated load requirements and interference effects (e.g. vibrations). As a rule, uncoupled foundations are built into the ground floor (or basement) for increased weight absorption. The uncoupling is achieved through a structural separation of the existing ceilings/floors. In multi-story buildings, the emphasis is placed on uncoupling the vibrations that are generated.

The load requirements in areas that are not used for production, assembly and storage tasks are generally low. Typical loading values range from 2 to 5 kN/m^2.

All matters regarding installation, mounting, alignment and insulation of machines and systems are part of foundation planning. (Rockstroh 1979, p. 21 ff)

Reasons for foundation planning:

- transfer of vibrations to workers or the surroundings should be prevented.
- the floor space of equipment with a relatively small footprint and a large mass needs to be increased (reduction in contact pressure).
- there is a need for a high level of machining accuracy to be achieved.
- the carrying capacity of floors in multi-story buildings is too low.

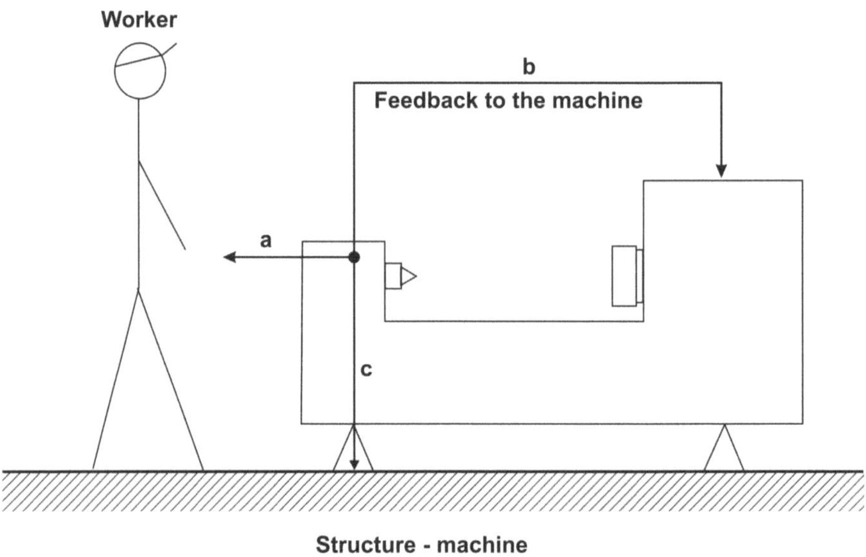

Fig. 3.79 Direction of action of vibrations

Direction of action towards the worker (a)

- Critical resonance frequencies in the case of vertical vibrations are:
- on the entire body 4-6 Hz
- on the brain 15-20 Hz
- on the eyes 40-60 Hz

In the case of horizontal vibrations, additional stresses arise at frequencies of 0.5 – 20 Hz. For the permissible vibration exposure times, refer to ISO standards 2631 and 2057.

Direction of action towards the machine system (b)

In this case vibrations can compromise the operating characteristics, bring about increased wear and tear through dynamic vibrations of up to 100 Hz, and cause material fatigue and a reduction in tool life. For assessment purposes, criteria are set for mechanical vibrations in VDI guideline 2056.

Direction of action towards the building and other systems (c)

Here the vibrations are transferred to parts of the building, other machines and plants via the supporting structure, foundations and subsoil. In principle there are two ways of preventing vibration transfer.

(1) Active suppression – prevention of transfer of vibrations to the surroundings
(2) Passive suppression – prevention of vibrations from the surroundings

Tasks of foundation planning and installation of the production facilities and other equipment are:

- absorption/transfer of the static base load
- absorption and transfer (in accordance with requirements) of the dynamic forces (vibrations) in operating mode
- fixing and stability of position in operating mode taking into account possible planned mobility of the equipment.

Forms of implementation are distinguished between those for

- fixed layouts
  - mounting elements without insulation effect (e.g. hooks, anchoring bolts etc.)
  - elements with active or passive vibration insulation (e.g. machine mounts, insulating packets with/without bolt insulation)
- flexible layouts
  - foundation-free installation using, among other things, supporting bolts, rubber bumpers (vibration insulated)

Types of mounting (see table 3.48)

Table 3.48 Types of mounting for machines and systems.

Type of mounting	Selection criteria
No mounting	Perpendicular force path of the working spindle
	High degree of stability of the machine's body
	Correct ratio of the base to the height of the machine
	Low natural frequency of the machine foundation system
Embedding	High degree of stability of the machine's body
	Vertical or horizontal force path
	Low occurrence of transverse forces
	Abnormal footprint in relation to the height of the machine
	Roughing with low cutting capacity
Bonding	Foundations must consist of unlayered material
	Machines that will be in the same position over a longer period of time
	Low cutting capacity
	Low natural frequency of the machine foundation system
Bolting	Roughing with high cutting capacity
	Finish machining with high-quality surface finish
	Increase in the footprint of the machine
	Large transverse forces
	High natural frequency of the machine foundation system
	Signs of imbalance of rotating parts

Note: Mounting only where necessary!

Fig. 3.80 shows types of foundation design by installation, alignment and vibration insulation type.

Installation types

- directly on the concrete floor
- on a support grid
- on block / strip foundations
- on oscillating foundations

Types of alignment

- no alignment
- with steel blocks
- with set screws
- with driving wedges

Types of vibration insulation

- no vibration insulation

- vibration insulation through interstices between foundation and floor
- vibration insulation through flexible layers between machine and substructure
- vibration insulation through spring elements between machine and substructure

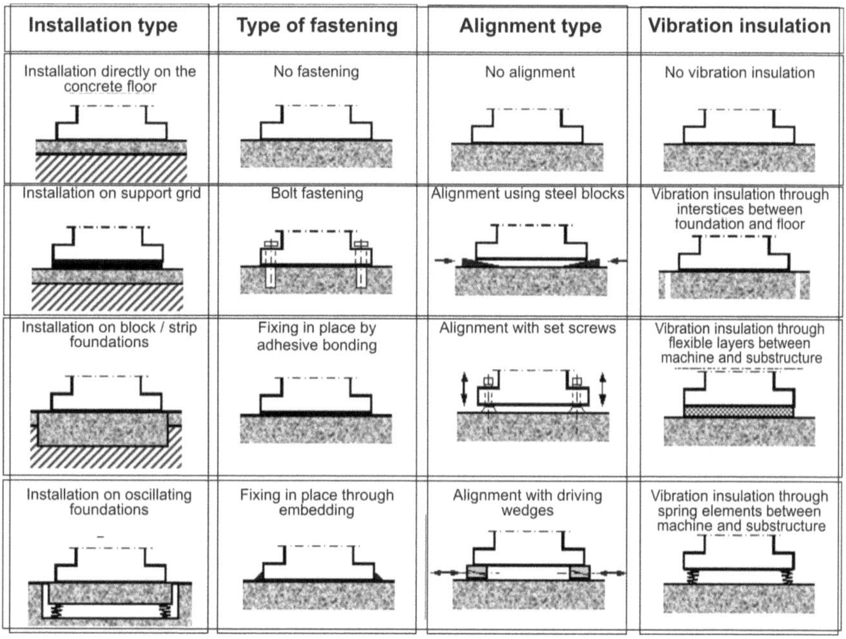

Fig. 3.80 Types of foundation design

A few selected examples of different forms of machine installation and foundation design are shown in fig. 3.81.

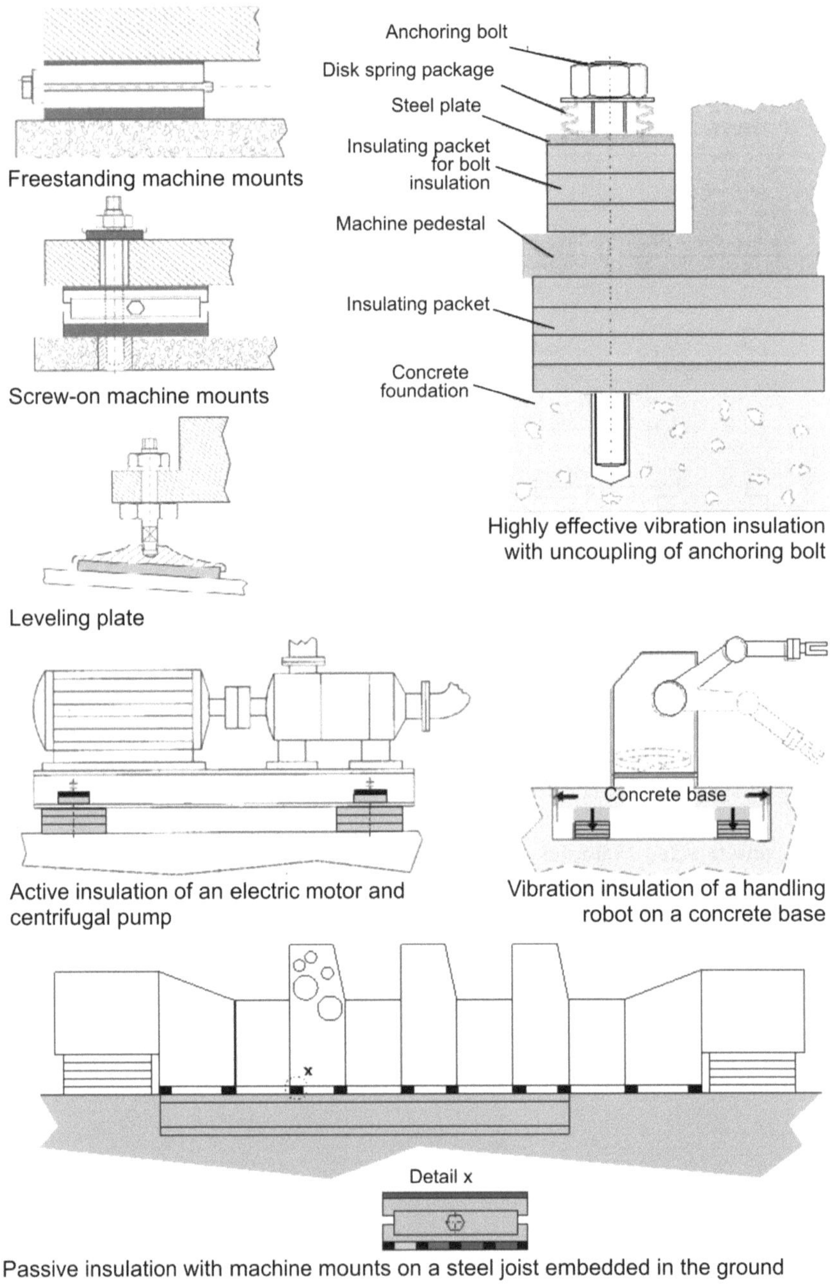

Fig. 3.81 Examples of machine and equipment foundation design (source: ISOLOC).

The outcome of the process of determining the types of foundation to be used is the foundation plan. It is part of the detailed layout planning and contains all information and measurements for installation, mounting, alignment and vibration insulation of the object (machine). fig. 3.82 shows an example of foundation planning. All measurements for the installation of cables, pipes and other connections for supply and disposal, and for the mounting and structural layout (openings, drill holes, channels, recesses) are set out in it.

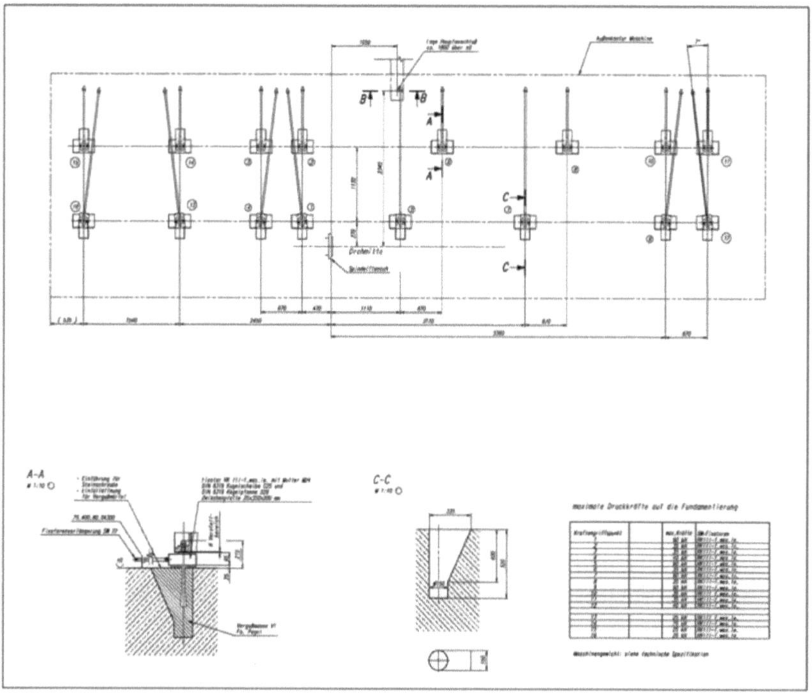

Fig. 3.82 Section from a detailed layout – foundations (source: NILES SIMMONS)

Note: the final machine installation plan can only be completed in terms of dimensions when the foundation plan with all structural details (coordinates and dimensions) has been fixed. First determine the dimensions for the foundation plan, then the machine installation plan.

e) Ergonomic workstation design

DIN 33406 is a useful tool for laying out workstations in the production area. Here, in view of the postures assumed while working, the workstations are categorized according to the following individual workstation types

- seated workstations
- standing workstations
- seated/standing workstations.

Wherever possible from a technical and task point of view, seated and seated/standing workstations are preferable, as they enable one-sided static muscle strain to be relieved / the strains arising from a one-sided body posture to be reduced.

Workstation measurements that are not dependent on the task include measurements governed to some extent by statutory regulations (see § 24 ArbStättV), and those that are determined anthropometrically to a certain degree. Measurements (not dependent on the type of workstation) that must be taken into account are:

- lateral clearance >= 1000 mm
- sagittal clearance >= 1000 mm

The dimensions are defined as follows (DIN 33406):

The *lateral clearance* is the distance between the longitudinal center plane of neighboring workstations. The *sagittal clearance* is the unobstructed width between the vertical reference plane and the closest point to the body of the workstation boundary behind the body. Task-dependent workstation measurements must be specified subject to the activity to be undertaken and the body height (see fig. 3.83).

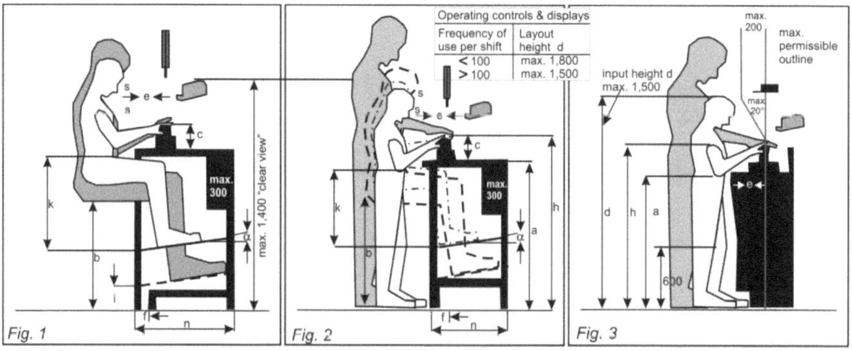

Term		Seated workstation (Fig. 1)	Seated/standing workstation (Fig. 2)	Standing workstation (Fig. 3)
Working position distance	e	0 to 325 A minimal distance to the front edge should be striven for		*Influence on:* arm posture, visual distance s, head inclination
Working height	h	See note for measurement c	Reference values for the working height (working area of the fingers is definitive) see Fig. 5	
Work surface height	a	min. 900 From a = 950 k = 520 to <u>750</u> is possible	900 to 1,080 *subject to:* body height equipment height workpiece height	*Influence on:* arm posture visual distance s head inclination
Work surface bottom edge to working position	c	keep as small as possible, large gaps cause awkward arm postures (posture work)		-
Knee space to working position	k_1	min. 250		-
	k_2	-		max. 250
Knee space: clearance width	k_3	800, in exceptional cases 600; It must be possible to stand up by pivoting the chair around		min. 600
Footrest: Free knee space Depth Inclination	 k n α	Installed in the workstation 520 to 750 (height adjustable), where a = 900 only k = 520 to 700 possible Target 550 (freedom of movement) 5° to 10°		-
Free foot space: clearance height	i	also for seat bracket min. 120		
Foot space to working position	f	max. 150		
Seating height (seat height)	b	adjustable 530 to 900, rolling until b = 650 permissible adjustable range min. 180, adapted to work area height		-
Visual distance 1)	s	when seated s is almost independent of the body size and can be adjusted via the seating height.	when standing s is strongly dependent on body size and can be influenced via the working height (see reference values Fig. 5)	

1) The visual distance is dependent on several factors. Reference values can be determined through comparisons with similar workstations or through testing. It should be possible to compare the following influences:

- Size and shape of the details that need to be identified
- Type of activity
- Color of object and surroundings
- Illumination
- individual visual faculty

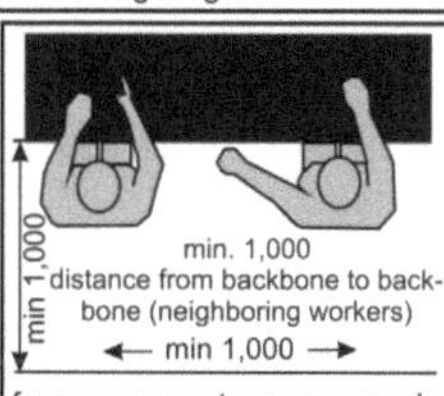

free movement area per workstation at least 1.5 m² (ArbStättV §24)

Fig. 4 View from above with knee space, back space, distance to from backbone to backbone and free movement area

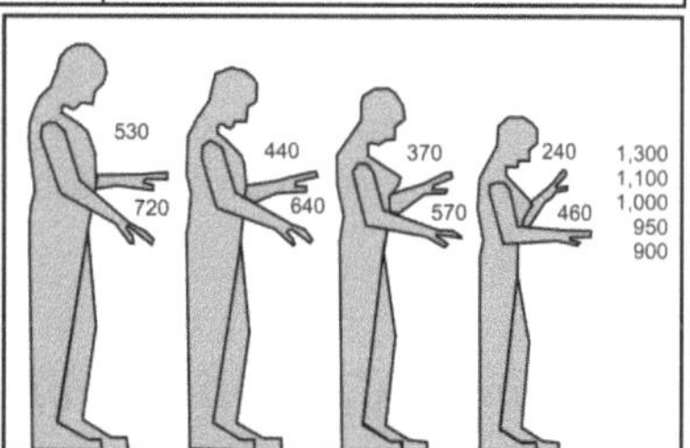

Fig. 5 Visual distances and lower arm position illustrated dependent on body size and working height

Fig. 3.83 Clearance measurements for seated and standing workstations

3.2.5.2 Detailed layout - example

The outcome of layout planning is the detailed layout, on which the implementation of the project is based (cf. fig. 3.84).

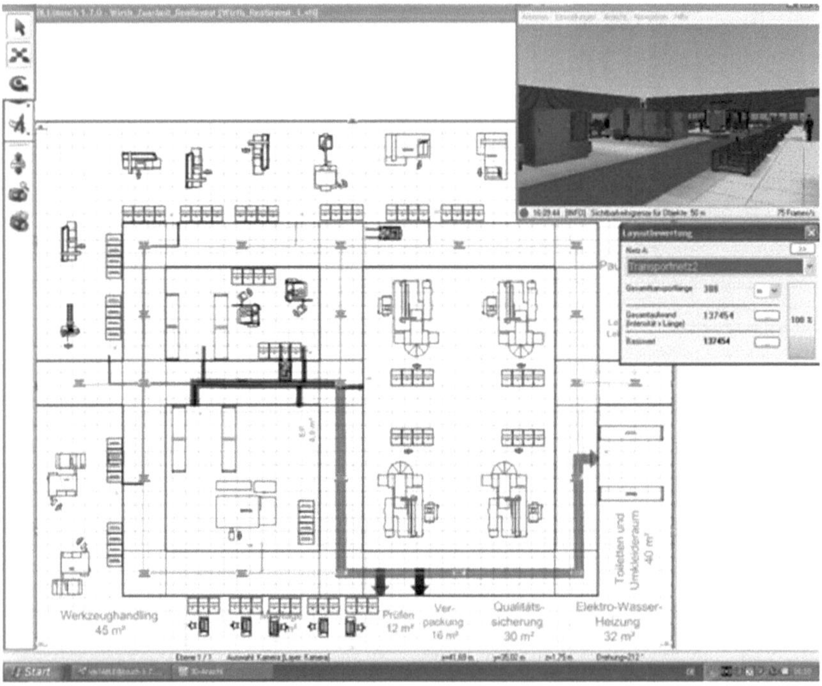

Fig. 3.84 Production section for pump impeller production (example of a detailed layout)

Note: in addition to dimensioning the layout of objects and pathways, technical and ergonomic design also includes workstation design tailored to people (workers) in terms of dimensions.

3.2.6 Complex planning project (example)

Parts manufacture and assembly – worm gear production (GUNT Company, Hamburg; Prof. PRÊT, Berlin). The example was worked through using the 0 + 5 + X planning model. In particular, this entailed the five specifications (01 – 05) regarding the project definition (complex I), the five project planning steps (5/1 – 5/5) for project development (complex II) and the profitability assessment specification (X6) for the project implementation (from complex III).

List of abbreviations

η_E	Degree of efficiency of the equipment	EM	Employee
PT	Payback time	f_W	Weighting factor
f_{CS}	Hourly set-up cost factor	f_{Cm}	Hourly manufacturing cost factor
C_S	Set-up costs	Cr_W	Criteria weight
C_m	Manufacturing costs	C_{Mat}	Material costs
CV_{t*}	Gap to the net present value in t* from zero (for amortization)	CV_{t*+1}	Gap to the net present value in t*+1 from zero (for amortization)
t*	Period during which for the last time there was a negative cumulative present value	D_W	Number of working days per year
T_{WFvj}	Target operating time of a worker	T_{need}	Time needed
t_u	Production time per unit	T_{Evj}	Target operating time of the equipment
t_s	Set-up time	T_{av}	Available time
GO	Goods outwards storage	GI	Goods inwards storage
W_C	Value equivalent for criteria weighting	X_{req}	Required number of units
X_L	Lot size	X_{opt}	Optimal lot size
N_{WF}	Number of workers	N_E	Amount (number) of equipment
N_E^*	Amount (number) of equipment (calculated)	N_L	Number of lots
IS	Interim storage	N_{Sh}	Number of shifts
$N_{TU,L}$	Transport units per lot		

3.2.6.1 Complex I - Project Definition

(01) Specification of input variables

Product program:	"M 000" worm gear
Materials:	GG-20, C45, 16MnCr5 BG, 42CrMo4 (semi-finished products cf. table 3.49)
Production technology and processes:	Parts manufacture and assembly
Number of units:	35,000-37,000 units/a
Production type:	Series production
Vertical integration:	Individual parts manufacture (cf. step 5/1) and assembly; purchase of standard parts; abrasive blasting, corrosion protection, hardening and hobbing - external cooperation
Plant and equipment information:	Processing machines (cf. 5/2 – (5)); one goods inwards store and one goods outwards store, an interim store; assembly stations
Logistics technology:	Mesh box pallets (800 x 1200 x 650 mm DIN 15142 / 1,000 kg), front loader
Working time:	Shift operation (7.5 h/shift), 225 d/a
Health, labor and environmental protection:	Laws, ordinances and guidelines
Requirements:	Use of an existing production building (50 x 40 x 6 m), reconfiguration; an office complex is annexed to the production building. (cf. fig. 3.85)

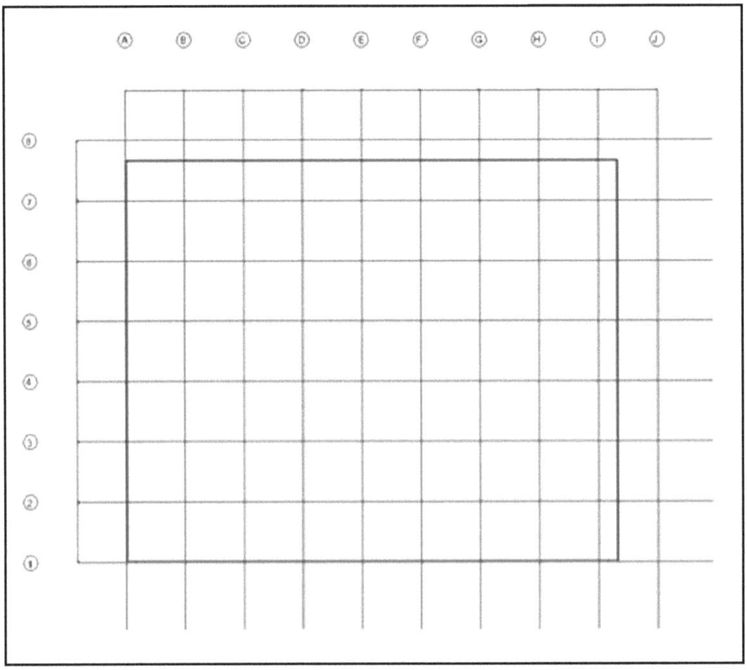

Fig. 3.85 Floor plan of existing production building

(02) Specification of the scope of analysis and corporate objectives

Corporate objectives:

1. favor modern, flow-based manufacturing principles
2. area and machinery arrangement tailored to material flow
3. logistically favorable transport route planning
4. minimal stocks (optimize storage and buffer quantity)
5. good access to machines and workstations (materials, personnel, information, utilities)
6. good floor space utilization (low floor space consumption)
7. good usability of residual floor space
8. humane and safe working conditions
9. implement multiple machine operation
10. overall transparency
11. ability to expand, flexibility and adaptability

a) Sales and marketing concept

Development of the market position

b) Product concept
The "M 000" worm gear is a fixed gear with only one fixed transmission ratio. It is used to adapt the torques and rotation speeds of a drive train to a load in accordance with requirements.

Typical uses for worm gears include traction drives for cranes and trolleys, escalators, augers, cable winches, lifting spindles or wiper drives on motor vehicles.

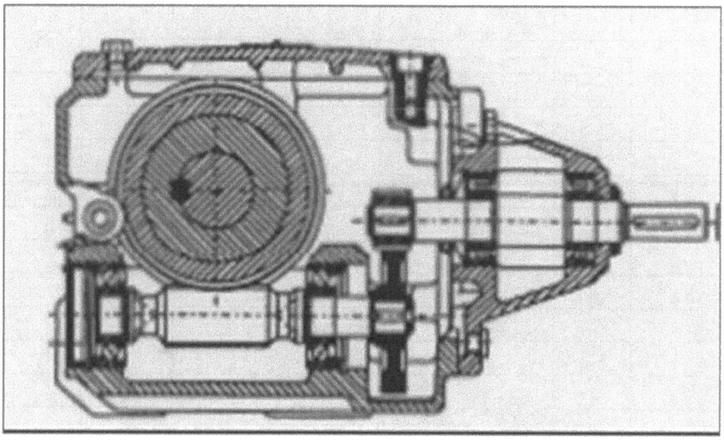

Fig. 3.86 "M 000" worm gear

c) Production concept:

- according to "Make or Buy" decision:
- in-house production according to parts list (cf. step 5/1)
- outsourced production/purchase parts according to parts list (cf. table 3.49)
- external cooperation for abrasive blasting, corrosion protection, hardening and hobbing work processes
- mechanical parts manufacture and assembly

Table 3.49 Worm gear and transmission housing parts list

1	2	3	4	5	6	7
Pos.	**Quantity**	**Unit**	**Name**	**Item code / Standard - short ref.**	**Purchase**	**Comment**
1	1	unit	Worm gear housing	M 001		GG-20
2	1	unit	Housing cover	M 002		GG-20
3	1	unit	Transmission housing	M 010		
4	1	unit	Worm gear	M 003		16 MnCr 5 BG
5	2	unit	Angular ball bearing	DIN 628 - 7304 B	X	
6	2	unit	Shim	DIN 988 - 42 x 52 x 2.5	X	
7	2	unit	Locking ring	DIN 472 - 52 x 2.0	X	

8	2	unit	Shim	DIN 988 - 42 x 52 x 0.1	X	
9	1	unit	Sealing cap	52 x 10	X	Busak + Luyken
10	1	unit	Feather key	DIN 6885 - B5 x 5 x 14	X	
11	1	unit	Drive wheel	M 004		16 MnCr 5 BG
12	1	unit	Locking ring	DIN 471 - 16 x 1	X	
13	1	unit	Output shaft	M 005		C 45
14	1	unit	Feather key	DIN 6885 - B 10 x 8 x 42	X	
15	1	unit	Worm wheel	M 006	X	
16	2	unit	Ball bearing	DIN 625 - 6207 Z	X	
17	2	unit	Shim	DIN 988 - 56 x 72 x 0.1	X	
18	2	unit	Locking ring	DIN 472 - 72 x 2.5	X	
19	1	unit	Shaft sealing ring	DIN 3760 - AS 35 x 72 x 12	X	
20	1	unit	Sealing cap	72 x 9	X	Busak + Luyken
21	1	unit	Seal	M 007	X	It Ö
22	6	unit	Fillister head screw	DIN 912 - M8 x 16	X	
23	6	unit	Screwed sealing plug	DIN 908 - M10 x 1	X	
24	7	unit	Sealing ring	DIN 7603 - A 10 x 13.5	X	It
25	1	unit	Vent screw	M 101	X	St
26	1	unit	Feather key	DIN 6885 - A 8 x 7 x 50	X	
27	1	unit	Engine seal	M 008	X	It Ö
28	4	unit	Spring washer	DIN 127 - A 8	X	
29	4	unit	Hexagon bolt	DIN 933 - M8 x 25	X	8.8
30	1	unit	Type plate	M 009	X	
31	2	unit	Grooved drive stud	DIN 1476 – 2.6 x 3	X	

d) Personnel concept:
Recruitment on the open labor market according to skills level

e) Profitability concept
Return on investment targets and costs

(03) Specification of basic planning cases

Basic case B – Reconfiguration of existing production facilities

(04) Specification of the planning scope

a) Planning phase: development/planning; core/conceptual planning
b) Planning objects: immovable items, moveable items, structures, personnel
c) Planning instruments: application of the “0 + 5 + X” planning model

(05) Specification of project design principles

- Approach according to the “top down” principle
- Use of flexibility and improvement principles (PDCA cycle), situation-driven planning

3.2.6.2 Complex II - project development

(5/1) Production program

Type of production program: definitive, restricted (cf. table 3.50)

Table 3.50 Production program – "worm gear" parts manufacture

Item	Drawing No.	Name	Raw material	Semi-finished product	Unfinished part weight [kg/unit]	Quantity [units/a]
1	M 001	Worm gear housing	GG-20	Cast part Model 183/97 (1)	7.5	36000
2	M 002	Housing cover	GG-20	Cast part Model 183/97 (2)	1.8	36000
3	M 011	Transmission housing	GG-20	Cast part Model 183/97 (3)	4.0	36000
4	M 012	Drive shaft	42CRMo4	Rd DIN 1013 30x169	0.9	36000
5	M 013	Drive pinion	16MnCr5 BG	Die forged part Rd DIN 1013 30 x 20	0.1	36000
6	M 003	Worm gear	16MnCr5 BG	Rd DIN 1013 35x159	1.2	36000
7	M 004	Drive wheel	16MnCr5 BG	Die forged part Rd DIN 1013 80 x 20	0.8	36000
8	M 005	Output shaft	C45	Rd DIN 1013 50x192	3.0	36000

(5/2) Determination of functions

(1) Product structure (cf. parts list table. 3.49)

(2) Manufacturing processes (parts production and assembly)

(3) Determination of manufacturing processes (operating sequence schematic and work schedules) (cf. fig. 3.87, table 3.51)

	Part produced in-house / Equipment	Worm gear housing	Housing cover	Transmis-sion housing	Drive shaft	Drive pinion	Worm gear	Drive wheel	Output shaft
1	(Abrasive blasting)	10	10	10		10		10	
2	(Corrosion protection)	20	20	20		20		20	
3	ABS 265				10		10		10
4	RAYO 165			30	20	30	20	30	20
5	H 100					40		40	
6	PE 150 C					50	30	50	
7	DMU 35	30	30	40					
8	CWK 500	40	40						
9	Deburring area	50	50			60	40	60	
10	(Carburizing)					70	50	70	
11	ZC 1				30	80	60	80	30
12	PF 150					90		90	
13	(Hobbing)						70		
14	Testing	60	60	50	40	100	80	100	40

Fig. 3.87 Operating sequence schematic – parts manufactured in-house (parts manufacture)

Table 3.51 Work schedules – assembly, testing, packing

Name of operating procedure		Equipment	Production time / unit t_u (s/unit)
50-10	Worm gear pre-assembly	Assembly area 1-3	55
50-20	Worm gear assembly	Assembly area 1-3	183
50-30	Drive shaft assembly incl. worm wheel	Assembly area 1-3	273
50-40	Sealing cap assembly	Assembly area 1-3	25
50-50	Sealing cap assembly with gasket	Assembly area 1-3	21
51-10	Drive shaft pre-assembly	Assembly area 4	55
51-20	Drive shaft assembly	Assembly area 4	83
51-30	Drive pinion assembly	Assembly area 4	49
52-10	Assembling of gear and drive	Assembly area 5	82

52-20	Filling of transmission	Assembly area 5	45
52-30	Assembly of vent screw with gasket	Assembly area 5	15
52-40	Housing cover assembly	Assembly area 5	42
60-10	Testing	Testing station	181
70-10	Type plate assembly	Packing area	38
70-20	Pack in film	Packing area	39
70-30	Print and enclose test report	Packing area	39
70-40	Pack in cardboard box	Packing area	35
70-50	Place on pallet	Packing area	35

Table 3.52 contains the (in-house production) workstations.

Table 3.52 Workstations – parts produced in-house

Part 1 - Worm gear housing M 001

Name of operating procedure	**Equipment**	**Set-up time ts [min/lot]**	**Production time per unit tu [min/unit]**
1-10 auto. abrasive blasting inside & out 1225	external	-	-
1-20 temp. corrosion protection 1225	external	-	-
1-30 milling of reference areas in jig (3 side processing), SPC	DMU 35	16	6
1-40 finish milling (5 side processing)	CWK 500	28	32
1-50 deburring	Deburring area	8	2
1-60 testing acc. to testing schedule	-	-	-

Part 2 - Housing cover M 002

Name of operating procedure	**Equipment**	**Set-up time ts [min/lot]**	**Production time per unit tu [min/unit]**
2-10 auto. abrasive blasting	external	-	-
2-20 temp. corrosion protection	external	-	-
2-30 face milling of face being built up to final machining allowance	DMU 35	12	2
2-40 drilling, facing and tapping of all drill holes, complete	CWK 500	30	2
2-50 deburring	Deburring area	8	1
2-60 testing acc. to testing schedule	-	-	-

Part 3 – Transmission housing M 011

Name of operating procedure	**Equipment**	**Set-up time ts [min/lot]**	**Production time per unit tu [min/unit]**

3-10 auto. abrasive blasting	external	-	-
3-20 temp. corrosion protection	external	-	-
3-30 turning complete	RAYO 165	22	10
3-40 drilling, facing and tapping of all drill holes, complete	DMU 35	20	4
3-50 testing acc. to testing schedule	-	-	-

Part 4 – Drive shaft M 012

Name of operating procedure	**Equipment**	**Set-up time ts [min/lot]**	**Production time per unit tu [min/unit]**
4-10 cutting 169 lg.	ABS 265	15	0.5
4-20 turning complete with grinding stock allowance	RAYO 165	25	5
4-30 grinding bearing and gasket seats	ZX 1	15	4
4-40 testing acc. to testing schedule	-	-	-

Part 5 – Drive pinion M 013

Name of operating procedure	**Equipment**	**Set-up time ts [min/lot]**	**Production time per unit tu [min/unit]**
5-10 auto. abrasive blasting	external	-	-
5-20 temp. corrosion protection	external	-	-
5-30 turning complete with grinding stock allowance	RAYO 165	18	4
5-40 broaching of channel	H 100	17	1
5-50 milling gear teeth, SPC	PE 150 C	24	4
5-60 deburring	Deburring area	5	1
5-70 carburizing	external	-	-
5-80 grinding ∅ 14M6 and plane surface	ZX 1	15	2
5-90 hob peeling of tooth flanks	PF 150	18	3.5
5-100 testing acc. to testing schedule	-	-	-

Part 6 – Worm gear M 003

Name of operating procedure	**Equipment**	**Set-up time ts [min/lot]**	**Production time per unit tu [min/unit]**
6-10 cutting 159 lg.	ABS 265	15	1
6-20 turning complete with grinding stock allowance	RAYO 165	25	5
6-30 milling worm gear teeth, SPC	PE 150 C	24	2
6-40 deburring	Deburring area	5	1
6-50 carburizing	external	-	-
6-60 grinding bearing seats	ZX 1	17	2
6-70 hobbing worm gear teeth	external	-	-
6-80 testing acc. to testing schedule	-	-	-

Part 7 – Drive wheel M 004

Name of operating procedure	**Equipment**	**Set-up time ts [min/lot]**	**Production time per unit tu [min/unit]**
7-10 auto. abrasive blasting	external	-	-
7-20 temp. corrosion protection	external	-	-
7-30 turning complete with grinding stock allowance	RAYO 165	25	5
7-40 broaching of channel	H 100	20	1.5

7-50 milling gear teeth	PE 150 C	24	4
7-60 deburring	Deburring area	5	1
7-70 carburizing	external	-	-
7-80 grinding internally and on front face	ZX 1	15	1.5
7-90 hob peeling of tooth flanks	PF 150 C	30	5
7-100 testing acc. to testing schedule	-	-	-

Part 8 – Output shaft M 005

Name of operating procedure	**Equipment**	**Set-up time ts [min/lot]**	**Production time per unit tu [min/unit]**
8-10 cutting 192 lg.	ABS 265	8	1
8-20 turning complete with grinding stock allowance	RAYO 165	25	7.5
8-30 grinding bearing seats	ZX 1	24	2.5
8-40 testing acc. to testing schedule	-	-	-

(4) Derivation of functional schematic

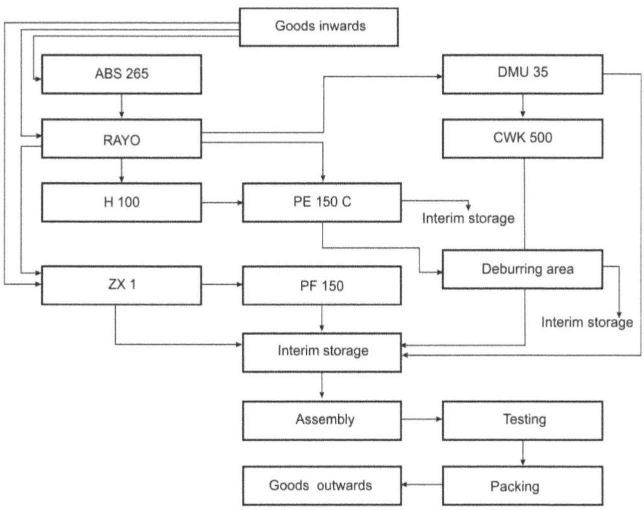

Fig. 3.88 Functional schematic (parts production and assembly)

(5) Pre-selection of equipment (technical and economic)
Technical:

No.	Name	Description	Manufacturer	Dimensions (mm)			Connected load (kW)	Weight (kg)
				Width W_{TE}	Depth D_{TE}	Height H_{TE}		
1	ABS 265	Band saw	Kurth GmbH	1,600	2,800	1,060	n.s.	n.s.
2	RAYO 165	Lathe	Pinacho	3,000	1,600	1,900	8.1	1,630
3	H 100	Broaching machine	STENHOJ/LLR Vertriebs GmbH	2,200	4,300	3,000	15	6,300
4	PE 150 C	NC hobbing machine	PFAUTER GmbH & Co.	2,200	4,300	3,000	48	7,500
5	DMU 35	Universal milling machine	Deckel-Maho	1,600	2,100	2,750	15	1,600
6	ZX 1	2-axis CNC cylindrical grinding machine with internal grinder	SCHAUDT GmbH	3,000	3,200	2,020	25	4,850
7	CWK 500	Processing center	Starrag Heckert AG	4,700	6,600	2,900	31	15,000
8	PF 150	CNC hobbing machine	PFAUTER GmbH & Co.	4,400	3,400	2,060	48	7,000

Fig. 3.89 Equipment characteristics

Economic (costs, lot sizes):

a) Formula for determining economic lot sizes according to ANDLER

$$X_{opt} = \sqrt{\frac{x_{tot} \times C_S \times 2}{C_m \times i_S}} \tag{3.69}$$

(C_S – Set-up costs; C_m – Manufacturing costs; i_S – Interest rate for storage)

b)

Table 3.53 Hourly cost rates

Item	Equipment	f_{CS}-Set-up [€/h]	f_{Cm}-Manufacturing [€/h]
1	ABS 265	18.90	15.90
2	RAYO	54.20	50.00
3	H 100	29.70	26.60
4	PE 150 C	57.30	52.10
5	DMU 35	38.30	34.30
6	CWK 500	48.60	44.00
7	Deburring area	18.40	15.30
8	ZX 1	35.30	31.20
9	PF 150	78.20	70.60

c)

Table 3.54 Determination of set-up and manufacturing costs

Item	Drawing No.	Name	t_s [min/lot]	f_{CS} [€/h]	C_S [€/lot]	t_u [min/unit]	f_{Cm} [€/h]	C_m [€/unit]
1	M 001	Worm gear housing	16	38.30	10.21	6.0	34.30	3.43
			28	48.60	22.68	32.0	44.00	23.47
			8	18.40	2.45	2.0	15.30	0.51
		Total			35.35			27.41
2	M 002	Housing cover			34.41			2.87
3	M 011	Transmission housing			32.64			10.62
4	M 012	Drive shaft			36.13			6.38
5	M 013	Drive pinion			81.41			12.66
6	M 003	Worm gear			61.76			7.46
7	M 004	Drive wheel			104.86			15.22
8	M 005	Output shaft			39.22			7.82

d)

Table 3.55 Determination of lot size X_L and number of lots N_L

Item	Drawing No.	Name	C_S [€/lot]	m [kg/unit]	Mat. price [€/kg]	C_{Mat} [€/unit]	C_M [€/unit]	M_C [€/unit]	X_{req} [units/a]	i_L [%]	X_{opt} [units/lot]	X_L [units/lot]	N_L [lot/a]
1	M 001	Worm gear housing	35.35			10.94	27.41	38.35	36000	8	911	450	80
2	M 002	Housing cover	34.41			3.50	2.87	6.37	36000	8	2,205	1,200	30
3	M 011	Transmission housing	32.64			5.93	10.62	16.55	36000	8	1,332	750	48
4	M 012	Drive shaft	36.13	0.9	1.15	1.04	6.38	7.42	36000	8	2,093	1,200	30
5	M 013	Drive pinion	81.41	0.1	1.00	0.10	12.66	12.76	36000	8	2,396	1,200	30
6	M 003	Worm gear	61.76	1.2	1.00	1.20	7.46	8.66	36000	8	2,533	1,200	30
7	M 004	Drive wheel	104.86			4.47	15.22	19.69	36000	8	2,189	1,200	30
8	M 005	Output shaft	39.22	3	0.80	2.40	7.82	10.22	36000	8	1,858	900	40

(to achieve shorter throughput times, X_L was set at approx. ½ X_{opt})

(5/3) Dimensioning

(5/3.1) Calculation of equipment quantities (parts manufacture)

Table 3.56 Calculation of amount of equipment for parts manufacture (shift variants using DMU 35 as an example)

Item	Equip-ment	Part no.	N_L [lots/a]	t_s [min/lot]	X_L [units/lot]	t_u [min/unit]	T_{need} [h/a]	D_W [d/a]	T_{Evj} [min/S]	η_{max}	N_{Sh} [1/d]	T_{av} [h/a]	N_E^* [unit]	N_E [unit]	η_E [%]
5	DMU 35	1	80	16	450	0	3621.3	248	480	0.90	1	1786	2.028		
		2	30	12	1200	2	1206.0	248	480	0.90	1	1786	0.675		
		3	48	20	750	4	2416.0	248	480	0.90	1	1786	1.353		
		Σ					7243					1786	4.06	4	101
5	DMU 35	1	80	16	450	6	3621.3	248	480	0.90	2	3571	1.014		
		2	30	12	1200	2	1206.0	248	480	0.90	2	3571	0.338		
		3	48	20	750	4	2416.0	248	480	0.90	2	3571	0.677		
	selected:	Σ					7243					3571	2.03	2	101
5	DMU 35	1	80	16	450	6	3621.3	248	480	0.90	3	5357	0.676		
		2	30	12	1200	2	1206.0	248	480	0.90	3	5357	0.225		
		3	48	20	750	4	2416.0	248	480	0.90	3	5357	0.451		
		Σ					7243					5357	1.35	2	68
1	ABS 265	Σ					1520				1	1627	0.93	1	93
2	RAYO 165	Σ					21981				3	5476	4.01	4	100
3	H100	Σ					1519				1	1706	0.89	1	89
4	PE 150 C	Σ					6036				2	3532	1.71	2	85
5	DMU 35	Σ					7243				2	3571	2.03	2	101
6	CWK 500	Σ					20452				3	5238	3.90	4	98
7	Deburring area	Σ					3622				2	3968	0.91	1	91
8	ZX 1	Σ					7247				2	3651	1.99	2	99
9	PF 150	Σ					5124				3	5357	0.96	1	96
	Grand total												17.3	18	96

(5/3.2) Calculation of the number of workers (parts manufacture)

Table 3.57 Calculation of the number of workers (machine operators)

Item	Equipment	Part no.	N_L [lots/a]	t_s [min/lot]	X_L [units/lot]	t_u [min/unit]	T_{need} [h/a]	D_W [d/a]	T_{WFvj} [min/S]	T_{av} [h/a]	N_{WF}
1	ABS 265	4	30	15	1200	0.5	307.5	225	450	1688	0.182
		6	30	15	1200	1	607.5	225	450	1688	0.360
		8	40	8	900	1	605.3	225	450	1688	0.359
		Σ					1520			1688	1.0
2	RAYO 165	Σ					21981			1688	14.0
3	H100	Σ					1519			1688	1.0
4	PE 150 C	Σ					6036			1688	4.0
5	DMU 35	Σ					7243			1688	4.0
6	CWK 500	Σ					20452			1688	13.0
7	Deburring area	Σ					3622			1688	2.0
8	ZX 1	Σ					7247			1688	4.0
9	PF 150	Σ					5124			1688	3.0
	Grand total										46

(5/3.3) Floor space dimensioning (parts manufacture and assembly)

The floor space is calculated according to the detailed floor space dimensioning method using plus factors.

Table 3.58 Calculation of floor space requirement (production area floor space)

Item	Equipment	Width W_{PA} [m]	Depth D_{PA} [m]	W_{PA}+0.8 [m]	D_{PA}+1.4 [m]	A_{WS} [m²]	Amount E	A_{WS}XE [m²]
1	ABS 265	1.6	2.8	2.4	4.2	10.1	1	10.1
2	RAYO 165	3.0	1.6	3.8	3.0	11.4	4	45.6
3	H100	2.2	4.3	3.0	5.7	17.1	1	17.1
4	PE 150 C	3.9	1.7	4.7	3.1	14.6	2	29.2
5	DMU 35	1.6	2.1	2.4	3.5	8.4	2	16.8
6	CWK 500	4.7	6.6	5.5	8.0	43.9	4	175.6
7	Deburring area	1.5	0.8	2.3	2.2	4.9	1	4.9
8	ZX 1	3.0	3.2	3.8	4.6	17.5	2	35.0
9	PF 150	4.4	3.4	5.2	4.8	25.0	1	25.0
10	Assembly	---------------		3.0	4.0	12.0	5	60.0
11	Testing	---------------		3.0	4.0	12.0	1	12.0
12	Packing	---------------		3.4	4.7	16.0	1	16.0
							Σ	447

Table 3.59 Calculation of floor space requirement (overall story floor space)

Σ Production station floor space A_{WS}		**447 m²**
Interim storage area A_I	(40% production area)	179 m²
Transport area A_T	(40% production area)	179 m²
Additional areas	(20% production area)	89 m²
Main workshop area A_{MA}		**894 m²**
Quality assurance	(15m² basic requirement {constant} + 7.5 m² per EM)	30 m²
Control and management	(2.5% of A_{MA})	22 m²
Tool handling	(5% of A_{MA})	45 m²
Toilets	(ASR37/1)	20 m²
Break room	(ASR 29/1-4)	30 m²
Power/water/heat	(1.8 m² per E)	32 m²
Secondary workshop area A_{SA}		**179 m²**
Net story area A_{NS}		**1073 m²**
Structure area A_S	6% of A_{NS}	**69 m²**
Total story area A_{TS}		**1142 m²**

(5/3.4) Determination of transport units per lot

Table 3.60 Determination of transport units per lot

Item	Drawing No.	Name	X_L [units/lot]	N_L [lot/a]	$N_{TU,L}$ Volumes [TU/lot]	$N_{TU,L}$ Weight [TU/lot]	$N_{TU,L}$ Adjustment [TU/lot]
1	M 001	Worm gear housing	450	80	6	4	6
2	M 002	Housing cover	1,200	30	1	3	3
3	M 011	Transmission housing	750	48	3	3	3
4	M 012	Drive shaft	1,200	30	1	2	2
5	M 013	Drive pinion	1,200	30	1	1	1
6	M 003	Worm gear	1,200	30	1	2	2
7	M 004	Drive wheel	1,200	30	1	1	1
8	M 005	Output shaft	900	40	2	3	3

The number of transport units per lot (e.g. containers/lot) is a function of the carrying capacity and the maximum volume of the selected container type.

(5/3.5) Dimensioning of costs (cf. step 5/2 – (5))

(5/4) Structuring (product flow)

(5/4.1) Flow chart (material flow), cf. Fig. 3.90

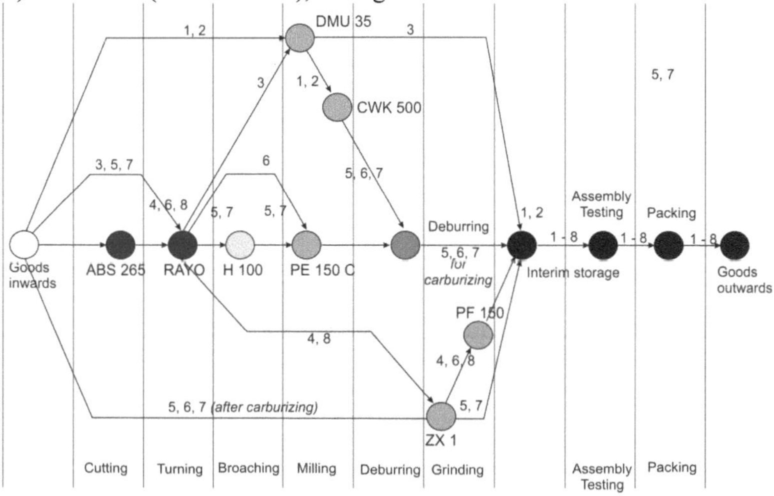

Fig. 3.90 Material flow (numbers 1 – 8 correspond to the relevant part number)

(5/4.2) Optimization of machine layout (after SCHMIGALLA)

a) Determination of transport intensities

Table 3.61 Transport intensities

Item	from	to	Part	N_L [lot/a]	$N_{TU,L}$ [TU/lot]	$N_L X N_{TU,L}$ [TU/a]	$\Sigma N_L X N_{TU,L}$ [TU/a]
1	GI	ABS 265	4	30	2	60	
2			6	30	2	60	
3			8	40	3	120	
							240
	GI	RAYO165					204
	GI	DMU 35					570
	GI	ZX 1					120
	ABS 265	RAYO 165					240
	RAYO165	H100					60
	RAYO 165	PE 150 C					60
	RAYO 165	DMU 35					144
	RAYO 165	ZX 1					180
	H100	PE 150 C					60
	PE 150 C	Deburring area					120
	DMU 35	CWK 500					570
	DMU 35	IS					144
	CWK 500	Deburring area					570
	Deburring area	IS					690
	ZX 1	PF 150					60
	ZX 1	IS					240
	PF 150	IS					60
	IS	Assembly					1,134

	Assembly	Inspection station						1,134
	Inspection station	Packing						1,134
	Packing	GO						1,134

b) Determination of sequence

Table 3.62 Transport matrix (absolute)

from \ to	GI	ABS 256	RAYO 165	H100	PE 150 C	DMU 35	CWK 500	Deburring area	ZX 1	PF	IS	Assembly	Testing	Packing	GO
GI		240	204			570			120						
ABS 265			240												
RAYO 165				60	60	144			180						
H100					60										
PE 150 C								120							
DMU 35							570				144				
CWK 500								570							
Deburring area											690				
ZX 1										60	240				
PF 150											60				
IS												1134			
Assembly													1134		
Testing														1134	
Packing															1134
GO															

Table 3.63 Transport matrix (relative)

from \ to	GI	ABS 256	RAYO 165	H100	PE 150 C	DMU 35	CWK 500	Deburring area	ZX 1	PF	IS	Assembly	Testing	Packing	GO
GI		0.21	0.18			0.5			0.1						
ABS 265			0.21												
RAYO 165				0.05	0.05	1.13			0.16						
H100					0.05										
PE 150 C								0.1							
DMU 35							0.5				0.13				
CWK 500								0.5							
Deburring area											0.61				
ZX 1										0.05	0.21				
PF 150											0.05				
IS												1			
Assembly													1		
Testing														1	
Packing															1
GO															

Table 3.64 Calculation of layout sequence

	GI	ABS 256	RAYO 165	H100	PE 150 C	DMU 35	CWK 500	Deburring area	ZX 1	PF 150	IS	Assembly	Testing	Packing	GO
GO															
Packing													1134		
	0	0	0	0	0	0	0	0	0	0	0	0	1134		
Testing												1134			
	0	0	0	0	0	0	0	0	0	0	0	1134			
Assembly											1134				
	0	0	0	0	0	0	0	0	0	0	1134				
IS						144		690	240	60					
	0	0	0	0	0	144	0	690	240	60					
Deburring area					120		570								
∑	0	0	0	0	120	144	570		240	60					
CWK 500						570									
∑	0	0	0	0	120	714			240	60					
DMU 35	570		144												
∑	570	0	144	0	120				240	60					
GI		240	204						120						
∑		240	348	0	120				360	60					
ZX 1			180							60					
∑		240	528	0	120					120					
RAYO 165		240		60	60										
∑		480		60	180					120					
ABS 256															
∑				60	180					120					
PE 150 C				60											
∑				120						120					
H100															
∑										120					
PF 150															

Layout sequence:
1. GO
2. Packing
3. Testing
4. Assembly
5. IS
6. Deburring area
7. CWK 500
8. DMU 35
9. GI
10. ZX 1
11. RAYO 165
12. ABS 256
13. PE 150 C
14. H100
15. PF 150

c) Machine installation (triangular network)

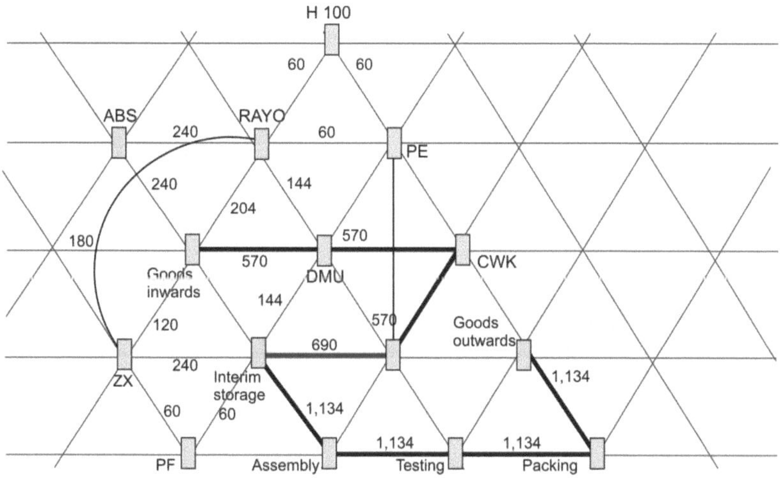

Fig. 3.91 Triangular network variant according to SCHMIGALLA's modified triangular approach as the ideal solution for machine installation

(5/5) Design

(5/5.1) Ideal layout (block layout)

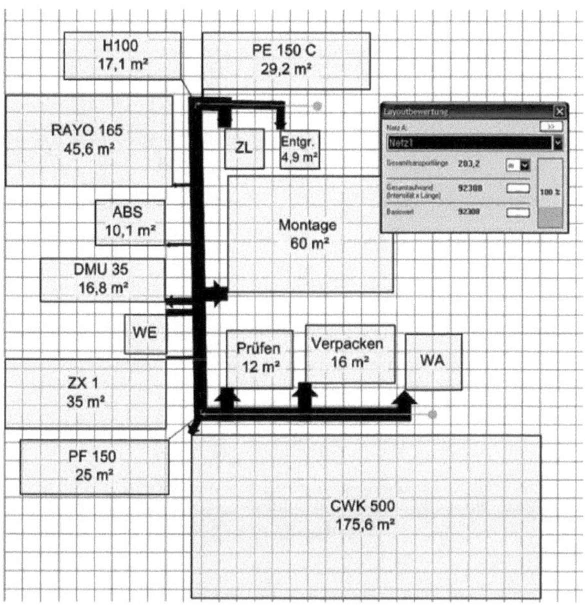

Fig. 3.92 Ideal layout (created with visTABLE.touch)

(5/5.2) Real layout variants

Variant 1

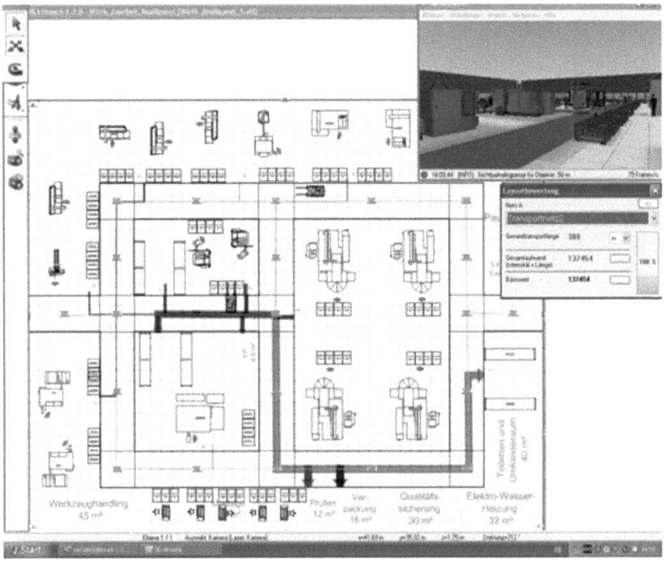

Fig. 3.93 Real layout – Variant 1

Variant 2

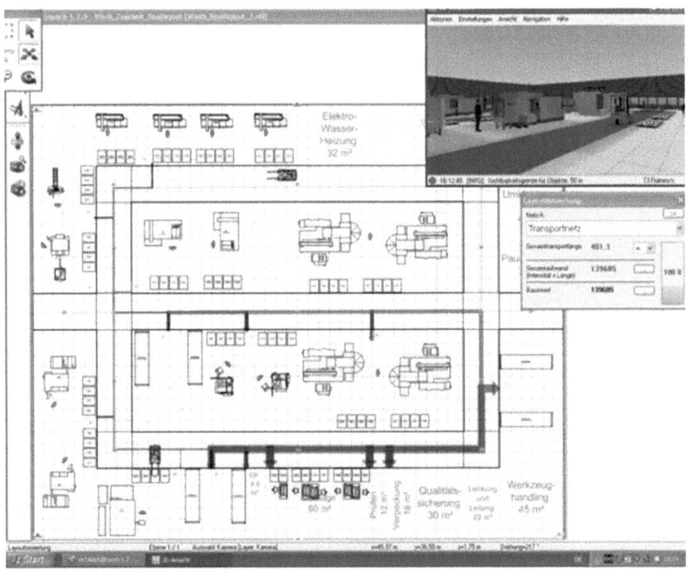

Fig. 3.94 Real layout – Variant 2

(5/5.3) Evaluation of real layout variants

Table 3.65 Cost-benefit analysis (<≙0; =≙0.5; >≙1)

Influencing factor	No.	1	2	3	4	5	6	7	8	W_C	Cr_W	Variant 1		Variant 2	
											%	f_W	$f_W \cdot Cr_W$	f_W	$f_W \cdot Cr_W$
Material flow-based layout of E	1		<	>	=	>	=	>	>	5	19	2.5	48	2	38
Floor space utilization	2	>		>	>	>	>	>	=	6.5	24	2	48	2.5	60
Residual floor space usability	3	<	<		<	=	=	=	<	1.5	6	2	12	2	12
Accessibility to workstations	4	=	<	>		>	>	>	>	5.5	20	4	80	4	80
Multi-machine operability	5	<	<	=	<		>	=	=	3	11	3.5	39	3	33
Ability to expand	6	=	<	=	<	<		=	=	2	7	1.5	11	0.5	4
Overall clarity	7	<	<	=	<	=	=		<	1.5	6	3	18	2	12
Transport effort	8	<	<	<	>	=	=	<		2	7	2	14	2.5	18
Σ (efficiency)										27	100		269		256

Variant 1 proves to be the better option according to the cost-benefit analysis.

(5/5.4) Detailed machine installation plan

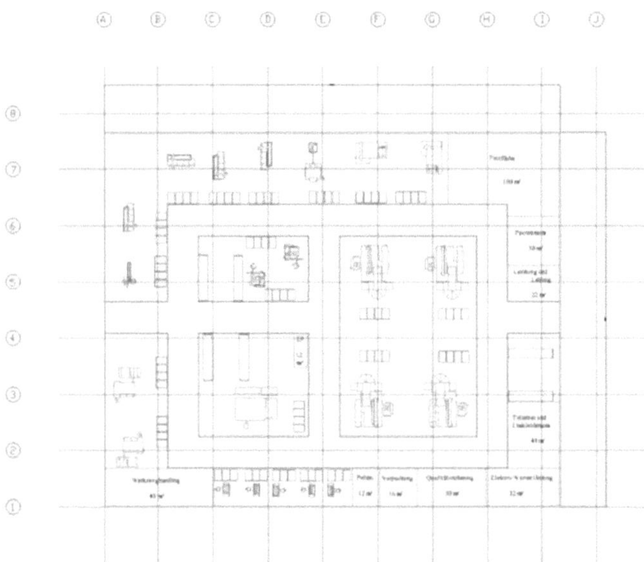

Fig. 3.95 Detailed machine installation plan

(x6) Profitability assessment (cf. Complex III)

Profitability is assessed by means of a dynamic investment appraisal. The "net present value" and "payback period" methods are used.

(1.) Net present value method (cf. chapter 6.1.2.1)

a) Capital investment

Table 3.66 Capital investment

Name		**Unit price**	**Quantity**	**Total Price**
Building	Low-rise building with standard requirements (incl. social rooms and office)		2000 m²	
	Reorganization costs	Approximate calculation		70,000.00 €
	Structural project design	Approximate calculation		25,000.00 €
	Inaccuracy margin	15%		14,250.00 €
				109,250.00 €
Electrical installation	Building distribution with assembly			14,000.00 €
	Overhead electrical power supply conduit	180 €/m	480	86,400.00 €
	Channel connection	390 €/m	120	46,800.00 €
	Machine connections	150.00 €	18	2,700.00 €
	Addition for incidentals			5,000.00 €
Water installation	Approximate calculation			7,500.00 €
Heating/ventilation	Approximate calculation			30,000.00 €
Project design	5%			4,675.10 €
				197,075.10 €
Lighting including installation	Lights	100.00 €	165	16,500.00 €
	Lamps	4.00 €	330	1,320.00 €
				17,820.00 €
Machinery	ABS 256	27,000.00 €	1	27,000.00 €
	RAYO 165	31,000.00 €	4	124,000.00 €
	H100	85,000.00 €	1	85,000.00 €
	PE 150 C	38,000.00 €	2	76,000.00 €
	DMU 35	62,000.00 €	2	124,000.00 €
	ZX 1	47,000.00 €	2	94,000.00 €
	CWK 500	93,000.00 €	4	372,000.00 €
	PF 150	55,000.00 €	1	55,000.00 €
	Assembly workstation	5,300.00 €	5	26,500.00 €
				938,500.00 €
STILL RX 70 fork lift truck		40,000.00 €	2	80,000.00 €
STILL HPT 22 pallet truck		300.00 €	6	1,800.00 €
Turnings container (V=0.5 m3; carrying capacity 1.0 t)		900.00 €	17	15,300.00 €
Costs for TUT	Mesh box pallets (800 x 1200 x 650 in	80.00 €	250	20,000.00 €

	accordance with DIN 15142)			
Heavy duty racking system		500 €/m²	39	19,500.00 €
Fire fighting equipment / First Aid	Charge - powder fire extinguisher	180.00 €	18	3,240.00 €
	First aid cupboard, large	240.00 €	5	1,200.00 €
	First aid cupboard, small	90.00 €	5	450.00 €
Office furniture & equipment	Desk	310.00 €	5	1,550.00 €
	Filing cabinet	180.00 €	5	900.00 €
	Roll front cabinet	165.00 €	5	825.00 €
	Chair	200.00 €	5	1,000.00 €
Social area furnishings	Changing room locker	160.00 €	40	6,400.00 €
	Urinal	400.00 €	5	2,000.00 €
	WC	300.00 €	6	1,800.00 €
	Single wash stand	600.00 €	2	1,200.00 €
	Double wash stand	1,150.00 €	4	4,600.00 €
	Chairs	30.00 €	40	1,200.00 €
	Table, round	210.00 €	3	630.00 €
	Table, square	200.00 €	8	1,600.00 €
Uncertainty margin		5%		8,259.75 €
			Total	**1,481,099.85 €**

b) Outgoing payments

Table 3.67 Material costs

			1st + 2nd years			3rd year (+10% of number of units)		
			DMC	MO (4% DMC)	MC	DMC	MO (4% DMC)	MC
Part name	Annual number of units	Material purchasing costs [€]	Direct material costs [€]	Material overheads [€]	Material costs [€]	Direct material costs [€]	Material overheads [€]	Material costs [€]
Worm gears	36.000	106.15 €	3,821,400.00 €	152,856.00 €	**3,974,256.00 €**	4,203,540.00 €	168,141.60 €	**4,371,681.60 €**

Table 3.68 Wage costs

Skill level	Workforce size in 1st + 2nd year	Workforce size in 3rd year	Average wage costs [€/h]	Working time per shift [h]	Working days per year [d/a]	Wage costs in the 1st / 2nd year [€/a]	Wage costs in the 3rd year [€/a]
Foremen	3	3	35	8	225	189,000.00 €	189,000.00 €
Logisticians	2	2	15	8	225	54,000.00 €	54,000.00 €
Machine operators	46	46	20	8	225	1,656,000.00 €	1,656,000.00 €
Quality assurance	2	2	20	8	225	72,000.00 €	72,000.00 €
Fitters	8	10	20	8	225	288,000.00 €	360,000.00 €
Unskilled labor	4	5	12.5	8	225	90,000.00 €	112,500.00 €
					Total	**2,349,000.00 €**	**2,443,500.00 €**

Table 3.69 Machine power costs

E name	Connected load [kW]	Working time in the 1st + 2nd planning years [h]	Costs per kWh [€]	Energy costs in the 1st + 2nd planning years [€]
ABS 265	6	3968	0.18	4,285.44 €
RAYO 165	8.1	11904	0.18	17,356.03 €
H100	15	3968	0.18	10,713.60 €
PE 150 C	48	7936	0.18	68,567.04 €
DMU 35	15	7936	0.18	21,427.20 €
ZX 1	25	11904	0.18	53,568.00 €
CWK 500	31	7936	0.18	44,282.88 €
PF 150	48	11904	0.18	102,850.56 €
	Total 1st + 2nd years			**323,050.75 €**
	Total 3rd year (1st + 2nd year + 10%)			**355,355.83 €**

Table 3.70 Manufacturing costs

		1st + 2nd years	3rd year
Direct manufacturing costs (DMC)	Energy costs	323,050.75 €	355,355.83 €
	Wage costs	2,349,000.00 €	2,443,500.00 €
	Total	**2,672,050.75 €**	**2,798,855.83 €**
Manufacturing overheads (MO)	10%	267,205.08 €	279,885.58 €
	Total	**2,939.255.83 €**	**3,078,741.41 €**

Table 3.71 Loan repayments

Loan for €2,500,000.00, term: 3 years; interest rate 6%		
Year	Borrowed capital [€]	Repayment [€]
0	2,500,000.00	-
1	2,650,000.00	750,000.00
2	2,014,000.00	1,500,000.00
3	544,840.00	544,840.00

c) Incoming payments

Table 3.72 Anticipated revenues

			1st + 2nd years	**3rd year (+10% of number of units)**
Part name	Annual number of units	Sales price [€]	Revenue [€]	Revenue [€]
Worm gear	36,000	235.25 €	8,469,000.00 €	9,315,900.00 €
		Total	**16,938,000.00 €**	**9,315,900.00 €**

Table 3.73 Depreciation and residual values

Name		Unit price	Quantity	Total price	Depreciation years	Residual value after 3 years
Machines	ABS 265	27,000.00 €	1	27,000.00 €		
	RAYO 165	31,000.00 €	4	124,000.00 €		
	H 100	85,000.00 €	1	85,000.00 €		
	PE 150 C	38,000.00 €	2	76,000.00 €		
	DMU 35	62,000.00 €	2	124,000.00 €		
	ZX 1	47,000.00 €	2	94,000.00 €		
	CWK 500	93,000.00 €	4	372,000.00 €		
	PF 150	55,000.00 €	1	55,000.00 €		
				957,000.00 €	10	669,900.00 €
Other (fork lift, racking system, equipment etc.)				**304,600.00 €**	10	213,220.00 €
					Total	**883,120.00 €**

Table 3.74 Annual incoming payments

Type	1st year	2nd year	3rd year
Revenue	8,496,000.00 €	8,496,000.00 €	9,315,900.00 €
Residual value			883,120.00 €
Total	**8,496,000.00 €**	**8,496,000.00 €**	**10.199,020.00 €**

d) Final value and investment appraisal – calculated according to the following formula

$$K_o = I_o + \sum_{t=o}^{u} \left(E_i - A_i\right) \bullet \left(1+i\right)^{-t} + \frac{L_n}{\left(1+i\right)} n \qquad (3.70)$$

Table 3.75 Payment series

Data			**t=0**	**t=1**	**t=2**	**t=3**
Outgoing payments	One-time outgoing payments [€]		1,481,099.85 €			
	Periodic payments	Material costs [€]		3,974,256.00 €	3,974,256.00 €	4,371,681.60 €
		Production costs [€]		2.939,255.83 €	2.939,255.83 €	3,078,741.41 €
		Loan repayment		750,000.00 €	1,500,000.00 €	544,840.00 €
		Total [€]	**1,481,099.85 €**	**7,663,511.83 €**	**8,413,511.83 €**	**7,995,263.01 €**
Service life [years]			3			
Incoming payments	Asset liquidation proceeds [€]					883,120.00 €
	Revenue [€]			8,469,000.00 €	8,469,000.00 €	10,199,020.00 €
		Total [€]		**8,469,000.00 €**	**8,469,000.00 €**	**11,082,140.00 €**

Required rate of return	6			
Net payment [€]	**1,481,099.85 €**	**805,488.18 €**	**55,488.18 €**	**3,086,876.99 €**
Addition of accrued interest			*1.06	58,817.47 €
Addition of accrued interest		*1.06²		905,046.51 €
Addition of accrued interest	*1.06³			- 1,764,013.62 €
Final value C_T [€]				2,286,727.35 €

Table 3.76 Annual profits

Year	1	2	3
Profit	**805,488.18 €**	**55,488.18 €**	**3,086,876.99 €**

(2.)Payback period method

Table 3.77 Payback period method

Point in time t	Net payment [€] N_t	Present value of the net payment [€] N_t*q^{-1}	Cumulative present value of the net payment [€]
0	- 1,481,099.85 €	- 1,569,965.84 €	- 1,569,965.84 €
1	805,488.18 €	853,817.47 €	716,148.38 €
2	55,488.18 €	58,817.47 €	657,330.91 €
3	3,086,876.99 €	3,272,089.61 €	2,614,758.70 €

$$PT \approx t^* + \frac{CV_{t^*}}{CV_{t^*} - CV_{t^*+1}} \tag{3.71}$$

(formula for calculating the payback period)

The investment object is profitable, as the payback period of 2.2 years is shorter than the duration of the project.

References

Cisek R (2005) Planung und Bewertung von Rekonfigurationsprozessen in Produktionssystemen, Utz, München
Conrad KJ (2002) Taschenbuch der Werkzeugmaschinen, Fachbuch, Leipzig
Forschungsberichte (2005) iwb/ Institut für Werkzeugmaschinen und Betriebswissenschaften, TU München
DIN 277 (2005) Grundflächen und Rauminhalte von Bauwerken im Hochbau, Beuth, Berlin
DIN 2860 (1990) – Montage- und Handhabetechnik, VDI, Düsseldorf
DIN 6789 (1990) – Dokumentationssystematik, Beuth, Berlin
DIN 8580 (2003) Justieren
DIN 8593 (1996) Fertigungsverfahren Fügen, Beuth Berlin

Dohms R (2001) Methodik zur Bewertung und Gestaltung wandlungsfähiger, dezentraler Produktionsstrukturen, Shaker, Aachen
Eversheim W, Schuh G (1999) (Hrsg) Produktion und Management 4 – Betrieb von Produktionssystemen, Springer, Berlin
Förster A (1983) Strukturierung von Teilflusssystemen der Fertigung im Maschinenbau, Dissertation B, TH Karl-Marx-Stadt
Götze U (2000) Kostenrechnung und Kostenmanagement, 2 Auflage, GUC, Chemnitz
Grundig CG (2000) Fabrikplanung, Hanser, München
Grundig CG (2006) Fabrikplanung: Planungssystematik - Methoden – Anwendungen, 2 aktualisierte Auflage, Hanser, München,
Hildebrand T, Mäding K, Günther U (2005) Plug + Produce – Gestaltungsstrategien für die wandlungsfähige Fabrik, IBF, TU Chemnitz
Hirsch A (2000) Werkzeugmaschinen, Vieweg, Braunschweig, Wiesbaden
IREGIA (2004) Leitfaden Flexibilisierung bestehender Fabrikstrukturen, IREGIA eV, Chemnitz
Kettner H, Schmidt, J, Greim HR (1984) Leitfaden der systematischen Fabrikplanung, Hanser, München
Kobylka A (2000) Simulationsbasierte Dimensionierung von Produktionssystemen mit definiertem Potential an Leistungsflexibilität, Diss, Wissenschaftliche Schriftenreihe, IBF, Heft 24, Chemnitz
Lotter B, Wiendahl HP (2006) Montage in der industriellen Produktion, Springer, Berlin
Rietz S (2001) Methodik zur zieladäquaten Bildung von Fertigungsgrobstrukturen, Barleben, Ges. für Innovative Systeme, Universität Magdeburg, 2000
REFA (1991) Methodenlehre der Planung und Steuerung, 6 Bände, Hanser, München
Rockstroh W (1973) Technologische Betriebsprojektierung, Technik, Berlin
Rockstroh W (1979) Gestaltung von Fertigungswerkstätten, Band 3, Technik, Berlin
Rockstroh W (1982) Projektierung von Fertigungswerkstätten, Technik, Berlin
Schmigalla H (1966) Methoden zur Vorausbestimmung des wirtschaftlichen räumlichen Strukturtyps und zur optimalen Gestaltung räumlicher Strukturen der spanenden Fertigung im Maschinenbau, Dissertation, TH Magdeburg
Schmigalla H (1995) Fabrikplanung, Begriffe und Zusammenhänge, Hanser, München
Schenk M, Wirth S (2004) Fabrikplanung und Fabrikbetrieb - Methoden für die wandlungsfähige und vernetzte Fabrik, Springer, München
Schulte H (2003) Planung von Fabrikanlagen, Vorlesungsskript, WBK Karlsruhe
TGL 13384 (1969) Blatt 1, Maschinenbau: Flächenbedarf für Arbeitsplätze in Werkmontagen, Bestimmung der Ausgangsdaten, verbindlich ab 1.1.1970
TGL 13384 (1969) Blatt 2, Maschinenbau: Flächenbedarf für Arbeitsplätze in Werkmontagen, Berechnung nach Arbeitsplatzarten, verbindlich ab 1.1.1970
VDI 3423 (2002) VDI-Richtlinie: Verfügbarkeit von Maschinen und Anlagen, VDI Handbuch Betriebstechnik, Beuth, Berlin
VDI 4400 (2004) Blatt 2 Logistikkennzahlen für die Produktion, Richtlinie Verein Deutscher Ingenieure, 12
Warnecke HJ (1992) Produktionsbetrieb, Band 1-3, Springer, Berlin
Wildemann H (1992) Die modulare Fabrik: Kundennahe Produktion durch Fertigungssegmentierung, Gfmt, St. Gallen
Wiendahl H, Scheffczyk HD (1998) Wandlungsfähige Fabrikstrukturen, in Werkstattstechnik 88, S 171 – 175
Wirth S (1989) (Hrsg) Flexible Fertigungssysteme, Technik, Berlin
Wirth S (2000) Vorlesungsskript Werkstoffe und Produktionssystemprojektierung, Technische Universität, IBF, Chemnitz
Woithe G (1965) Projektierung von Betriebsanlagen des Maschinenbaus, Lehrberichte, TH, Magdeburg

3.3 Project implementation (complex III)

This covers the planning activities detailed in figure 3.96 together with the relevant basic conditions and outcomes from the project development stage (complex II). It includes execution planning (execution project) and implementation.

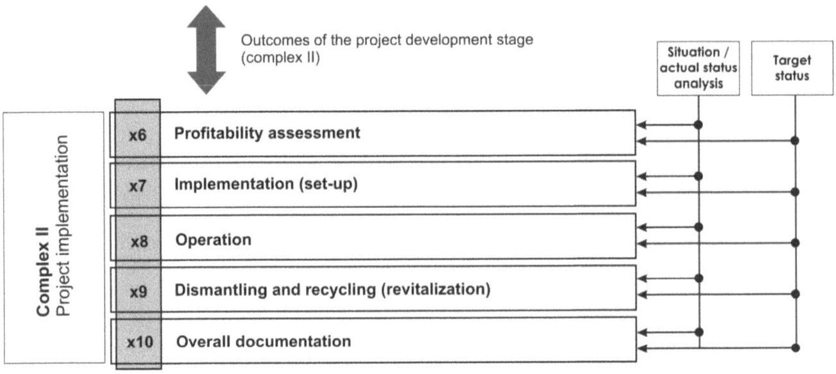

Fig. 3.96 "0 + 5 + X" planning model project implementation

The project is implemented on the basis of the profitability assessment and decisions (basic policy decisions) with the realization / set-up of the production facilities. This involves the implementation planning / implementation project from the material and technical realization of all objects through to operation, including testing and commissioning. (Schenk & Wirth 2004)

Table 3.78 contains a checklist for conducting a rough assessment of project implementation activities. This checklist is useful as it contains all of the most important statements and information for the project implementation. From experience, permanent monitoring and reevaluation of the activities in line with the changed situation will be required.

Table 3.78 Checklist – 0 + 5 + X Planning Model - Project implementation

❒	5	**Project development**			
			a) Planning project		
		❍	available	❍	not available
		❍	available to some extent		
			b) Implementation project		
		❍	available	❍	not available
		❍	available to some extent		
			c) Approvals		
			d) Requirement specifications		
			e) Technical specifications		
❒	X6	**Profitability assessment**			
			a) Static methods		
		❍	Cost comparison method	❍	Accounting rate of return method
		❍	Rate of profit method	❍	Payback period method
			b) Dynamic methods		
		❍	Net present value analysis	❍	Overhead calculation
		❍	Annuity calculation		
			c) Lease- buy		

		❍	Lease	❍	Buy
		❍	Combination		
❐	X7	**Implementation (project planning, control and monitoring)**			
			a) Approvals		
		❍	Occupational safety		
		❍	Regulations (machines/plant/equipment)		
		❍	Regulations, clearances and layout (plant/walls/doors)		
		❍	Environmental protection (immissions/water/air/noise)		
			b) Equipment and plant implementation		
		❍	Request for quotation	❍	Tender
		❍	Offer	❍	Order release
		❍	Purchase order	❍	Supply contract
		❍	Purchasing contract	❍	Contract for work
			c) Technical specifications		
		❍	Workforce	❍	Equipment
		❍	Technical requirements – building structure	❍	Technical requirements – trades
		❍	Technical requirements – operating facility	❍	Technical requirements – plants
			d) Design work and plant construction		
		❍	In-house activity	❍	Outsourced activity
			e) Delivery/Commissioning		
		❍	Feasibility study	❍	Order preparation
		❍	Assembly	❍	Function test
		❍	Acceptance protocol		
			c) Project conclusion/evaluation		
		❍	Performance parameters	❍	Documentation
		❍	Training	❍	Maintenance
❐	X8	**Operation**			
			b) Start-up/run-down planning		
		❍	Start-up	❍	Ramp-up
		❍	Normal operation		
			e) Operation and utilization		
		❍	Production planning/control	❍	Monitoring
		❍	Servicing/maintenance	❍	Adjustments
		❍	Continuous improvements (CIP)		
❐	X9	**Dismantling - Recycling**			
		❍	Remediation	❍	Decommissioning
		❍	Revitalizing	❍	Reuse
		❍	Further use	❍	Recycling
❐	X10	**Documentation**			
		❍	Task definition/specifications/decisions		
		❍	Project planning, control and monitoring		
		❍	Project documentation (planning/implementation)		
		❍	Consulting documents/records/decision documents		
		❍	Approvals/correspondence/tenders		
		❍	Profitability assessments/status reports		

For this purpose, binding orders and agreements must be implemented and concluded with each individual production facility outfitter / system supplier for every item (machinery / plant / installations / trades) identified in the planning project (complex II). Cost and technical specifications must be strictly regulated. The planning and implementation process must be managed both in detail and as a whole with the equipment suppliers. Fig. 3.97 shows the process from the point of view of the supplier.

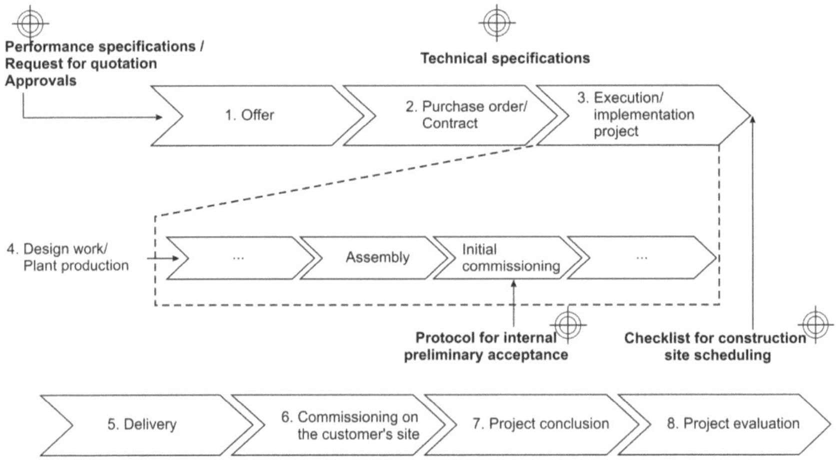

Fig. 3.97 Client-contractor process (production facility outfitter/supplier)

Depending on the situation, new tenders and additional bid comparisons and evaluations are both common and expedient. The operational planning and implementation of the process should be executed as a closed loop process as follows:

(1) Request for quotation/tender/offer (requirement specification)

(2) Purchase order/contract

(3) Execution/implementation project (technical specifications)

(4) Design work/plant production in supplier enterprises

(5) Delivery

(6) Commissioning

(7) Project conclusion and

(8) Project evaluation

3.3.1 Profitability assessments and project evaluation (x6)

While during the project development (complex II) predominantly variant evaluations and selective profitability assessments are carried out, here a specific profitability assessment for individual and all planning phases as a whole should be carried out by means of an investment appraisal. To do this, static and dynamic methods and lease-buy methods used in investment appraisal are applied. Examples of the investment assessment for procedures a), b) and c) are summarized in chapter 6.1.

a) Static methods

- cost comparison method
- accounting rate of return method
- rate of profit method
- payback period method

b) Dynamic methods

- net present value method, see complex planning project (point 3.2.6.2, x6)
- internal rate of return method
- capital recovery method

c) Lease-buy methods

3.3.2. Realization (x7)

Realizing the project is the objective and task of project implementation. Permits and approvals are required for this and must be obtained. These relate, for example, to building regulations, safety, environmental protection and health protection (hygiene). The safety of employees (occupational safety) and the protection of the public are of particular importance. Various ordinances, rules and regulations exist to help safeguard this. Approvals and permits are a part of the project implementation and belong to the execution project. A selection of required acceptance inspections and regulations is summarized below. (Schulte 2003, p. 132 ff)

a) Occupational safety – in Germany, approvals required from:

- Office for Labor Protection and Safety *(Amt für Arbeitsschutz und Sicherheitstechnik)*
- TÜV (the Technical Inspection Agency)
- Employers' liability insurance association (*Berufsgenossenschaft*)
- the fire department

b) Regulations affecting machines/plant/equipment:

- UVV (German accident prevention regulations)
- Individual accident prevention regulations of the industrial employers' liability insurance associations (VBG regulations)
- Guidelines, safety rules, principles and data sheets issued by the industrial employers' liability insurance associations (ZH1 regulations)
- German standards (DIN)
- technical rules (TR)

- VDI guidelines (VDI)
- VDE guidelines (VDE)
- FEM recommendations (FEM)
- CE mark (CE – Communautés Européennes)

c) Regulations affecting clearance gaps and the layout of systems/walls/doors etc.

- German Trade, Commerce and Industry Regulations (GewO § 120a)
- Regulations governing the workplace (ArbStättV)
- Workplace guidelines (ASR)
- German standards (DIN)
- Workplace requirements (standard reference work Josef Krämer 5 volumes)

a) Environmental protection – acceptance inspections required by:

- Office for Immission Control *(Amt für Immissionsschutz)*
- Environment agency, regional water authorities, water protection authorities
- District chimney sweep
- Federal Water Act (WHG § 19)
- safety data sheets (DIN 52900)
- Federal immission protection regulations (BImSchV)
- Technical instructions on air *(TA Luft)*
- Technical instructions on noise *(TA Lärm)*

3.3.2.1 Request for quotation/tender/offer

a) *Requests for quotation* can be made by the CL or the CO to fine tune their objectives for planning and realization.

b) *Tender documents* relate to the main coordinating contractor, general contractors, groups of trades and individual trades. Maintaining a list of suppliers for the selection of firms has proved useful for the *award procedure. Public and limited tenders and negotiated procedures or awards* on the basis of requirement specifications are customary for this.

c) *Offers* are, as a general rule, provided free of charge to the awarding authority (CL) by the bidder (CO). Tender documents should cover the following points:

- general information
  (e.g. project information, contacts, guidelines for tender preparation, list of abbreviations)
- project description
  (e.g. task definition, performance data and general conditions, process description, interfaces)

- tender specifications
  (e.g. what is required of the plant, components, data about the plant, control, acceptance)
- prices and acknowledgement
  (e.g. price composition, acknowledgement of tender)
- contractual conditions
  (e.g. general contract conditions such as contractual basis and scope of services, project development, technical / commercial conditions, purchasing agreement conditions, project-specific conditions)
- appendix
  (e.g. construction site regulations, acceptance protocol, tender specification items, requirement specifications, performance specifications, technical specifications, accompanying plans)

Questionnaires and checklists as part of the documentation have proven useful for facilitating comparison of the detailed incoming offers. In the case of standard products from catalogues, a request for quotation is often sufficient.

d) *Evaluation of offers* and *order release*

The supplier compiles an offer for the scope of services on the basis of the request for tender. If the target requirements do not fulfill the desired purpose, a bidder comparison will be necessary. To do this, specified knock-out criteria (e.g. profitability, costs, equipment) must be considered.

3.3.2.2 Purchase order/contract

a) *Purchase order* - is placed by the CL (investor/operator) on the basis of requests for quotation, tenders and offers. It initiates the customer order.
b) *Contract* – constitutes the binding legal basis for the order. It is concluded on the basis of a technical specification.

When drafting a contract, domestic and international (e.g. EU) regulations (AGBG, BGB, HGB) should be taken into account. Contracts for labor and services, work performance, purchasing and service level agreements should be considered.

When drafting a contract, particular attention should be paid to the clauses governing liability, guarantee and termination. The following fig. 3.98 shows the major criteria and consequences that can be regulated in the contract with regard to supply and performance.

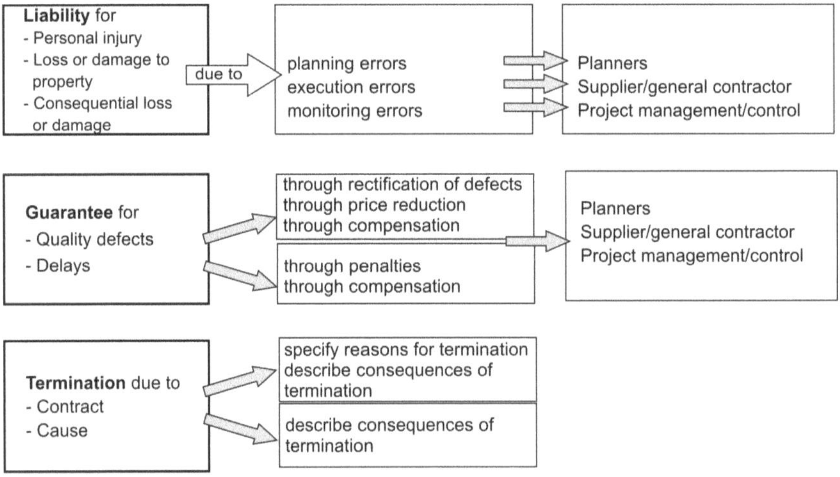

Fig. 3.98 Drafting of contract – liability, guarantee and termination

3.3.2.3 Technical specifications / execution project

The contents of the technical specifications must be reflected in the execution project. For the CO (supplier/outfitter), all realization documents must be prepared, coordinated with regard to content and supplied in accordance with the contract.

This relates primarily to facilities and equipment for:

a) Personnel/workforce: changing, washing and social facilities
b) Machinery and equipment: manufacturing, assembly equipment including jigs and fixtures, tools and testing equipment, logistics equipment including auxiliary THS equipment
c) Technical systems: structure (building), technical building services, heating, ventilation, air conditioning, lighting, supply and disposal facilities for technical utilities, safety equipment and interference protection
d) Operating materials: equipment for transport, staging, storage of liquid and gaseous substances

The execution project must contain a technical specification for every item of equipment. A general requirement and technical specification is shown below in the form of a checklist. The technical and requirement specifications for material flow and automation systems are contained in VDI 2519 and VDI/VDE 3694 respectively.

NB: agreements/contracts that fix the rights and responsibilities of the partners should be concluded for complexes I to III either separately or as a

whole. This applies for purchase orders, supply and purchase of products (services) and plant in accordance with the technical specifications.
The subject matter of the contract includes: contract contents, contracted services (service parameters), deadlines, costs (price) and location as well as clauses governing warranty, liability, guarantee and termination. In addition, provisions regarding, for example, safety and labor protection, as well as constraints should be agreed upon.

Table 3.79 contains a thorough checklist for the three project definition, development and implementation complexes which can be used as a tool for monitoring situation-driven changes in the planning and execution process.

Table 3.79 Checklist – requirement/technical specifications (example of areas of emphasis)

0	**Presentation of client**							
	(Enterprise, sector, market position, production program, performance program, contact information, contact person/cooperation partners)							
1	**Complex I (project definition)**							
1.1	ACTUAL status							
	□	Description of technical and organizational processes						
		□	Plant description					
			○	characteristic data	○	operational dependencies		
			○	process-related interrelationships				
		□	Description of process flows in normal operation					
			○	product/material flow	○	information flow	○	document flow
			○	energy flow	○	value flow	○	work flow
		□	Description of process flows in irregular operation					
			○	useful operation	○	operating procedure in the event of partial system breakdown		
	□	Description of technical components						
		□	Available equipment/plant for machining and processing					
		□	Power		□	Existing control systems		
		□	Operating materials		□	Existing material flow components		
	□	Organization (work areas, responsibilities, competences)						
		□	Organization for enterprise/division/section/group					
		□	Organizational structure					
			○	layout organization	○	leadership structure (work assignments)		
			○	engineering	○	services	○	safety
			○	planning	○	production	○	quality
		□	Process flow organization					
			○	flow sequences				
		□	Operational organization (operating modes/downtime)					
		□	Work instructions					

		○ specifications/controls/ reports	○ procedural instructions
		○ safety requirements	○ procedures/manufacturing instructions/ formulations
	□ Reporting (work and production reports)		
□ Transport, handling and storage processes (THS processes)			
	□ Transport		
		○ Sankey diagram	○ commodity to be conveyed (shape, dimensions, weight)
		○ transport utilities	○ material flow matrix (source/destination matrix)
		○ daily/seasonal fluctuations in volume flows	
	□ Storage/buffering		
		○ storage utilities	○ type of storage (racking, flow-through, block storage etc.)
		○ storage strategy (equal distribution, ABC distribution etc.)	
		○ static storage data (number of storage areas)	
		○ dynamic storage data (number of transfers in and out of storage)	
	□ Loading and unloading/handling/transfer		
		○ frequencies	○ processing times/cycle times
		○ fluctuation range of processing times (min/max values)	
	□ Testing/control/quality assurance		
		○ testing time	○ fluctuation range of testing time (min/max values)
		○ quality stages	○ quality standards (ISO)
1.2	TARGET status		
	□ Brief description of the task		
	□ Classification and description of task		
	□ Description of workflows		
	□ Data representation and quantity structure		
	□ Future considerations		
		○ description of future expansions and expansion stages	
		○ indication of the required capacity reserves	
		○ changes to products	○ changes to the manufacturing program

□ Specification of enterprise objectives			
	□ Market position objectives	□ Return on investment objectives	□ Financial aims
	□ Social aims	□ Prestige aims	□ Strategic objectives
□ Performance objectives			
	□ Increase in output	□ Reduction in consumption and emissions	
	□ Quality improvement	□ Reduction of emissions	
	□ Plant capacity utilization	□ Improvement in process control	
	□ Plant network	□ Improvement of reliability	

		□	Safety improvements			
		□	Improvement of flexibility			
	□	Profitability				
		□	Rationalization	□	Reduction in plant equipment	
		□	Cost reduction	□	Minimization of investment	
	□	Concepts				
		□	Sales, marketing	□	Product	□ Personnel
		□	Production	□	Profitability	□ Flexibility
1.3	Specification of basic planning cases					
		□	New building	□	Reorganization	□ Expansion
		□	Deconstruction	□	Revitalization	□ Recycling
	□	Due to…				
		□	Market changes	□	Technology/methods	□ Obligations (Authorities)
		□	Safety requirements			□ Rationalization
		□	Restructuring of the organization			□ Other
1.4	Specification of the planning scope					
	□	Planning phases				
		○	Development/planning	○	Set-up	○ Start-up
		○	Operation	○	Dismantling	○ Core/conceptual planning
		○	Objective planning	○	Execution planning	○ Detailed/system planning
	□	Planning objects				
		○	Personnel	○	Structure	○ Moveable objects / plants
		○	Property/buildings			
	□	Planning instruments				
		○	Models	○	Methods	○ Tools
		○	Theories			
1.5	Specification of project design principles					
		○	As a whole	○	Stages	○ Top down
		○	Bottom up	○	Variants	○ Flexibility
		○	Improvement	○	Situation	○ Synergy
		○	Profitability			
2	**Complex II (project development)**					
2.1	Production and performance program					
	□	Production program scheduling				
	□	Planning data (deadlines, etc.)				
		○	available	○	not available	○ available to some extent
	□	Operating variables (production types, level of detail, production program type)				
		○	available	○	not available	○ available to some

							extent
	□	Process flows					
		○	available	○	not available	○	available to some extent
		○	material flow (quantity, processing time, throughput time)				
		○	information flow (scanners, RFID, data transfer)				
		○	energy flow (supply, information regarding machine consumption)				
		○	value flow (feedback into the system)				
	□	Preparation of production program					
		□	Product structure				
		○	available	○	not available	○	available to some extent
		□	Type representatives				
		○	available	○	not available	○	available to some extent
		○	variable/flexible				
		□	Make	□	Buy		
2.2	Determination of functions						
	□	Product structure/manufacturing structure (technical information for the entire system)					
		□	availability	□	robustness, fault tolerance		
		□	consumption data	□	performance data (response times, throughput)		
	□	Manufacturing/assembly process					
		□	manufacturing process stages	□	manufacturing process operations		
		□	workflow schematic	□	functional schematic		
	□	Processes (technical data regarding material flow and components)					
		□	protection classes	□	electromagnetic data		
		□	mechanical, electrical and climatic data				
	□	Manufacturing mode					
		□	manufacture of parts	□	assembly	□	combination
	□	Selection of equipment (technical, economic)					
		□	technical/functional (parts production and assembly)				
		○	special purpose machinery	○	transfer line		
		○	processing center	○	standard machinery		
		○	general purpose machinery	○	manufacturing/assembly system		
		○	manufacturing/assembly cells				
		□	technical/functional (THS technology)				
		○	transport	○	handling	○	storage
		□	economic, ecological				
		○	water consumption	○	energy consumption	○	recyclability
		○	reusability	○	investment/acquisition costs		

		○	operating costs				
2.3	Dimensioning						
	□	Equipment					
		□	individual	□	multiple	□	complex plants and systems
		□	amount of equipment (parts manufacture)				
		○	product, workpiece-related components				
		○	equipment-related components				
		□	amount of equipment (assembly)				
		○	product/assembly unit-related components				
		○	equipment-related components				
		□	number of THS facilities				
		○	product – material to be conveyed	○	material to be conveyed - conveyor	○	conveyor - conveyor
		○	conveyor – conveying system	○	conveying system – conveying system		
		○	conveying system – storage system				
		○	conveying system – control system	○	conveying system – personnel		
		○	control system – personnel				
	□	Workforce					
		□	number	□	skills level	□	group/ team structure
		□	individual operation	□	multiple machine operation		
	□	Methods					
		□	static	□	dynamic		
	□	Times					
		□	utilization times	□	order times	□	time calculation
	□	Floor space					
		□	approximate floor space classification	□	detailed floor space classification		
		□	floor space assessment	□	determination of total floor space		
		□	structured determination of floor space				
2.4	Structuring						
	□	technical and organizational structural variants					
		□	operational structure	□	supply-chain network		
	□	Logistics structure					
		□	procurement	□	production	□	sales & marketing
	□	Process flow principles					
		□	JIT	□	JIS	□	KANBAN
		□	CONWIP				
		□	process flow organization	□	segments	□	centers (cost, service)

	□	Production structure					
		□	spatial	□	chronological		
	□	Structural form					
		□	transport-oriented	□	object-oriented	□	combination
	□	Manufacturing/assembly forms					
		□	manufacture of parts	□	assembly	□	combination
2.5	Design						
	□	Layout					
		□	approximate layout	□	detailed layout	□	ideal layout
		□	real layout				
	□	restrictions					
		□	labor protection guide-lines	□	statutory provisions		
	□	Arrangement of objects relative to the transport route					
		□	in parallel	□	at right angles	□	diagonal
	□	Dimensional design					
		□	minimum clearances	□	safety clearances		
	□	Foundations					
		□	foundations	○	block foundations	○	oscillating foun-dations
		□	foundationless	○	rigid installation	○	flexible installation
	□	Ergonomic design					
		□	seated workstations	□	standing workstations	□	combined workstations
		□	single-operator work-stations	□	multiple-operator workstations		
	□	Quality characteristics					
		□	operational reliability	□	fitness for purpose	□	user-friendliness
		□	maintainability				
	□	Quality assurance					
		□	measures (organizational, administrative, technical)				
		□	methods, quality control plans (standards to be applied, audit and review plan, testing plan)				
		□	interim acceptances	□	tools, utilities		
	□	Proof of quality					
		□	disclosure of quality-control plans				
		□	reports, process audit records in accordance with quality-control plans				
3	**Complex III (project implementation)**						
3.1	Profitability assessment						
	□	financial interrelationships					

			○	equipment economies				
			○	reimbursement of machine costs	○	enterprise-specific dependencies		
		□	Static methods					
			○	cost comparison method	○	accounting rate of return method		
			○	rate of profit method	○	payback period method		
		□	Dynamic methods					
			○	net present value analysis	○	cost-plus method	○	capital recovery method
		□	Lease-buy methods					
			○	lease	○	buy	○	combination
3.2	Implementation (set-up)							
	□	Project organization						
			□	personnel	□	responsibilities	□	work locations/ rooms
			□	working environment	□	working times		
	□	Project execution						
			□	project planning				
			○	activity plan	○	milestones		
			□	project control (approval procedures)				
			□	project monitoring (deadlines, costs, quality)				
			○	project meetings	○	progress reports		
	□	General requirements						
			□	documentation	□	change management	□	error tracking
			□	maintaining project history				
	□	Approvals						
			□	occupational safety				
			□	regulations (machines/plant/equipment)				
			□	regulations (clearances and layout of systems/walls/doors)				
			□	environmental protection (immissions/water/air/noise)				
			□	energy efficiency				
	□	Equipment and plant implementation basis						
			□	request for quotation	□	tender	□	offer
			□	order release	□	purchase order	□	supply contract
			□	purchasing contract	□	contract for services		
	□	Technical specifications (client/contractor requirements)						
			□	workforce	□	building structure technical requirements		
			□	equipment	□	manufacturing facility technical requirements (building)		
			□	technical requirements	□	other requirements		
			□	trades technical requirements				
	□	Design work and plant production						

		□	in-house activity	□	outsourced activity	
	□	Supply installation (contractor)				
		□	assembly	□	installation site preparation	
		□	acceptance protocol	□	function test	
		□	feasibility study			
	□	Project conclusion/evaluation				
		□	performance parameters	□	documentation	
		□	training	□	maintenance/servicing	
3.3	Operation					
	□	Assembly				
		□	operator's requirements (type, scope, restrictions, e.g. in day-to-day operation)			
		□	installation guidelines	□	installation and assembly planning	
		□	assembly execution	□	provision of equipment and personnel	
		□	software implementation	□	installation locations for the equipment	
	□	Commissioning				
		□	operator's requirements (type, scope, restrictions)			
		□	commissioning planning	□	switching on the plant	□ manual/automatic operation
		□	takeover of data inventories			
		□	provision of equipment and personnel			
	□	Test operation, acceptances				
		□	operating conditions (live operation, simulated operation, shift operation, duration, interruptions)			
		□	requirements of the operator (plant provision)			
		□	test operation/acceptance planning (partial acceptance)			
		□	conditions of acceptance (criteria, fault tolerance, duration, availability)			
		□	provision of equipment, software and personnel			
		□	acceptance execution			
	□	Training (plan, costs, skills and qualifications)				
		□	product/technology training for			
		○	operators	○	control station personnel	○ supervisors
		○	maintenance technicians			
	□	Operating schedule (determination of organizational process flows and conditions for the following types of operation)				
		□	normal operation (working methods, operating methods, operating times)			
		□	start-up, restart (reaction to breakdowns, breakdown plans, secondary processing and malfunction time)			
		□	emergency operation	□	interrupted operation (criteria, development, duration, general conditions)	
	□	Maintenance				
		□	planning and strategy (in-house / outsourced activity)			

		□	requirements				
		○	reaction times	○	skills level of personnel		
		□	measures in the event of faults/breakdowns				
		○	diagnosis	○	utilities	○	elimination
		○	replacement parts inventory				
		□	preventative measures (scope, frequency, general conditions)				
3.4	Recycling						
		□	remediation	□	decommissioning	□	reuse
		□	revitalization	□	further use	□	recycling
3.5	Documentation						
		□	documentation requirements from the point of view of the operator, the plant supervisors and maintenance				
		□	list of documents supplied with deliveries				
		□	scope of individual documents				
		□	structure of documents (norms, standards, language)				
		□	type of document generation	□	storage of documents (medium, location)		
		□	updating of documents				
	□	Content (technical solutions)					
		Brief description of the solution					
		Classification and description of the technical solution					
		□	structural plan of the technical solution				
		□	description of the individual function (the detailing of the functions may require further stratification)				
	□	Description of the technical solutions for regular operation (normal operation, start-up and restart and for irregular operation (interrupted operation, emergency operation)					
	□	Profitability and performance parameters					

3.3.2.4 Construction design / plant production

The equipment and plant are built on the basis of the technical specification and arrangements between the CL and CO. The production of the plant that is ordered requires further subcontractors (SCO), with whom the outfitter/supplier must initiate and conclude independent contracts and orders. This applies for both the design and manufacture of the plant. Lists of questions for detailed offers are created for this purpose. An example for the production of an assembly plant is shown by the checklist in Table 3.80.

Table 3.80 Assembly checklist

☐	**Basic information**					
	Are there specifications regarding required deadlines?					
	☐	Submission of the offer	☐	yes	☐	no
	☐	Awarding of order	☐	yes	☐	no
	☐	Customer pre-acceptance at the contractor's site	☐	yes	☐	no
	☐	Delivery	☐	yes	☐	no
	☐	Commissioning	☐	yes	☐	no
	☐	Final acceptance / production start	☐	yes	☐	no
	Are there specifications regarding payment conditions?					
	☐	Are the payment conditions acceptable?	☐	yes	☐	no
	Are there specifications regarding the client's regulations?					
	☐	Equipment specifications/company-specific standards/guidelines (please indicate revision status)	☐	yes	☐	no
	☐	Confidentiality agreements (please indicate revision status)	☐	yes	☐	no
	Are there specifications regarding external regulations?					
	☐	Laws and guidelines (liability ordinances, labor protection ordinances, employment laws)	☐	yes	☐	no
	☐	Employers' liability insurance association rules (safety regulations / accident prevention regulations)	☐	yes	☐	no
	☐	Environmental regulations (taking specific federal state regulations into account)	☐	yes	☐	no
	☐	Standards, state of the art, patents	☐	yes	☐	no
	Is there information about the product description?					
	☐	Are assembly drawings including process parameters (forces, torques) available?	☐	yes	☐	no
	☐	Are component assembly drawings available and complete? (incl. dimensions, tolerances, surface characteristics, mechanical characteristics, materials information, fits)	☐	yes	☐	no
	☐	Are assembly parts lists available and complete? (incl. quantities, materials)	☐	yes	☐	no
	☐	Is there information regarding the provision of sample parts? (please indicate quantities and deadlines)	☐	yes	☐	no
☐	**Plant concept and performance parameters**					
	Technological concept					
	☐	Is there sufficient information on the required technologies/processes? (component parts, principle workflow [assembling, pressing on/in, welding, testing, etc.])	☐	yes	☐	no
	Is there information on performance data?					
	☐	Main and secondary processing times	☐	yes	☐	no
	☐	Technical availabilities (individual and overall availabilities)	☐	yes	☐	no
	☐	Process capabilities (measurement for stability and reproducibility) (please indicate tolerances!)	☐	yes	☐	no
	☐	Output (number of units/unit of time)	☐	yes	☐	no
	☐	Planned use period (living environment)	☐	yes	☐	no
	Is there information on general and mechanical design features?					
		Plant installation size (max. floor space requirement, room height)	☐	yes	☐	no
	☐	Customer-specific restrictions regarding the handling and interlinking system (linear unit, circularity, robot system, storage)	☐	yes	☐	no
	☐	Customer-specific restrictions regarding feeder technol-	☐	yes	☐	no

		ogy/periphery (organizational condition of the incoming parts: ordered, unsystematic [palletizer, vibratory hopper conveyor])				
	☐	Flexible design of the plan due to product variants (type change) / changed customer requirements necessary	☐	yes	☐	no
	☐	Language versions (multilingual labeling, monitor and screen displays, terminals, consoles [operating, control, driver's consoles] etc.)	☐	yes	☐	no
	☐	Raw materials (deviations from standard)	☐	yes	☐	no
	☐	Color schemes/paintwork (incl. surface characteristics [galvanizing, nickel plating])	☐	yes	☐	no
	Is there information on the interfaces to be taken into account for utility connections?					
	☐	Electricity network (are there any peculiarities?)	☐	yes	☐	no
	☐	Compressed air network (are there any peculiarities?)	☐	yes	☐	no
	☐	Hydraulic network required (please indicate parameters!)	☐	yes	☐	no
	☐	Extraction system required (please indicate parameters!)	☐	yes	☐	no
	☐	Cold water supply required (please indicate parameters!)	☐	yes	☐	no
	Is there information on the control and electrotechnology (process automation)?					
	☐	Degree of automation (1. hybrid plant or 2. fully automated)	☐	yes	☐	no
	☐	Operating concept and display structure	☐	yes	☐	no
	☐	Component lists for supplier parts for the automation hardware	☐	yes	☐	no
	☐	Types of operation (incl. functions)	☐	yes	☐	no
	Is there information regarding the pneumatic and hydraulic systems?					
	☐	Customer-specific restrictions / special features (hydraulic reservoir, filtration, pumps, cylinders, valves, hydraulic pipes, mixing, special applications)	☐	yes	☐	no
	☐	Component lists for pneumatic system supplier parts	☐	yes	☐	no
	☐	Component lists for hydraulic system supplier parts	☐	yes	☐	no
	Is there information on the safety concept and ergonomics?					
	☐	Customer-specific restrictions regarding the safety concept (danger areas and protective equipment [enclosures, safety shut-off mats, light barriers], integration of emergency stop circuits/buttons etc.)	☐	yes	☐	no
	☐	Permissible noise emissions (max. sound pressure level)	☐	yes	☐	no
	☐	Ergonomic design (structural dimensions, catwalk grating, etc.)	☐	yes	☐	no
☐	**Quality assurance scope of work**					
	Is there information on quality assurance?					
	☐	Special requirements regarding the scope of quality assurance work (adjustment, calibration and master parts)	☐	yes	☐	no
	☐	Functionalities of the quality assurance measures (separate outward transfer)	☐	yes	☐	no
	☐	Measurement and testing processes (testing parameters, conditions of testing, testing sequence, operating figures to be achieved incl. quantification)	☐	yes	☐	no
	☐	Requirement for proof of testing equipment capabilities? (please indicate tolerances!)	☐	yes	☐	no
☐	**Scope of information technology and communication interface work (MDA/PDA/traceability system)**					
	Is there information regarding the need for an MDA/PDA/traceability system?					
	Are there customer-specific requirements (ONLY RELEVANT IF CONTAINED IN THE SCOPE OF PERFORMANCE [see 4.1])?					

	☐	Data acquisition and processing (scope of data acquisition, processing functions, functions of data output, information on file formats)	☐	yes	☐	no
	☐	Data management (database system, access to data, transactions and data security, archiving)	☐	yes	☐	no
	☐	Software (system software, application software)	☐	yes	☐	no
	☐	Transfer of source codes to the customer (software source code)	☐	yes	☐	no
	☐	Hardware (scope of information-related hardware, special features with regard to the hardware environment, equipment list)	☐	yes	☐	no
	☐	Communication interfaces (interfaces between man and machine / machine and machine [computers on the same level or on a higher level])	☐	yes	☐	no
	☐	Availability of the information technology / data acquisition system	☐	yes	☐	no
☐	**Scope of documentation, maintenance and operation**					
	Is there any information on the documentation?					
	☐	Type of documentation (1. standard or 2. customer-specific documentation)	☐	yes	☐	no
	☐	Scope of special features of documentation	☐	yes	☐	no
	☐	Language versions of the documentation (relevant for export)	☐	yes	☐	no
	Is there any information regarding installation and commissioning on the customer's site?					
	☐	Customer-specific requirements regarding the transportation of the plant	☐	yes	☐	no
	☐	Required utilities for the installation (means of transport, fork lift, crane systems)	☐	yes	☐	no
	☐	Installation conditions (gate dimensions, transport routes, building heights)	☐	yes	☐	no
	☐	Operator conditions (ambient temperature, max. rel. humidity, if nec. clean room requirements, vibrationless installation location, absolute avoidance of aggressive vapors and gases, etc.)	☐	yes	☐	no
	Is there information regarding training and orientation?					
	☐	Scope of training (target groups, number and duration, range of subjects)	☐	yes	☐	no
	Is there information on guarantees and services?					
	☐	Special requirements (availability period of service staff / supply of services, replacement parts storage, logistics concept)	☐	yes	☐	no
	☐	Is remote diagnosis or service necessary for the fulfillment of the guarantee / supply of services?	☐	yes	☐	no
	☐	Is an additional service premium required due to export of the plant? (inclusion in calculation)	☐	yes	☐	no

The initial commissioning of items (equipment/plants) takes place for function testing at the supplier's (manufacturer's) site. It is then dismantled to make it suitable for transport and delivered to the client/ordering party.

3.3.2.5 Delivery/Commissioning

Delivery (in accordance with the contract) and commissioning (individual and overall) of the equipment/plants and facilities form a single unit. Prior to final delivery/commissioning, all important prerequisites for the viability of the project in all of its complexity are examined and tested by the client (CL) and contractor (CO) according to:

a) technological,
b) human resource,
c) economic,
d) financial,
e) regulatory and
f) market requirements

The interaction between the CL and the CO during the process of implementation, i.e. who is responsible for what and when, is demonstrated in fig. 3.99. It is part of Machinery Directive 2006/42/EC, which prescribes the correct documentation, safety apparatus and characteristics.

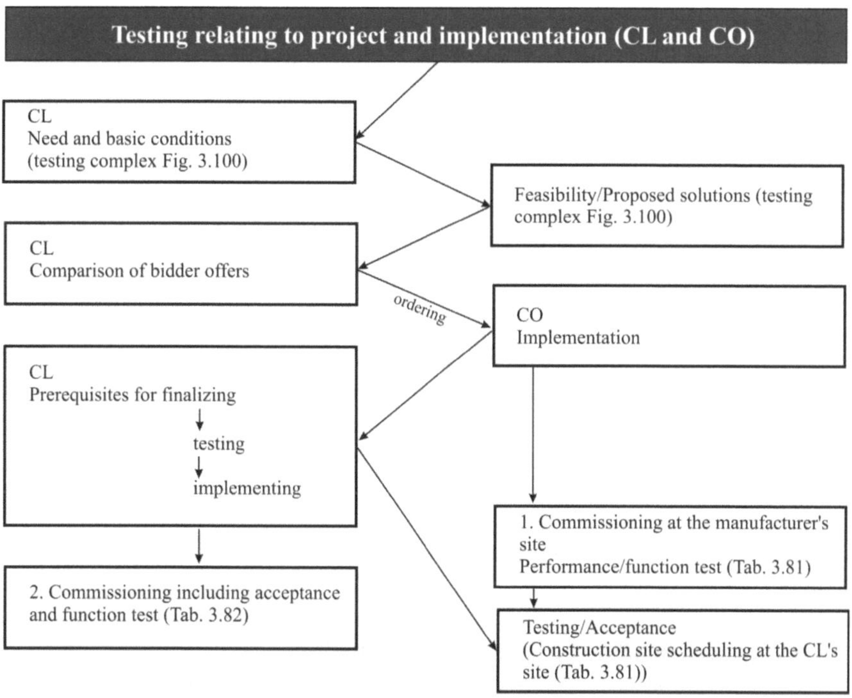

Fig. 3.99 Interaction between the CL and the CO

3.3.2.6 Implementation testing complexes

In order to prepare for implementation, the preconditions detailed in fig. 3.100 should be checked by the CL and CO.

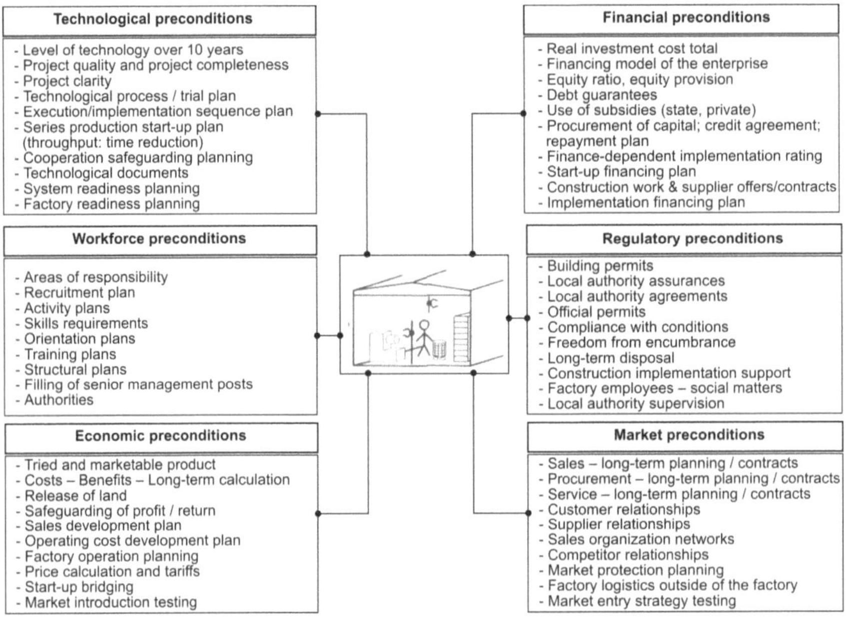

Fig. 3.100 Implementation testing complexes (after Helbing 2007)

The following project management tasks must be tackled:

- status analysis of existing projects and sub-projects (completeness, quality compliance, profitability, customer acceptance, approvals)
- specification of implementation stages (e.g. production technology/structure)
- checking and confirmation of financing (finance model)
- implementation plans (deadlines, milestones)
- fixing of the organizational form for control of the implementation process (manager, groups, suppliers)
- specification of the contents and times for implementation, trialing and start-up in the implementation schedule
- conditions of award of contract, tender document contents, proviso conditions, exclusion conditions to be governed contractually
- secure obligation to cooperate of the professional trades groups and outfitters
- control financing of implementation progress on the basis of the financing plan
- stipulate capacities for operational planning for the elimination of faults, changes, adjustments and adaptability.

- fix acceptance conditions and warranty entitlements

Once all of the conditions and measures have been satisfactorily checked, the installation site must be prepared in accordance with table 3.81. After the plant has been installed on the customer's site, the plant acceptance procedure takes place by means of a function test in accordance with table 3.82. Compliance with the operational reliability and plant acceptance parameters must be confirmed by the CL and CO in the form of an acceptance test record with binding signatures. Training programs, maintenance and change management services are also fixed.

Table 3.81 Installation site preparation checklist

☐	**Interfaces**					
	Power supply network					
	☐	Voltage	☐	not relevant	☐	clarified
	☐	Frequency	☐	not relevant	☐	clarified
	☐	Power	☐	not relevant	☐	clarified
	☐	Comments	☐	not relevant	☐	clarified
	Compressed air system					
	☐	Nominal diameter (supply and disposal)	☐	not relevant	☐	clarified
	☐	Air pressure	☐	not relevant	☐	clarified
	☐	Air consumption	☐	not relevant	☐	clarified
	☐	Comments	☐	not relevant	☐	clarified
	Hydraulic system					
	☐	Nominal diameter (supply and disposal)	☐	not relevant	☐	clarified
	☐	Pressure	☐	not relevant	☐	clarified
	☐	Max. flow rate	☐	not relevant	☐	clarified
	☐	Comments	☐	not relevant	☐	clarified
	Extraction system					
	☐	Nominal diameter (supply and disposal)	☐	not relevant	☐	clarified
	☐	Volume/unit of time	☐	not relevant	☐	clarified
	☐	Comments	☐	not relevant	☐	clarified
	Cold water supply					
	☐	Nominal diameter (supply and disposal)	☐	not relevant	☐	clarified
	☐	Water consumption	☐	not relevant	☐	clarified
	☐	Temperature	☐	not relevant	☐	clarified
	☐	Comments	☐	not relevant	☐	clarified
☐	**Communication network interfaces**					
	Technological concept					
	☐	Integration in available operating network	☐	not relevant	☐	clarified
	☐	Coupling with data servers or ERP systems	☐	not relevant	☐	clarified
☐	**Installation and Commissioning**					
	Special transport features					
	Auxiliary equipment for installation					
	☐	Hoists	☐	not relevant	☐	clarified
	☐	Max. carrying capacity	☐	not relevant	☐	clarified
	☐	Driven floor conveying equipment	☐	not relevant	☐	clarified

	☐	Max. carrying capacity	☐	not relevant	☐	clarified
	☐	Hand-operated floor conveying equipment	☐	not relevant	☐	clarified
	☐	Max. carrying capacity	☐	not relevant	☐	clarified
	Conditions of installation					
	☐	Door sizes (height x width)	☐	not relevant	☐	clarified
	☐	Transport route width	☐	not relevant	☐	clarified
	☐	Layout plan for unusable areas (pillars, ducts, etc.)	☐	not relevant	☐	clarified
	☐	Space requirement for assembling the plant	☐	not relevant	☐	clarified
	Operator conditions					
	☐	Ambient temperature	☐	not relevant	☐	clarified
	☐	Relative humidity	☐	not relevant	☐	clarified
	☐	Clean room class	☐	not relevant	☐	clarified
	☐	Prevention of local heating of machinery and control elements	☐	not relevant	☐	clarified
	☐	Absolute avoidance of aggressive vapors and gases	☐	not relevant	☐	clarified
	☐	Electromagnetic compatibility in accordance with DIN EN 50140 / VDE 0847 for routing of control cables	☐	not relevant	☐	clarified
	☐	Max. floor load	☐	not relevant	☐	clarified
	☐	Floor quality, non-slip concrete floor, max. gradient 1:2,000	☐	not relevant	☐	clarified
	☐	Foundations according to foundation plan	☐	not relevant	☐	clarified
	☐	Vibrationless installation location (poss. vibration insulation)	☐	not relevant	☐	clarified
	Test parts for acceptance					
	☐	Deadline	☐	not relevant	☐	clarified
	☐	Quantity	☐	not relevant	☐	clarified
	Required documents/approvals					
☐	**Commissioning personnel**					
	Client's commissioning personnel					
	☐	Required personnel (operators, maintenance personnel)	☐	not relevant	☐	clarified
	Contractor's commissioning personnel					
	☐	Seating possibilities for special commissioning activities (programming)	☐	not relevant	☐	clarified
	☐	Possibilities for storage of documents, tools, measuring devices and other utilities (cabinets)	☐	not relevant	☐	clarified
	☐	Labor protection measures to be taken (PPE), with respect to emissions (noise, turnings & chips, dust, vapors)	☐	not relevant	☐	clarified
	☐	Labor protection measures to be taken in the case of commissioning outside the EU (e.g. vaccinations)	☐	not relevant	☐	clarified
☐	**Plant layout**					
	Plant layout with identification of:					
	☐	Machine footprint area	☐	not relevant	☐	clarified
	☐	Operating area/operating controls	☐	not relevant	☐	clarified
	☐	Repair areas	☐	not relevant	☐	clarified
	☐	Areas in hazardous zones	☐	not relevant	☐	clarified

	☐	Auxiliary areas for storage and ancillary equipment	☐	not relevant	☐	clarified
	☐	Transport areas	☐	not relevant	☐	clarified
	☐	Connections for supply of utilities	☐	not relevant	☐	clarified
☐	**Hazardous substances, packaging and disposal**					
	Required storage areas for hazardous substances (e.g. hydraulic fluid)					
	☐	Responsibilities for disposal (plant packaging, hazardous substances packaging)	☐	not relevant	☐	not relevant

Table 3.82 Plant acceptance and function test checklist

Fault simulation									
	What happens if no parts are installed? Tested assembly components:								
			☐	yes	☐	no	☐	n/a	
			☐	yes	☐	no	☐	n/a	
			☐	yes	☐	no	☐	n/a	
			☐	yes	☐	no	☐	n/a	
	Too many parts are installed in one location? Tested assembly components:								
			☐	yes	☐	no	☐	n/a	
			☐	yes	☐	no	☐	n/a	
			☐	yes	☐	no	☐	n/a	
			☐	yes	☐	no	☐	n/a	
	Defective parts are installed? Tested assembly components:								
			☐	yes	☐	no	☐	n/a	
			☐	yes	☐	no	☐	n/a	
			☐	yes	☐	no	☐	n/a	
			☐	yes	☐	no	☐	n/a	
	Mix-up-proof storage of compliant and defective parts?								
			☐	yes	☐	no	☐	n/a	
			☐	yes	☐	no	☐	n/a	
			☐	yes	☐	no	☐	n/a	
			☐	yes	☐	no	☐	n/a	
Ergonomics									
	Working heights OK (workstation, parts supply, parts storage)		☐	yes	☐	no	☐	n/a	
	No difficulty in inserting/removing parts		☐	yes	☐	no	☐	n/a	
	Manual pre-, interim- or final assembly possible without difficulty.		☐	yes	☐	no	☐	n/a	
	Working possible without bent back		☐	yes	☐	no	☐	n/a	
	Tools easy to handle (screws, pliers)		☐	yes	☐	no	☐	n/a	
	Manipulation of rotatable components, adjusting wheels, stopcocks etc. all possible without difficulty (adequate freedom of movement)		☐	yes	☐	no	☐	n/a	
	Sufficient space for depositing components before and after the work step		☐	yes	☐	no	☐	n/a	
	Sufficient operating space available for working (standing, seated, arms, legs, head)		☐	yes	☐	no	☐	n/a	
	Terminal boxes, HQE conveyor belt, hydraulic unit, filter, wearing parts etc. are sufficiently accessible for servicing and repair		☐	yes	☐	no	☐	n/a	
	When carrying and lifting loads > 15 kg, a hoist must be available		☐	yes	☐	no	☐	n/a	
			☐	yes	☐	no	☐	n/a	

		☐	yes	☐	no	☐	n/a
		☐	yes	☐	no	☐	n/a
		☐	yes	☐	no	☐	n/a
Accumulation of dirt							
	Any accumulated dirt or waste arising is absorbed and collected	☐	yes	☐	no	☐	n/a
	Plant is easy to clean	☐	yes	☐	no	☐	n/a
	Unnecessary accumulation of dirt is prevented	☐	yes	☐	no	☐	n/a
		☐	yes	☐	no	☐	n/a
		☐	yes	☐	no	☐	n/a
		☐	yes	☐	no	☐	n/a
Control							
	Monitoring and display of:						
	Testing of operating consumables (grease, oil)	☐	yes	☐	no	☐	n/a
	Compressed air (too low and too high pressure reported)	☐	yes	☐	no	☐	n/a
	Protected area	☐	yes	☐	no	☐	n/a
	Control voltage	☐	yes	☐	no	☐	n/a
	Visibility of fault signals adequate	☐	yes	☐	no	☐	n/a
	Data storage medium with program copy available	☐	yes	☐	no	☐	n/a
	Structure of OP functional, transparent, comprehensible	☐	yes	☐	no	☐	n/a
	Structure of OP consistent	☐	yes	☐	no	☐	n/a
	Signal lamps OK	☐	yes	☐	no	☐	n/a
	All signal lamps can be seen at once (high enough)	☐	yes	☐	no	☐	n/a
	Lighting of illuminated buttons OK	☐	yes	☐	no	☐	n/a
	Lamp test available and OK	☐	yes	☐	no	☐	n/a
	User guidance on OP well executed (buttons, menu)	☐	yes	☐	no	☐	n/a
		☐	yes	☐	no	☐	n/a
		☐	yes	☐	no	☐	n/a
		☐	yes	☐	no	☐	n/a
		☐	yes	☐	no	☐	n/a
Details from the requirement specification and general information							
		☐	yes	☐	no	☐	n/a
		☐	yes	☐	no	☐	n/a
		☐	yes	☐	no	☐	n/a
		☐	yes	☐	no	☐	n/a
		☐	yes	☐	no	☐	n/a
		☐	yes	☐	no	☐	n/a
		☐	yes	☐	no	☐	n/a

3.3.2.7 Project conclusion / project evaluation

The project conclusion documents all of the goods and services supplied that were agreed in the technical and other specifications. Amendments, alterations and other specifications are part of the documentation (cf. 3.3.5.).

The project evaluation is carried out by both the CL and the CO. This is with the aim of recording potential improvements and consequences for future project planning and implementation.

3.3.3 Operation (x8)

The operation of production facilities is scheduled and implemented by means of the trial run, commissioning and start-up planning.

The production facilities are utilized via production planning and control, monitoring (service and maintenance), and change and adaptation planning and implementation during day to day operation. During operation, situation-driven, operational project design and implementation planning is required, which makes efficient technical and organizational alignment of the order (in accordance with fig. 1.4) possible. The flexible adaptation of the project on the basis of CIP recommendations, production and supply changes etc. is a permanent task of the enterprise's management.

3.3.4 Dismantling and recycling (x9)

These processes are based on utilization and recycling planning and implementation. The commitment to precautionary, sustainable, environmentally compatible economic activity includes the reuse of resources such as equipment, plants and buildings.

This is done in the following order: remediation, decommissioning, reuse, further use and recycling. This also includes the reconditioning of brownfield sites (revitalization).

3.3.5 Documentation (x10)

This includes the most important records and documentation for the entire production facility (all equipment, plants and service systems) relating to the contents, process flows, individual projects, decisions and approvals. In addition to the information from the client, contractor and cooperation partners, constituent parts of the documentation are:

a) Cooperation partners' documentation regarding project planning / execution / implementation

b) Task definition, specifications and decisions
c) Real layout for technological processes and facilities, for structural plans, material flows, supply and disposal, for installation, changes and adaptations as blue print documentation
d) Procedures and processes relating to the overall project and organization
e) Solutions developed (ideas, calculations, sketches, drawings, models, charts and sub-projects)
f) Advisory and protocol documentation (client/contractor)
g) Conditions, permits, decisions by authorities
h) Task definitions, tenders, offers, evaluation and selection of offers
i) Documents, correspondence, photographic material (incl. in digital form)
j) Reports (trial run, start-up), acceptance, evidential and handover records
k) Approval documentation (plant documentation in accordance with European Machinery Directive – 2006/42/EC)
l) Profitability assessment/status reports

References

European Machinery Directive (2009) 2006/42/EC
Helbing K (2007) Fabrikprojektierung, book manuscript, Rostock
Schenk M, Wirth S (2004) Fabrikplanung und Fabrikbetrieb - Methoden für die wandlungsfähige und vernetzte Fabrik, Springer, München
Schulte H (2003) Planung von Fabrikanlagen, lecture notes, WBK Karlsruhe
VDI 2519 (2001) Vorgehensweise bei Erstellen von Lasten- und Pflichtenheften, Düsseldorf
VDI/VDE 3694 Lastenheft/Pflichtenheft für den Einsatz von Automatisierungssystemen, Düsseldorf

4 Material flow and logistics

4.1 Material flow

The manufacturing processes within a production facility also require integral transport, handling and storage processes, which are known as logistics processes.

Logistics is the planning, implementation and control

- of the movement and positioning of people and/or goods and
- the associated supporting activities.

Logistics ensures the

- supply with commodities (goods, materials and operating materials), and also with specific information and information carriers
- disposal of substances (waste, residues, operating materials) and the recycling of objects as well as
- changes to the location of objects (physical objects and information).

Logistics processes represent a significant cost factor. Smooth logistics processes ensure and influence both the production facility's efficiency (e.g. utilization of production systems) and its performance (e.g. output and short, transparent throughput of orders).

Logistics is characterized by the global consideration of material flows, which are apparent through movement (e.g. a moving fork lift), and inventory (e.g. parts in storage), plus information and value flows.

Fig. 4.1 shows the central position of logistics in the enterprise including in control and manufacturing.

M. Schenk et al., *Factory Planning Manual*, DOI 10.1007/978-3-642-03635-4_4,

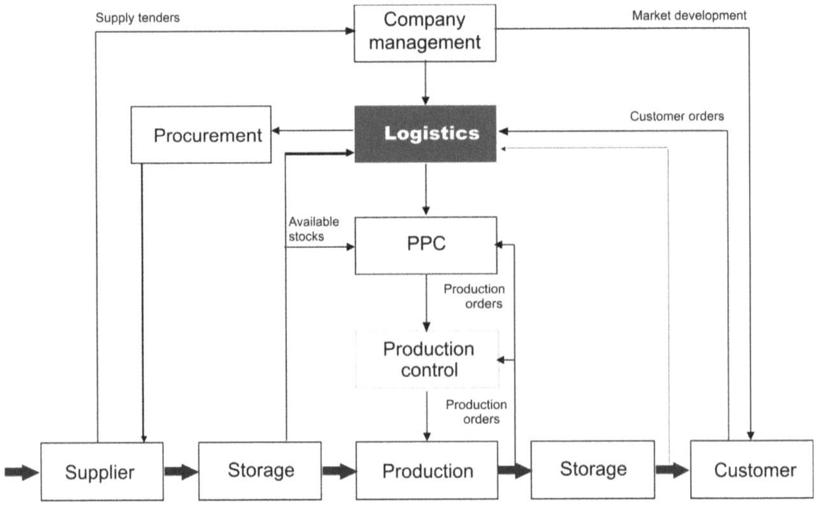

Fig. 4.1 Central position of logistics in the enterprise between markets and production (Krampe 2007, p. 141)

The organizational synchronization of the throughput of an order, which must be managed in terms of logistics, is shown in fig. 4.2.

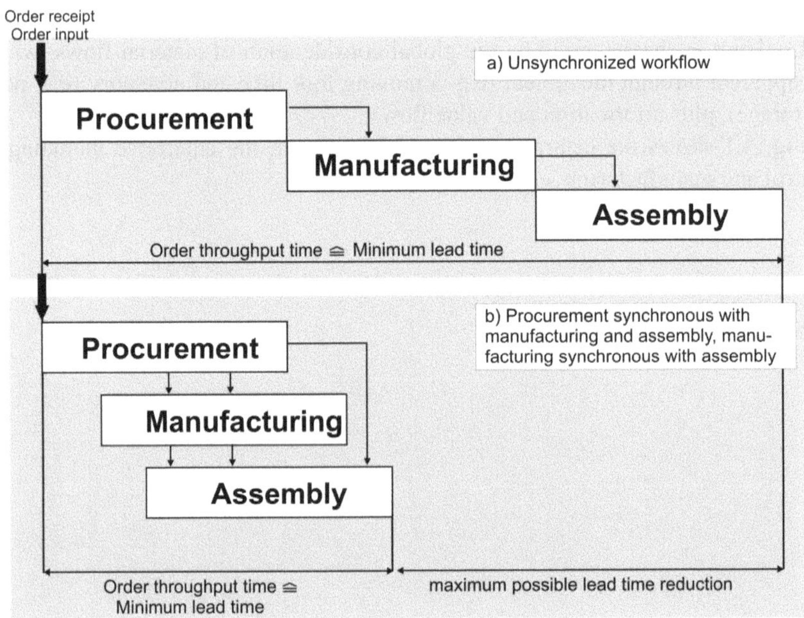

Fig. 4.2 Synchronization of procurement, manufacturing and assembly (Jünemann 1989)

The Supply Chain Management (SCM) method has proven valuable for efficiently structuring the value added chain for production and logistics processes. It enables the integration of all partners in the value added chain, the process-oriented structuring of all material and information flows and the use of modern IT solutions for process control (cf. chapter 6.4).

Note: material, energy, information, personnel and capital flows together with their systems and equipment and connection to one another must be planned and implemented for every production facility according to the logistics principles of value added chain design.

4.2 Logistics product, process and system

The cognitive model assists in logistics planning by segmenting "Logistics" into the logistics product, process, system and facilities. This makes it possible to simplify the processing tasks and at the same time provides the opportunity for control in respect of the completeness of the planning of material flows and logistics (cf. table 4.1 for more on this).

Table 4.1 Logistics perspectives (Ziems 2007)

Categories	Physical perspective	Information-related perspective
Logistics product	Commodities, e.g. goods, parts, waste, animals people	Information, e.g. orders inventory information accounting identification number
Logistics processes	Material flow Storage Transport, handling Collecting/distributing Picking/sorting Packaging/repackaging Labeling/identifying	Information and communication flow Capturing Processing Storing/administering Circulating/utilizing
Logistics systems/facilities	Material flow systems/facilities Man and machine e.g. Communications technology Storage technology Conveying technology Packaging technology	Information and communication systems Man and machine e.g. Identification technology Control technology Control center technology Communications technology

Note: there are different ways of illustrating logistics products, processes and systems, which should be employed according to their practicability in the specific case.

4.2.1 Logistics product

As can be seen from table 4.1, the logistics product can be either a product or a piece of information. Table 4.2 contains a nomenclature for describing a logistics product and incorporates both outcome and process aspects.

Table 4.2 Description of the logistics product as the 6+1 "rights" of logistics

The 6+1 "rights"		Question	Requires the...
1	The right object	WHAT	identification; assessment of logistics characteristics
2	In the right quantity	HOW MUCH	counting, weighing etc.
3	In the right place	WHERE, WHERE TO	recording the location conditions on site
4	At the right time	WHEN, HOW OFTEN, HOW LONG	recording the times, calculating the duration, complying with tolerances
Process aspects:			
5	At the right cost	HOW (low cost)	assessment of the logistics process with respect to effectiveness and efficiency
6	Of the right quality	HOW + (quality-compliant)	assessment of the logistics process with respect to safeguarding the integrity of the objects
+ 1	Environmentally-compatible	HOW ++ (environmentally sound)	assessment of the logistics process with respect to ensuring environmental compatibility

During the execution of the logistics processes, logistics objects may display a different or changing external appearance. (cf. fig. 4.3). In addition, objects may also be nested (e.g. parts in packaging, packaging in a container, container on a pallet, pallets in a box container).

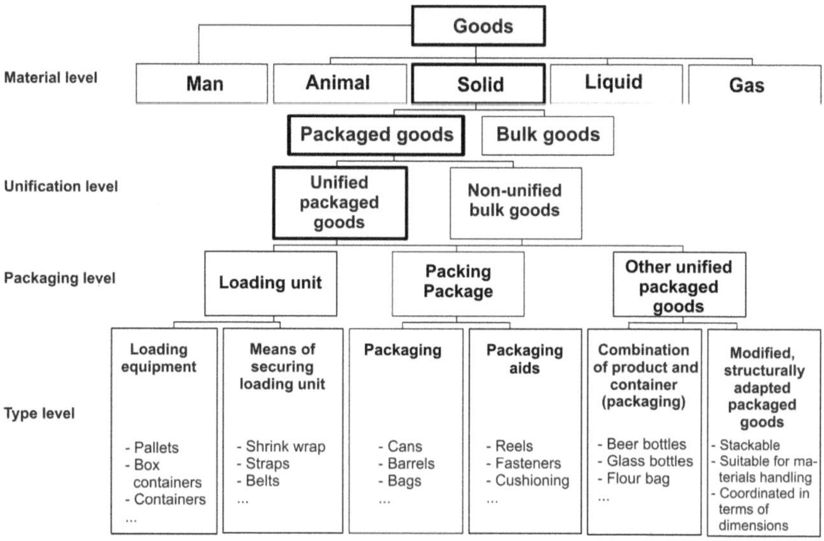

Fig. 4.3 Goods classification (based on Grossman 1990)

Major features of an object description, which are derived from table 4.2, are contained in table 4.3.

Table 4.3 Features of object description in logistics (based on Ziems 2008)

	Characteristic	Abbreviation	Description	
The right object	Sort	S	Sort description	
			– Identification number	
			– Sort characteristics (among others physical, chemical properties, value)	
			Range	
			– Breadth of range	
			– Depth of range	
In the right quantity	Quantity	Q	Mass	m in kg
			Volume	v in m^3
			Number	n in units
			Density	d in kg/m^3, units m^3
In the right place	Place	P	Place name	
			Place coordinates P (P_x, P_y, P_z)	
			Conditions on site	

At the right time	Time	t	Point in time	
			– Date	xx.xx.xxxx
			– Time	xx:xx
			– Tolerance	-xxx min; + xxx min

As well as the object descriptions in table 4.3, information on the costs, services or special conditions (e.g. product requires verification or to be kept cool; product is valuable) should also be added in order to obtain a complete description of the logistics product.

4.2.2 Logistics process

The logistics process includes transformation tasks together with their characteristics and material flow operations (table 4.4), e.g. transporting, handling, storing, sorting, identifying, labeling, packaging.

Table 4.4 Operand transformations in the physical material flow

	Characteristics						
					(ancillary)		
Transformation task	**Time**	**Place**	**Quantity**	**Configuration**	**Appearance**	**Information**	**Operations**
Safekeeping	Δ t						Storage
Relocation	Δ t	Δ P					Transportation
							Handling
Change of quantity	Δ t	Δ P	Δ Q				Collecting
							Distributing
Change of range	Δ t	Δ P	Δ Q	Δ C			Sorting
							Assorting
Change of appearance	Δ t				Δ A		Packing
							Repackaging
							Unpacking
Change in the form of information	Δ t					Δ I	Labeling
							Identifying
Changes to units of goods with respect to	Environment		Content Relationships		Process characteristics		

Similarly to the operation sequence in table 4.4, the following applies for the information flow: saving (storing), relaying (transporting), capturing (collecting), distributing (loading & unloading), sorting, integrating, filtering, processing, labeling, identifying. In this way, any logistics process can be assembled as if from individual building blocks.

A process description is often used to document logistics flows, procedures or work instructions. Selected aspects of a process description are detailed in table 4.5, which can also be used as a checklist.

Table 4.5 Main points of a process description (cf. VDM, 2002)

Descriptive detail	Means of implementation
Process identification	– How is this process characterized? – What elements does the process contain? – Where does the process begin and end? – Where does the process stand in the system?
Process owner	– Who is responsible for describing and developing the process?
Process participants	– Who is responsible for individual tasks within the process? – What function does this person have?
Process objective	– What is the role of the process? – What benefits does the process have for internal and external customers? – What benefits does the process have for the enterprise? – How can the objectives be measured and tracked?
Process customers (internal/external)	– Who benefits from the outcomes of this process? This can be the person responsible for a subsequent process, legislators, purchasers, operators, users of the product etc.
Process inputs	– What initiates the process? – What is required by the process in order to successfully implement it? This can be information, documents, products, specified cycles, etc.
Rules in the process	– What is the input (materials or information)? – What specifications and rules are there for the process? These can be means, methods, information, criteria for external services, applicable procedures, guidelines, codes of practice, procedural instructions, work instructions etc. – What influence do these rules have on the process?
Process outcome	– What is the outcome (output) of the process flow? This can be a product, a service, a decision, information etc. – How is this outcome tested?
Process verification and documentation	– What information, documents and records are needed for the process? – What information, documents and records are generated by the process?
Key process control data	– What parameters are used to control the process? (deadline, time or cost parameters)
Interaction with other processes	– Which other processes have an influence on this process? How? – Which other processes are influenced by this process? How?
Process supplier	– Who is responsible for performing the required preparatory work for this process? This can be the person responsible for the preceding process, legislators, customers, the company, etc.

The following features in particular must be described in order to illustrate a logistics process:

- potential (generation of and demand for goods)
- topologies (paths, distances, routes)
- flows (movements, inventories, blockages, impediments)
- resources (floor space, volumes, personnel, costs, time)
- restrictions

a) Depiction of material flow

A range of commonly-used models exists in order to illustrate material flows (cf. table 4.6).

Table 4.6 Types of material flow models

	Tangible	Abstract	
		Graphical	Mathematical
static (unchanging)	Architectural model	Graph Map Block diagram Flow diagram (Sankey diagram)	Connection matrix (Adjacency matrix) Transport matrix Formula
dynamic (changing)	Mobile functional model Model railroad Flight simulator	Vehicle navigation system Animation Interactive animation Influenceable	Analytical: y = f(t) Statistical: simulation model (program)

- Schematic graphic depictions

These can be distinguished according to the correctness of position, proportions, and in terms of quantity and type as follows (tab. 4.7):

Table 4.7 Distinguishing characteristics of schematic graphic depictions

<table>
<tr><th>Characteristic</th><th colspan="4">Properties</th></tr>
<tr><td>Location of sites</td><td colspan="2">correct location</td><td colspan="2">incorrect location</td></tr>
<tr><td>Distances</td><td colspan="2">to scale</td><td colspan="2">not to scale</td></tr>
<tr><td>Flow object quantities</td><td>not depicted</td><td colspan="2">scale flow width</td><td>figures</td></tr>
<tr><td>Flow object types</td><td colspan="2">not distinguished</td><td colspan="2">distinguished</td></tr>
</table>

Fig. 4.4 contains customary forms of illustrating logistics processes.

<table>
<tr><th>Example</th><th>Name and features</th></tr>
<tr><td>⇨△○⇨△○⇨+⇨+⇨□⇨△⇨
Process flow plan
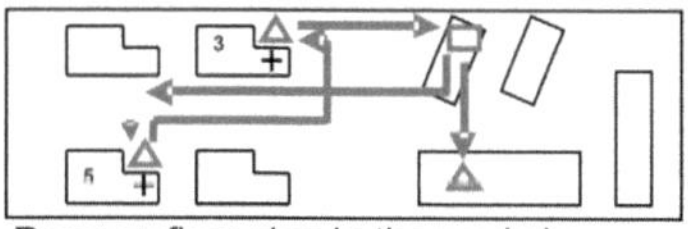
Process flow plan in the workshop layout</td><td>Material flow process symbols
(internationally recommended (FEM))
<table>
<tr><td colspan="4">Provision processes</td><td colspan="2">Other processes</td></tr>
<tr><td>▷</td><td>⇨</td><td>○</td><td>△</td><td>+</td><td>□</td></tr>
<tr><td colspan="2">Transport</td><td>Handling</td><td>Storage</td><td>Proces-sing</td><td>Testing</td></tr>
</table>
Location of sites: correct position
Distances: not to scale
Flow object volumes: not shown
Flow object types: not differentiated</td></tr>
<tr><td>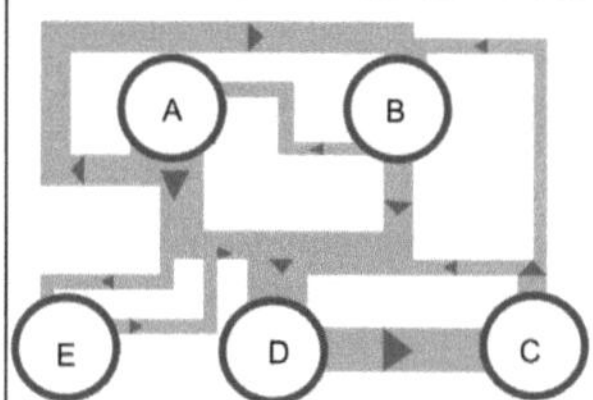</td><td>Sankey diagram
The Sankey diagram shows material flow relationships and also their direction using arrows. Intensities can be seen by the thickness of the arrows.
Location of sites: not in the correct position
Distances: not to scale
Flow object volumes: scale flow width
Flow object types: differentiated</td></tr>
<tr><td>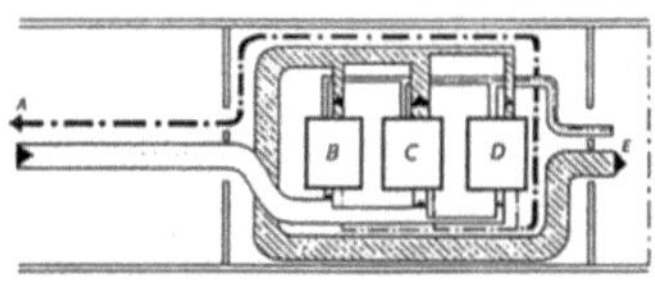</td><td>Sankey diagram showing correct positions
Location of sites: correct position
Distances: not to scale
Flow object volumes: scale flow width
Flow object types: differentiated</td></tr>
<tr><td>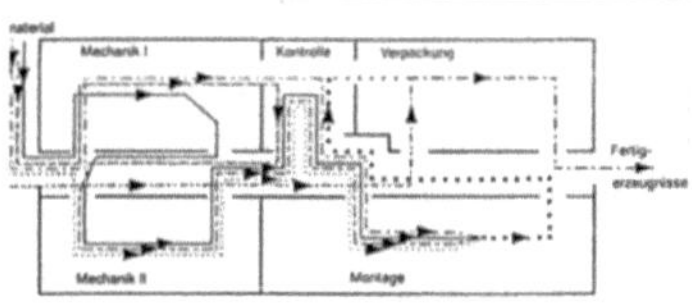</td><td>Floor layout plan with material flow lines
Location of sites: correct position
Distances: not to scale
Flow object volumes: not shown
Flow object types: differentiated</td></tr>
</table>

Fig. 4.4 Model and characterization of forms of material flow illustration (Part 1)

Example	Name and features
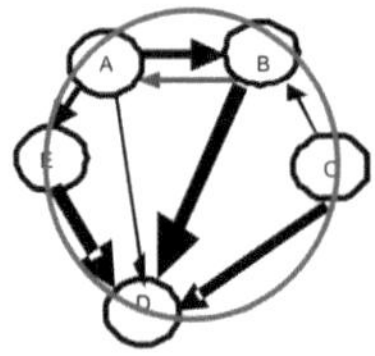	**Circle diagram** Location of sites: not in the correct position Distances: not to scale Flow object volumes: shown Flow object types: differentiated
	**Simulation model** Used for dynamic evaluation with regard to parameters such as throughput time and inventory. Location of sites: not in the correct position Distances: not to scale Flow object volumes: shown Flow object types: differentiated
	**VR Model** Used for realistic depiction. Can be enhanced with simulation and animation. Location of sites: correct position Distances: to scale Flow object volumes: shown Flow object types: differentiated
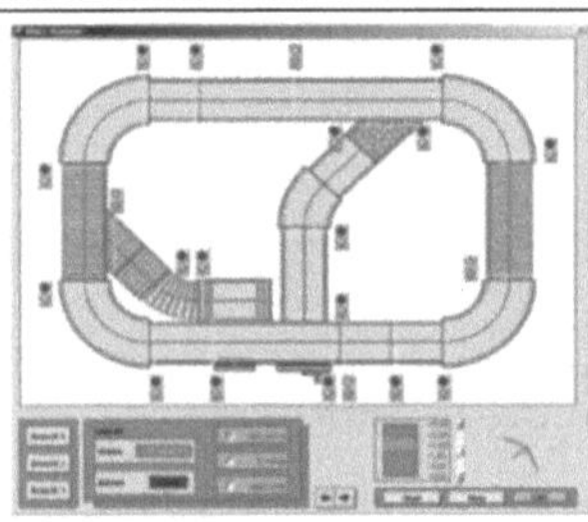	**Control center in 2D** Used for to scale 2-D depiction. Can be enhanced through visualization of situations, simulation and animation. Location of sites: correct position Distances: to scale Flow object volumes: shown Flow object types: differentiated

Fig. 4.5 Model and characterization of forms of material flow illustration (Part 2)

- Transport matrix structure

The transport matrix is especially well-suited for calculations. For this reason it is explained in more detail below. Fig. 4.6 shows the general structure of a transport matrix:

- the sources (locations in which a product is generated) are entered in the lines and
- the sinks (locations in which there is a demand for a product) are entered in the columns.

- the data field contains the information for the relationship between a particular source and sink.

Depending on the way in which the transport matrix is used, different situations can be illustrated. (cf. table 4.8). Separate matrices are required for flow volumes of different flow objects.

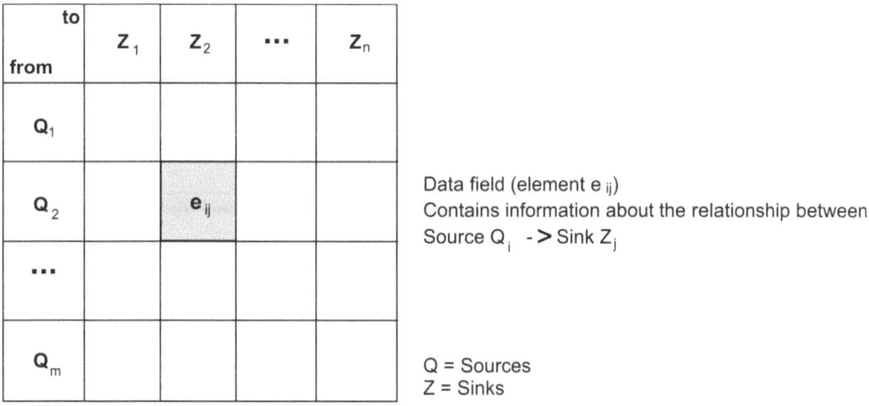

Fig. 4.6 General structure of a transport matrix

Table 4.8 Use of the transport matrix

Name and content			Unit
Evaluation matrix	Distance matrix How far are source Q_i and sink Z_j away from one another?	$e_{ij} = s_{ij}$	[m, km, miles]
	Duration matrix How long does a journey between source Q_i and sink Z_j last?	$e_{ij} = t_{ij}$	[d, h, min, sec]
	Transport work matrix What transport effort is executed on the way from the source Q_i to the sink Z_j?	$e_{ij} = B_{ij}$	[tkm]
	Load matrix What is the size of the load on route from the source Q_i to the sink Z_j (e.g. how many vehicles, journeys)?	$e_{ij} = F_{ij}$	[Number of vehicles, number of journeys]
	Cost matrix What costs are incurred for the transport from the source Q_i to the sink Z_j?	$e_{ij} = K_{ij}$	[$, €]
Flow matrix	Quantity matrix How many units must be transported from the source Q_i to the sink Z_j?	$e_{ij} = M_{ij}$	[t, kg, m^3, l, number of load units, number of loads]
	Capacity matrix How much transport capacity is available on the route between the source Q_i and the sink Z_j?	$e_{ij} = k_{ij}$	[t, kg, m^3, l, number of load units, number of loads]
Network matrix	Connection matrix Does a direct connection exist between the source Q_i and the sink Z_j (1=yes, 0=no)?	$e_{ij} = h_{ij}$	[0] or [1]
	Relationship matrix Does a transport relationship exist between the source Q_i and the sink Z_j (1=yes, 0=no)?	$e_{ij} = r_{ij}$	[0] or [1]
	Traffic lane matrix How many traffic lanes exist on the route between the source Q_i and the sink Z_j?	$e_{ij} = FS_{ij}$	[Number]
	Limitation matrix Are there limitations arising through tonnages or heights on the route from the source Q_i to the sink Z_j (1=yes, 0=no)?	$e_{ij} = r_{ij}$	[t, m]

b) Depiction of information flow

The following fig. 4.7 contains major forms of illustration for information and business processes.

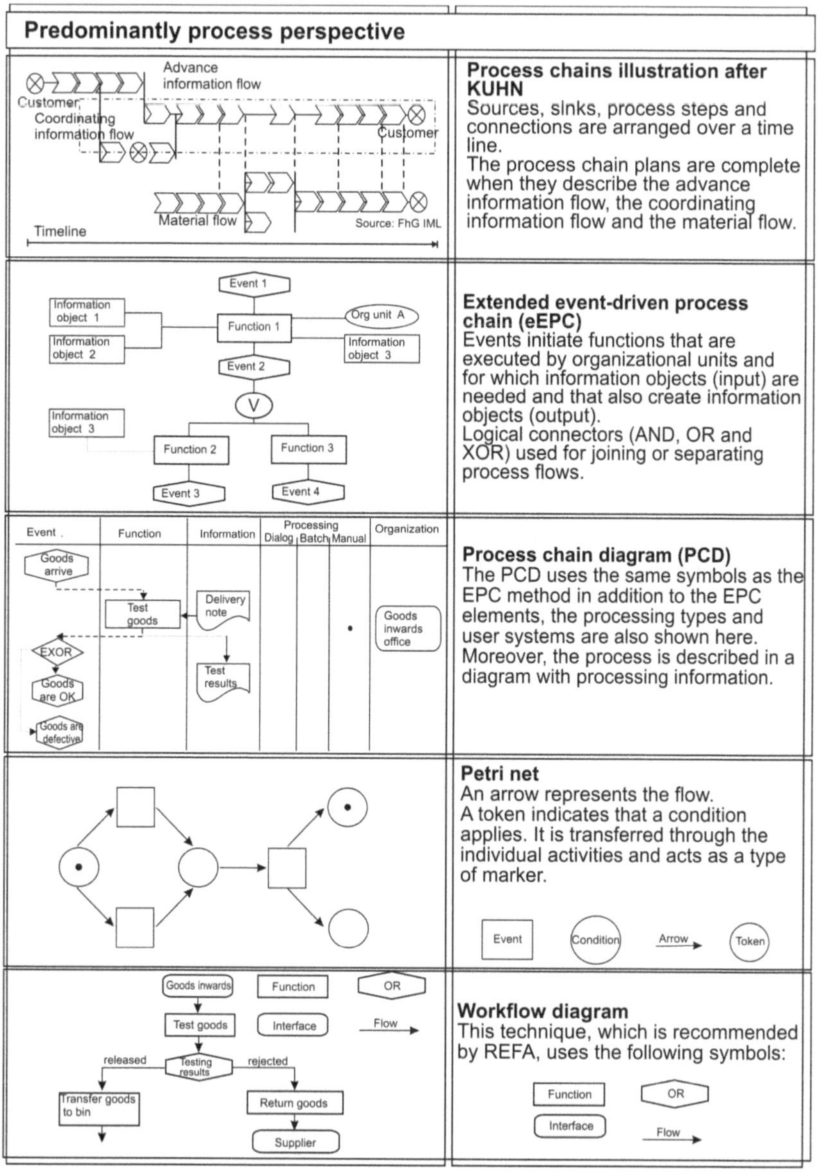

Fig. 4.7 Examples of forms of illustration for information flows and business processes.

4.2.3 Logistics system

Here the general system model provides support that stems from systems theory. As is known, systems theory describes the structure, connections and behavior of systems. The following must be defined for the installations being planned (cf. table 4.9):

Table 4.9 Description of the installations being planned as a system

Characteristic	Example
Planning output parameters (known or to be defined by the planner):	
System designation	Workshop
System boundary	Is formed by the structural envelope (the building) (of course this may, however, also be freely defined).
System inputs	Everything that enters the building in the material or information flows
System outputs	Everything that leaves the building in the material or information flows
System surroundings (system environment)	Identification of the influences and restrictions arising from the environment
Planning outcomes:	
System elements and sub-systems	Outcome of planning: Selection and dimensioning of the elements of the logistics system Formation of sub-systems
System structure	Outcome of planning: formation of structures through technical and/or organizational connection of individual elements
Relationships	Outcome of planning: definition and evaluation of the relationships between the elements

The systems can be described from different perspectives. Fig. 4.8. shows a few selected perspectives.

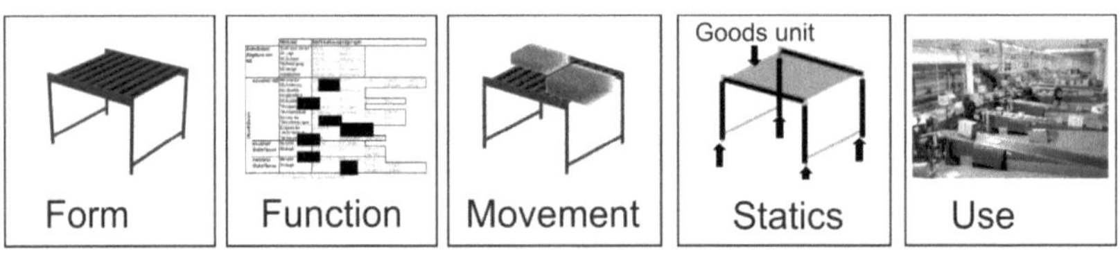

Fig. 4.8 System perspectives

In addition, the following data is collected from other perspectives and required for the different planning steps:

- cost-related data for investment lists or investment accounting,
- databases for subsequent simulations,
- connection requirements of the technical equipment as the basis for connection plans,
- disposal-related data for the visualization of disposal process sources, sinks and methods,

- capacity-related data as the basis for capacity planning and for the dimensioning of equipment and components,
- descriptive texts such as, for example, servicing information, ordering information, etc.

4.2.4 Logistics relationships

The subdivision into product, system and process means that it is also necessary to consider the relationships between these aspects. (cf. table. 4.10).

Table 4.10 Logistics broken down into logistics product, logistics system and logistics process (Illés 2007)

Description aspect	Logistics product	Logistics system	Logistics process
		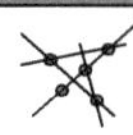	
Logistics product 	•	**R1**	**R2**
Logistics system 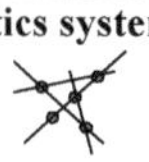	R1	•	R3
Logistics process 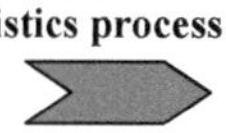	R2	**R3**	•

R1 Relationship between logistics product and logistics system:

The logistics product defines the requirements of the logistics system. The logistics system must be dimensioned and structured in such a way that it is in a position, functionally and in terms of capacity, to fulfill the requirements/logistics performance expected of it. This affects the number of system elements (e.g. the amount of rack space in a high-rack warehouse) or the performance of the individual system elements (e.g. the speed of a transport conveyor) as well as, for example, the reliability of the system elements (e.g. technical availability) or the formation of structures through technical connections or the formation of organizational units.

R2 Relationship between logistics product and logistics process:

The logistics product also defines the requirements of the logistics process. The logistics process must be designed in such a way that it is potentially in a position to fulfill the requirements/logistics performance expected of it. At the same time the processes should, of course, also be effectively and efficiently designed. To this end, the operations should be examined in terms of necessity (waste) and the manner of their execution.

R3 Relationship between logistics system and logistics process:
This relationship covers the coordination of configuration (elements and structure) and sequence (process) and thus the classic "scheduling dilemma." This requires the planner to constantly strive to achieve a compromise between high capacity utilization (system basis) and throughput times (process basis).

4.3 Material flow and logistics planning

4.3.1 Planning steps and planning tasks

This entails the planning of material flows (logistics chains) and business processes. Here too the material flow planning steps are:

- performance program coordination (qualitative determination of the logistics commodity including the logistics utilities),
- determination of functions (logistics technology (processes) and THS equipment/systems),
- dimensioning (amount of logistics equipment, floor space and workers),
- structuring (layout of the logistics resources in combination with the process flows and production facilities based on the material flow calculation) and
- design (layout of the entire production facility with different forms of material flow).

The information flow is planned in the same way (see chapter 6.3.2).

The material flow inside the building can take different forms. There is no single form that guarantees optimal material flow in all cases. Important basic forms are contained in the following table (table 4.11).

Table 4.11 Common basic forms of material flow in buildings (Kettner 1984, p. 160 ff)

Basic form	Example
	**Linear** Typical for manufacturing and for assembly work-flows with fixed flow sequence
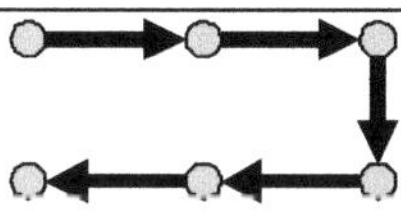	**U-shaped** Typical for picking
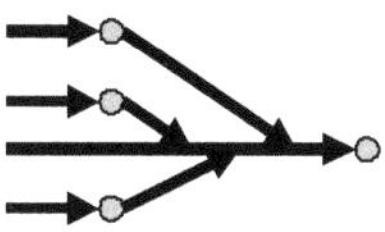	**Converging** Typical for assembly with connected pre-assembly stages
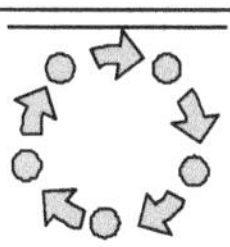	**Ring-shaped** Typical for collection rounds
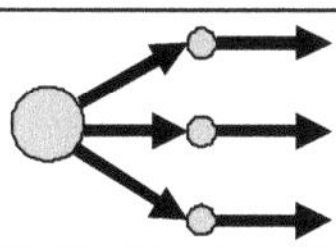	**Diverging** Typical for disassembly
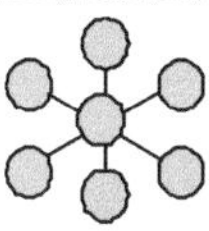	**Star-shaped** Typical for assembly/disassembly
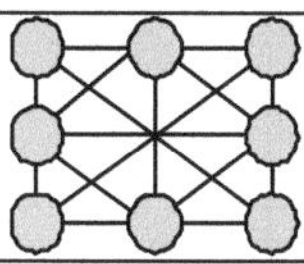	**Network-shaped** Typical for flexible, frequently-changing workflow sequences

The following planning rules apply for the planning, especially for the material flow (cf. table 4.12 as a checklist):

Table 4.12 Checklist of important planning rules for designing material flows

Aspect	Planning recommendation / rule	Observed?
Parameters	– form practical units (e.g. procurement unit = manufacturing unit = transport unit = storage unit = packing and shipping unit)	□
	– classify parameters to facilitate the planning task => reduce planning effort	□
Workflows	– undertake clear definition of workflows (collect and bring systems, control centers, stations)	□
	– provide measuring points for the identification and monitoring of progress	□
	– guarantee short throughput times	□
	– plan main transports (materials) and secondary transports (e.g. waste) in an equally effective and efficient way	□
	– limit handling processes to a minimum	□
	– minimize control costs	□
System/plants		
Transport route	– materials should flow in a linear fashion	□
	– the material flow should only link a small number of areas	□
	– plan short transport routes	□
	– do not plan any route intersections	□
	– do not plan movements in opposite directions	□
Floor space	– allot minimal floor space and room occupation while securing the ability to function	□
Engineering	– minimize the variety of types of technology used => favorable conditions for capacity utilization, for redundancy, for optimization of maintenance, low stocks for replacement parts	□
Logistics equipment	– variety of types (see above). Limit technology	□
	– take automation into account	□
	– achieve high capacity utilization of equipment and plants	□
Logistics utilities	– select as small as possible	□
	– variety of types (see above). Limit technology	
Costs	– minimize investment costs	□
	– minimize overall costs taking operating costs for n years into account	□
People	– minimize number of required workers	□
	– design workstation and workflows ergonomically	□

The problem and situation analysis forms the starting point for material flow planning plus a precise specification of tasks and objectives.

The goal of logistics planning for a given planning task is (fig. 4.9):

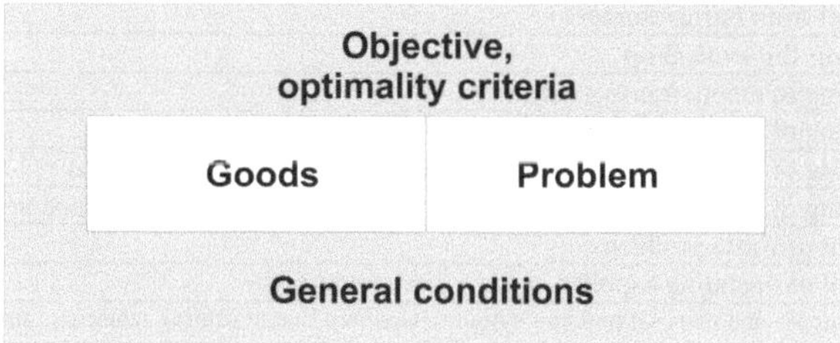

Fig. 4.9 Material flow planning task

- to develop (find, design) suitable solutions for problems,
- to demonstrate their suitability,
- to formulate all specifications for procurement and construction, assembly, commissioning, operational management, servicing/maintenance, operating instructions, organization and disposal for a given planning task (fig. 4.9).

To do this, a varied range of eligible propositions should consciously be included in the planning. Planning execution requires the ability to select operations and connect them as well as to select conveying and storage equipment as planning building blocks for logistics facilities, an assessment of the conditions of use and areas of usefulness of logistics components, a command of the techniques of dimensioning, layout and performance assessment as well as the ability to define functional ordering and procurement tasks. These planning tasks should therefore be carried out by factory planners, material flow technicians, conveying technicians or logisticians.

The following checklists contain important material and information flows that need to be planned (cf. tables 4.13 and 4.14). They answer the question of "what" should be planned.

Table 4.13 Personnel and material flow checklists

Personnel flows (company employees, suppliers, visitors)	
Material flows:	
Supply in the workshop	
☐	Supply processes with materials (basic materials, individual parts, component assemblies, additional materials, purchase parts, standard parts, finished products) from outside into the workshop
☐	Supply processes with tools, jigs and fixtures, replacement parts
☐	Supply processes with operating supplies and auxiliary materials
☐	Provision of empty transport utilities
☐	Provision of empty storage utilities
☐	Flows of personnel into the workshop
Transfer within the workshop	
☐	Transfer of materials (basic materials, individual parts, component assemblies, additional materials, finished products) from workstation to workstation in technological sequence and if necessary also in closed loop processes
☐	Transfer from the workstation to buffer storage

☐	Removal from buffer storage
Disposal from the workshop	
☐	Recycling of empty transport utilities
☐	Recycling of empty storage utilities
☐	Recycling of used tools, used jigs and fixtures, used replacement parts
☐	Recycling of materials (individual parts, assembly components, finished products)
☐	Disposal of waste products
☐	Disposal of operating supplies and auxiliary materials
Flow systems of supply and disposal/building systems (auxiliary manufacturing materials, air (supply air, exhaust air), water (drinking, process and waste water), electrical power (drives, heating, data processing), compressed air, technical gases, ambient air (air conditioning), water vapor, warm water (heating), etc.) – cf. Appendix 6.13	

Table 4.14 Information flow checklist

Information flows:	
Supply in the workshop	
☐	Supply of information in the workshop
Transfer within the workshop	
☐	Transfer of information from workstation to workstation
Disposal from the workshop	
☐	Recording of information in the workshop (esp. progress and availability information)
☐	Feedback of information (e.g. KANBAN post) from the workshop

In order to clarify the complexity and scope of the tasks within the subsequent material flow planning, important tasks that a logistician must deal with when designing and optimizing the material flow are enumerated below in the form of a summary list. The planning tasks are classified according to logistics product, logistics process and logistics system and summarized in table 4.15.

Table 4.15 Examples of planning tasks

Logistics product/logistics service	
	– category formation according to shared logistical features – volume planning (primary and secondary demand) – calculation of lots – determination of the average arrival time interval – calculation of arrival rate – procurement planning, determination of resource requirements
Logistics process	
(1)	Material flow:
	– definition of the flows – analysis of the flow rate (continuous or intermittent flow rate) – calculation of inflow and outflow capacity – calculation of average duration in the system – calculation of probabilities for dynamic systems (arrival time, service time) – calculation of working time / performance time – calculation of service time (overall time – repair times) – conducting of functional analyses (branches, single server) – analysis of system parameters (applicable requirements & general conditions) – calculation of system load (technical processes having an influence from the outside,

	geometric conditions) – calculation of material flow indicators (capacity utilization, capacity reserves) – calculation of performance indicators (throughput, flow intensity, volume, speed) – calculation of maximum capacity – calculation of shipping time (supply time/loading time) – determination of the routes between the source and destination locations – transport optimization, route scheduling – optimization of empty running – increasing of loading capacity – calculation of traffic density – elimination of points of conflict – planning of transport system
(2)	Information flow:
	– definition of the flows and coordination with material flows – definition of measurement points in the information flow – selection of an information system (ERP system) – checking of the process for Kanban suitability – specification of logistics principles (e.g. supply, storage and picking principles) – selection of identification system – selection of suppliers and service providers – scheduling – setting up of simulation model and running simulation – planning and setting-up of information and energy flows – planning of the systems for data storage
(3)	Working cycle:
	– decision as to whether single cycles, dual cycles or multiple cycles will be implemented – calculation of the starting point for the conveying equipment's working cycle – duration calculation for individual and multiple cycles
(4)	Picking:
	– determination of the procedure used (man to goods, goods to man) – calculation of basic times for picking: → receipt of order and empty container → picking + handover of paperwork – calculation of travel times – calculation of sorting times – selection of picking procedure – determination of optimal number of lines for picking orders
(5)	Storage:
	– selection of basic principle – specification of storage sequence (e.g. FiFo, LiFo, etc.) – determination of storage organization – calculation of storage indicators (inflow, stock)
Logistics system	
(1)	Equipment/operating terminal:
	– determination/optimization of location – planning of machinery layout – calculation of flow intensity and throughput – checking of ground conditions (inside/outside; covered/open) for use of optimum conveying equipment – optimization of ground conditions (floor covering, ramps) – setting up of transfer points – planning of disposal

	– calculation of maximum capacity of equipment – calculation of amount of equipment required – calculation of the service rate of the control apparatus – dimensioning of waiting rooms – calculation of set-up times – calculation of average waiting time in the queue
(2)	Means of conveyance/transport
	– analysis of working cycle of conveying equipment – calculation of working cycle time → calculation of the time that the conveying equipment needs to collect and deliver an item → calculation of the duration of loaded and empty running – calculation of the time that the conveying equipment needs to reach full speed – calculation of conveying equipment's start-up and slow-down times – calculation of stopping distance of the conveying equipment – calculation of capacity of the conveying equipment – calculation of the amount of conveying equipment (e.g. in the case of shuttle connection) – trade off between route length, costs (capacity), amount and type of conveying equipment required (combined traffic) e.g. use of fork lift trucks + conveyor belt – calculation of amount of transport equipment required – selection of conveying equipment
(3)	Storage/Buffering:
	– where should the goods waiting to be transported by the conveying equipment be stored? – where will the goods be stored (buffered) in the interim? – division of the warehouse into storage areas – dimensioning of the warehouse – storage in high-rack warehouse: → calculation of the applicable movement time for each rack, calculation of selection cycle, determination of order of racks approached → determination of the size of store and the number of storage areas (capacity of the warehouse) – if necessary, dimensioning of interim storage areas – determination of optimal storage zoning – calculation of required buffer storage, dimension
(4)	Route:
	– calculation and optimization of route widths – calculation of route length and travel time – calculation of the route loading
(5)	Loading utilities:
	– calculation of the empty equipment to be provided – selection of utilities (for transport, storage, picking) – calculation of optimal loading unit
(6)	Overall solution:
	– calculation of maximum capacity of the entire system – parallel or series connection of operating controls – determination of location of goods inwards and goods outwards storage – verification of system reliability → determination of the availability of the internal transport system taking malfunction indicators into account → determination of the achievable operating times and delivery rate – calculation and optimization of reliability and availability of the system

In order to create universal processes, all interfaces must be taken into consideration. It is important that the logistics processes within the company should not be optimized in isolation, but instead that they be integrated smoothly into the external material flow. This means that at this point questions should be asked such as:

- what is known regarding the provision of inputs (material, individual parts, component assemblies)?
- which means of conveyance/transport will be used to make delivery?
- which transport utilities will be used?
- how should the delivery take place?
  - JIT (just in time)
  - JIS (just in sequence)
  - cross docking
  - via a goods inwards store
- how will the finished products be shipped?
- how should they be made available?

An analysis of the material flow over all of the processing stages necessary for manufacture, including staging and disposal areas for materials, utilities and operating supplies, plus buffering and storage, constitutes the starting point for the internal part.

Requirements regarding the identification of parts and batches in order to completely verify which part was installed in which product, by whom and when, are also to be included, so that they can be implemented subsequently in an appropriate manner.

In order to be able to assess the correctness of the planning, concrete material flow planning objectives must be derived and defined from the overall plans. Variants and the preferred solution should all be evaluated on the basis of these objectives.

NB: material flow planning approaches:

(1) Problem and situation analysis plus precise definition of the project tasks, setting of objectives and determination of the performance program.

In this step the requirements are specified in concrete terms and objectives defined. This step helps to clarify and fix the degree of planning freedom.

(2) Determination of functions (functional concept)

In this step all logistics processes are formulated as a process chain (logistics chain) taking the twin aims of effectiveness and efficiency into account, the material flow is calculated and the logistics facilities (equipment) are selected.

(3) Dimensioning
In this step the basic resources and capacities are calculated. To do this, specific elements must be selected and the quantity/size of the logistics facilities must be determined.
(4) Structuring
This step is used to arrange the logistics facilities that have been selected and quantified.

(5) Design
The layout is created in conjunction with structuring (structural concept).

(6) Verification of functions using a model
This step involves the functional verification of the planning; simulation models and virtual representations are commonly used for this today.

4.3.2 Transport

The largest logistics cost elements arise as a result of the quantities, volumes and mass of the objects that must be moved and stored. By contrast, the transfer, processing and supply of information is associated with comparatively low costs. The result is the principle of the dominance of material flows.

As well as the fundamental transport process flows that have to be fixed, the results of transport planning must define:

- the type and amount of transport equipment,
- the type and amount of transport utilities,
- staging and disposal areas,
- the type and location of transport routes,
- the type, size and principle location of buffers,
- the skills level and number of workers and
- the type and scope of all utilities

Solutions for the associated information flow are developed in coordination with this. Once the production sequence (approximate technology) and material flows in terms of direction, intensity and chronological progression are known, it is possible to make a preliminary decision regarding the choice of transport equipment.

The same applies to the initial decision regarding the choice of a continuous or intermittent mode of operation. In order to make both initial decisions, the following data must be compiled (table 4.16):

Table 4.16 Data describing the transport task and solution

Data describing transport task	Content examples
Commodity to be conveyed	Type, number, mass, dimensions, logistical characteristics
Transport time	Overall transport time and individual times for pick-up, travelling, braking and delivery
Frequency of transport	Frequency per type of transport
Transport utilities	Type, characteristics, dimensions, costs
Means of transport	Type, dimensions, carrying capacity, speed, costs
Transport route	Length, type, nature

The following process applies for the selection of the means of transport = rough selection as part of the instrumental concept:

1. Formulation of the transport task

The transport task is determined by:

- the type and characteristics of the commodity to be transported (product, transport unit with utilities)
- transport volume in transport units
- transport route, loaded and empty sections
- transport time
- special transport regulations (safety, quality, positioning accuracy)
- throughput volume

2. Selection of conveying principle (continuous or discontinuous) and specification of transport levels (overhead, ground-based) and crane track installation (line, area, room)

The preliminary decision regarding the equipment's positioning relative to the floor -

- under-floor transport (e.g. conveyor belts)
- floor-based transport (e.g. fork lifts)
- alternating floor transports (e.g. elevators in the case of multi-level production)
- overhead transport (crane systems)

- has great significance for the structural design, routing of service ducts and overall arrangement of the objects (layout planning). This decision has, for example, an influence on the floor space areas, the floor load capacity, the room heights, ceiling load capacity and pillar spacing, and thus has a significant influence on the construction costs.

3. Specify transport levels (overhead, ground-based)
   In approximate terms, the amount of transport equipment required can be determined via the transport capacity or transport intensity.

Transport capacity:

$$QT = (mT \bullet s) / t \qquad [tkm/a;\ tm/h] \qquad (4.1)$$

4. Determine crane track installation (lines, areas)

Loading and unloading and handling procedures must be kept to a minimum, as their effect does not enhance value.

The ergonomic, gender-dependent limits for manual handling procedures limit the size of the transport utilities and the weight to be handled. These limits can be overcome by partial or total automation. Manipulators and robots must be provided for this purpose and integrated where practical. (Martin 2004)

5. Calculate transport equipment load (permissible/maximum load). Fix load parameters (floor and ceiling load).

4.3.3 Picking

"Picking" is the compiling of certain sets of items from an overall range of goods provided based on information about requirements (orders). As a result, the condition of the items is transformed from a storage-specific condition to a consumption-specific one." (VDI 3590)

Picking is necessary for parts that are to be manufactured or installed or component assemblies, as well as for tools, measuring and testing equipment, special jigs and fixtures, replacement parts, packaging and other items that need to be provided. The picking processes should always be planned before the storage processes in order to enable cost-effective overall solutions. Otherwise storage restrictions often prevent optimal picking processes.

a) Single and multi-stage picking

A choice needs to be made between single-stage and multi-stage picking. Multi-stage picking includes consolidation, which means the process of physically bringing together parts for a delivery from different (spatial) areas.

Picking can be characterized by means of the type of provision, the removal and delivery, and the process strategy. This classification is illustrated by fig. 4.10 and explained in more detail below (table 4.17). VDI guideline 3590 sheet 3 (VDI 3590-3, p. 4 ff.) also lists different properties of the features of the information and material flows in a morphological box. It provides a further overview of the widely-diversified configuration options that are discussed in more detail at this point.

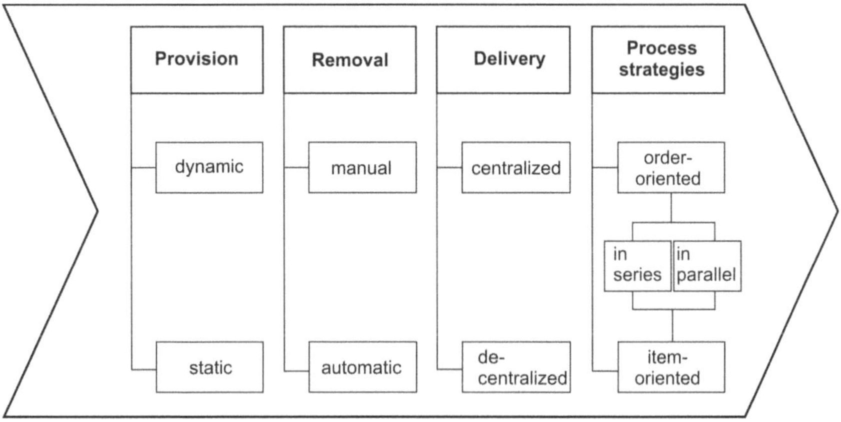

Fig. 4.10 Types of picking, differentiated according to provision, removal, delivery and process strategies

In many cases picking vehicles are utilized to assist in efficient execution of the picking process and support the workers. Various different kinds of picking vehicles are used, among which are order picking trolleys, order picking trucks (horizontal/vertical), order picking stackers and storage and retrieval units for high-rack warehouses. Furthermore, picking is in some cases carried out automatically by order-picking robots. To link all picking areas into a single unit, transport, conveying and sorting systems are required. They assist in replenishing supplies and providing empty containers, plus collection, removal, consolidation, distribution and sorting.

All forms of continuous conveyors and driverless transport systems are used to supply the picking area with pallets and heavy general cargo, which are combined to form a functioning, complex system with the help of junctions, branches and vertical conveyors.

Both continuous and discontinuous conveyors are used to transport small containers (cf. Gudehus 2000, p. 137). In two-stage picking systems, overhead circular conveyors or continuous elevators (paternosters), for example, are used as collection systems.

Distribution into the second picking stage is often carried out using sorting equipment (= sorters). These can be tilt tray, cross belt or sliding shoe sorters, for example. They distribute the goods to the various packing station terminals that are allocated to the individual orders. The order units can subsequently also be distributed to the goods outwards store using sorters in accordance with their various different destinations.

b) Planning of conveying and sorting facilities

When planning, attention should be paid to ensuring that the required performance is achieved in compliance with both processes and deadlines. Specifically, bottlenecks must be configured in such a way that the desired throughput can still be achieved at the bottleneck. There are several possible variations for this (table. 4.17).

Table 4.17 Possible picking variations (MH95 p. 327 ff, Gudehus 2004, p. 700 ff., HG08 p. 121 ff.)

Provision	
Dynamic provision synonym: "goods to man"	The goods are brought to a fixed picking area. The items that need to be picked are removed there. Application: e.g. for carousels and automatic small parts stores
Static provision synonym: "man to goods"	The picker follows instructions to go to a certain storage area where the goods to be picked can be found. He removes the goods and collects the various individual items in a single area. – the picker goes on foot and takes a non-driven trolley with him – the picker drives (horizontally only) with a mobile picker or a horizontal order picking truck to the staging area. – the picker uses a vertical order picking truck that is able to move both horizontally and vertically. – the picker uses a storage and retrieval unit that is able to move both horizontally and vertically at the same time.
Removal	
Manual removal	The materials are removed manually by a picker from a rack or a storage / transport container. – manual picking without technical support – mechanical picking with human assistance
Automatic removal	This removal principle assumes dynamic provision. The materials are removed automatically from the rack or storage / transport container. – automatic picking without human involvement (picking robots) – automatic extraction, with which an extractor device transports the items from a continuous flow channel in to an order bin or onto a conveyor.
Delivery	
Centralized delivery	Once the goods have been picked, they are transferred to a central location, for example the consolidation area.
Decentralized delivery	The picked materials are collected in a decentralized location and then transported to the next station. The decentralized delivery takes place in the immediate vicinity of the picking workstation.
Process strategy	
Task-oriented process flow strategy	Different materials are picked using a pick list for a certain order only. This process is also known as single-stage picking. It can be done in series or in parallel. (see below)
Item-oriented process flow strategy	Quantities of a certain number of materials are removed for different orders. They are assigned to the individual orders afterwards. This process is also known as two-stage/multi-stage picking. It can be done in series or in parallel. (see below)
Process flow principle	
In series	The sub-orders are picked one after the other.
In parallel	Several sub-orders are processed at once.

c) Other picking activities
Other activities need to be undertaken as part of picking. Examples of these are:

- counting, measuring, weighing, packaging
- formation of sets
- labeling and tagging
- quick and easy to perform pre-assembly and preparation activities

d) Units (merchandise)
The units are differentiated according to VDI 3590 sheet 2 (VDI 3590-2, p. 3) into:

- storage unit – unit in which the items to be picked are stored.
- transport unit – unit with which the provisioning system is supplied
- feeding unit – unit with which the respective storage location is replenished
- removal unit – quantity of an item to be picked that is removed in one movement
- collection unit – quantity of same or different items, that arises through removal
- shipping unit – unit that is formed by consolidating collection units

e) Operating strategies
In addition to the technology used and the selected picking principle, the successful operation of a picking system depends very heavily on the operating strategies used. Both performance and costs are influenced by them. Gudehus 2000, p. 151 ff, classifies the operating strategies as follows:

- occupancy strategies
  (e.g. fixed or free picking location arrangement, fixed or free reserve location arrangement, fast movers, picking station sequence optimized for packing, station occupancy optimized for picking)
- processing strategies
  (e.g. real time processing, batch processing (time or quantity-based), batches can be defined as either fixed (fixed batch) or flexible (flowing batch))
- movement strategies (see below)
- removal strategies
  (e.g. FIFO, clearing of batches already broken into, quantity adjustment)
- replenishment strategies
- empties strategy
  (e.g. the provision of empties in batches or order-wise; removal of empties by the picker or separate disposal)

Fig. 4.11 contains sample movement strategies that are commonly used for manual picking processes:

- serpentine strategy
- branch and pick strategy (without repeating aisles)

- branch and pick strategy (aisles are repeated)

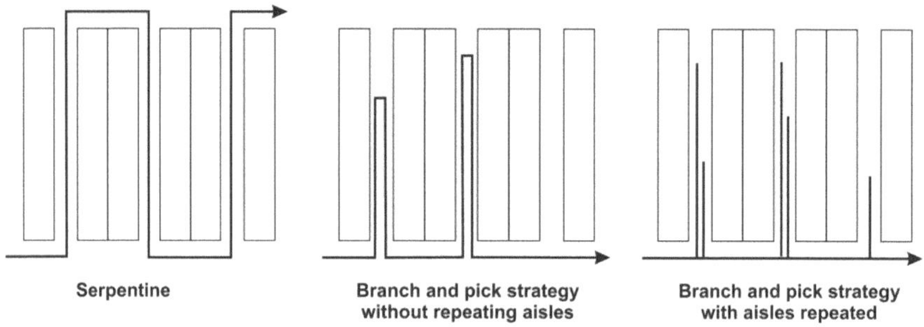

Fig. 4.11 Possible movement strategies (after Gudehus 2000, p. 159)

In the case of technology-supported or automatic picking processes, the following strategies are used:

- single-cycle strategy
- dual-cycle strategy
- aisle change strategy
- feed strategy

This list only contains the most frequently used strategies. Special strategies exist for particular tasks such as stock transfers, for example. In order to implement the actual picking material flow, information flows are required that prepare, control and close the process. They include the

- entry of orders
- order processing
- order transmission

f) Dissemination of information in the picking process

The picking process is carried out using prepared information. In a classic picking procedure this information is presented as a pick list. In the case of more modern forms, it is possible, for example, to transform the information into light signals (pick to light) to support the picking process. Removal and delivery of the items must then be confirmed. With this last step in the information process the procedure is reported as being complete and the picking of a new order can begin. The following technical options are available for issuing information for the picker:

- picking lists
- pick-by-monitor screen display
- pick-by-light and put-to-light
- pick-by-voice

To process / enter information, bar code scanners or RFID readers are used, depending on the available technology. It is possible to combine these input devices with simultaneous issue of information using personal digital assistants (PDAs). A complete description of the picking process material flow also includes the return transport of loading units that have been broken into and the supply and disposal of cmptics.

Important parameters for the evaluation of the picking process are:

The *picking time,* which describes the whole of the picking process, i.e. for the assembling of an order. It is made up of the following:

- basic time: administrative activities, collecting containers, delivery
- travel time: time taken to negotiate the route between the removal locations
- grab time: removal, collection, conveying, delivery
- dead time: searching, counting, checking, labeling

The *picking output* describes the number of picked parts, goods or units in a certain time period.

The *picking quality* is the expression in percentage terms of the accuracy of the goods picked.

4.3.4 Storage

Larger stores and buffer stocks can be dimensioned and designed separately. Over and above this there are also buffer stocks within workshops, which are required for time decoupling. Such buffers can be categorized as stationary buffers (e.g. magazines) and flowing (dynamic) buffers (e.g. circular conveyors))

According to VDI 2411, storage is any planned depositing of workpieces within the material flow. A store therefore describes a room, area or unit, in which objects can be placed. In a store, different activities take place, which can be roughly subdivided into five categories:

- in-storage preparation (preserving, packing, palletizing, labeling)
- transfer to bin (loading of transport units, transport to the storage location)
- storing (bridging of the storage time, possibly maintenance, checking of stocks, relocation)
- transfer from bin (picking, consolidating, packing, labeling)
- in-storage post-processing (unpacking, repacking or packing procedures, cleaning, dust removal)

a) Storage functions
The functions of a warehouse are:

- safeguarding of the material supply
- safeguarding of the ability to deliver for internal and external customers
- balancing of time and quantity-based fluctuations in demand

- sorting and preparing of materials and customer orders
- maturation and refining of goods
- speculating on price fluctuations, shortages, strikes or seasonal deliveries

(cf. Bichler 1997, p. 155 f)

b) Goods inwards and goods outwards

Goods inwards and goods outwards are the warehouse's two interfaces to the surrounding environment. The performance and function capability of the warehouse is substantially dependent on the design and dimensioning of these two interfaces. Incorrect planning or organization can result in serious malfunctions and bottlenecks. Goods inwards and outwards consist of ramps, gates, staging areas and other functional areas.

The following possibilities exist for arranging their layout:

- separate goods inwards and goods outwards
- adjacent goods inwards and goods outwards on one side of the building (possibility of demand-dependent use of the gate modules)
- combined goods inwards and goods outwards (alternate or demand-driven use)

c) Storage facilities

According to Jünemann (1989), in addition to simple floor storage, these can be classified according to the following major types (see fig. 4.12):

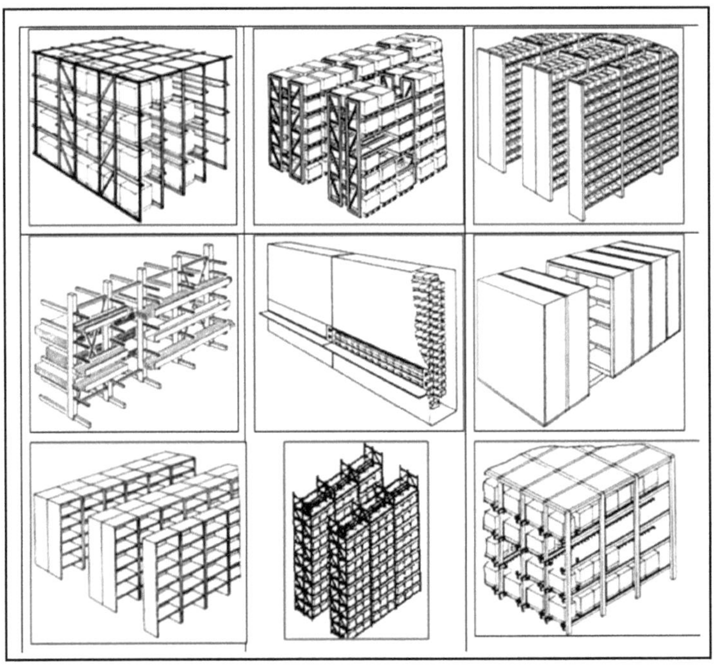

Fig. 4.12 Storage facilities (Jünemann 1989, p. 153 ff)

Storage elevators are used predominantly for the picking process. (fig. 4.13)

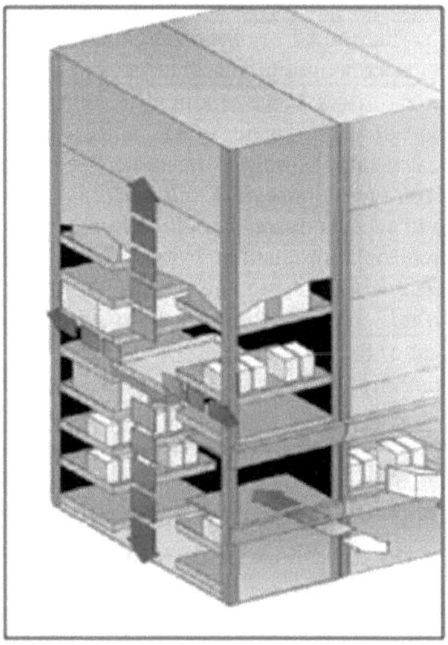

Fig. 4.13 Storage elevator (source: Megamat 2001)

d) Storage operating strategies

Storage operating strategies influence the performance and costs of the storage system. In the case of new projects, the right selection will not only enable savings to be made in terms of investment, but also reduce the subsequent operating costs. For existing storage facilities, an optimal operating strategy will lead to an increase in throughput and the utilization of rooms and floor space.

e) Occupancy strategies

In addition to the transport costs within the warehouse, the degree of utilization of the storage area and the to-bin and from-bin transfer times will also be influenced by the form of warehouse organization selected. (cf. table 4.18)

Table 4.18 Major storage occupancy strategies

Fixed storage locations synonyms: systematic, rigid	Each storage unit has its own fixed location. This is dependent on the planned inventory level. Materials with a high handling frequency should be allocated to a storage location that guarantees a short transport route to the transfer points.
Free storage locations synonym: chaotic	A storage unit is allocated any free space. Storage location administration is necessary in order to obtain a precise overview of the suitable free storage areas. The available storage space must be able to hold the item on the basis of its size and shorter transport routes should be implemented for storage units that are handled more frequently.
Zoning	Division of the warehouse into different sections. These zones are suitable for certain groups of articles or defined types of loading units.

Fast movers	Items that are handled quickly are stored close to the entrance/exit, meaning that the average distance travelled by the storage equipment is reduced.
Even distribution strategy	The stock of an item is distributed evenly throughout the warehouse in order to safeguard maximum security of access.
Space adaptation	Objective: greater degree of floor space / room utilization. Small storage areas are allocated small storage units and large storage areas are allocated large storage units and a large inventory. In the case of ***single-item*** or ***single-batch occupancy***, storage areas that are able to accept several storage units are only allocated the same article or a production batch. The opposite, ***mixed-item occupancy***, enables the storage of different items in one storage location, if this consists of several storage positions.
Minimize storage areas for units that have already been broken into	Objective: Improvement of the fill level Loading units from partially filled storage areas are always released from stock first.

f) Movement strategies

The sequence of the to-bin and from-bin transfers to be carried out by the conveying system and storage appliances is specified on the basis of the movement strategies.

References

Bichler K (1997) Beschaffungs- und Lagerwirtschaft. Gabler, 7 Auflage

Grossmann G (1990) chapter 1; page 16 in Krampe u.v.a. Wissensspeicher für Technologen – Transport, Umschlag, Lagerung. 1. Auflage, Fachbuchverlag, Leipzig, ISBN3-343-00451-0

Gudehus, T (2000) Logistik 2 – Netzwerke, Systeme und Lieferketten. Springer, Berlin

Gudehus T (2004) Logistik: Grundlagen Strategien Anwendungen. Springer, 2 Auflage

Illés B, Glistau E, Coello Machado N (2007) Logistik und Qualitätsmanagement. 1 Auflage, Miskolc, ISBN 978-963-87738-1-4.

Gleißner H, Femerling C (2008) Logistik: Grundlagen – Übungen – Fallbeispiele. Gabler, 1 Auflage

Jünemann R (1989) Materialfluss und Logistik, Springer, Berlin

Jünemann R (1990) Transport-, Lager- und Kommissioniersysteme für die 90er Jahre. TÜV Rheinland, Köln

Kettner H, Schmidt I, Greim HR (1984) Leitfaden der systematischen Fabrikplanung. Hanser, München

Krampe H, Lucke HJ (2007) Grundlagen der Logistik. Huss, München, 3 Auflage

Megamat (2005) Megalift FSE brochure.

Martin H (1998) Transport- und Lagerlogistik – Planung, Aufbau und Steuerung von Transport- und Lagersystemen. Braunschweig, Wiesbaden, Vieweg, 2 Auflage

Martin H, Römisch P, Weidlich A (2004) Materialflusstechnik. Konstruktion und Berechnung von Transport-, Umschlag- und Lagermitteln. Braunschweig, Wiesbaden, Vieweg, 8 Auflage

VDI 2411 Begriffe und Erläuterungen im Förderwesen

VDI 3590-1 (1994) VDI 3590 Blatt 1: Kommissioniersysteme – Grundlagen. Berlin, Beuth

VDI 3590-2 (2002) VDI 3590 Blatt 2: Kommissioniersysteme – Systemfindung. Berlin, Beuth

VDI 3590-3 (2002) VDI 3590 Blatt 3: Kommissioniersysteme – Praxisbeispiele. Berlin, Beuth

VDI 3657 (1993) VDI 3657: Ergonomische Gestaltung von Kommissionierarbeitsplätzen. Berlin, Beuth, 1993

VDI 4415 (1999) VDI 4415: Automatisierte Kommissionierung. Berlin, Beuth

VDMA Nachrichten (2002) H.7, p. 66

Ziems D (2008) Vorlesungsskript Technische Logistik I und II. Otto-von-Guericke-Universität Magdeburg

5 Situation-driven case studies

The following section contains case studies illustrating situations that customers were unable to plan for in advance, i.e. situation-driven occurrences. This also applies for the selection and determination of production buildings and structural solutions, which had to occur at a time when the technological processes had not yet been fully determined. The same is also valid for case studies regarding production solutions for parts manufacture and assembly, as well as for integrated organizational solutions.

5.1 Production buildings and structural solutions

The production building has a pivotal role. It has an inward effect on the production facilities and an outward one on the infrastructure (fig. 5.1).

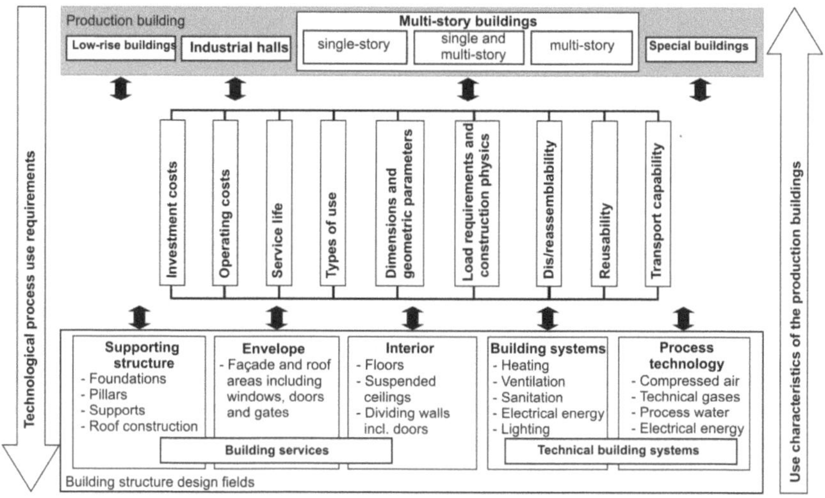

Fig. 5.1 Influencing factors on the building characteristics of factories (Hildebrand 2005, p. 114)

Production buildings are selected by comparing the parameters for building requirements with the building's characteristics (Wirth & Hildebrand 2001; IREGIA 2004).

Use requirements – qualitative and quantitative building-related criteria and their parameters, which result from the technological processes, logistics workflows and machinery, and plant for the manufacture of goods. They determine the requirements in terms of the arrangement and adaptability of the building.

Use characteristics are qualitative and quantitative building-related criteria and parameters with which the building is equipped. They are dependent on the type of

M. Schenk et al., *Factory Planning Manual*, DOI 10.1007/978-3-642-03635-4_5,

use (industrial, administrative buildings) and the building categories (low-rise buildings, industrial hall-type buildings, multi-story buildings).

The use characteristics should be taken into account during the planning stage

- on the one hand through machinery and plant "mobility-enhancing" parameters (production mobility reserves)
- and on the other hand building parameters (structural mobility reserves)

While the mobility potential refers predominantly to the technical equipment in the building, the flexibility potential of the building itself can be developed through structural measures - taking economic restrictions into account. (IREGIA 2004; Zäh, Bayerer 2004)

Note: the building structure acts as the central interface inwards to production and outwards to the infrastructure. It must be possible to efficiently assign the building requirements (resulting from the production process) and characteristics (resulting from the building parameters) in terms of their main parameters. The interface design is a result of the juxtaposition of the requirements placed on the building with the building's characteristics, which either exist already or must be satisfied.

5.1.1 Building requirements

The general use requirements can be consolidated into six requirement groups.

(1) Technological requirements: technological viability, expandability, changeability, flexible machine installation, load capacity feasibility, supply and disposal possibilities, working conditions and ergonomics.

(2) Spatial requirements: pillar-free rooms/areas, optimal room geometry (pillar spacing/dimensions between center lines), modifiable internal walls, lighting, building openings, optimal installation and aisle widths, barrier-free rooms, fire compartments.

(3) Protection requirements: climate, emissions, noise, fire, heat, radiation, health, labor, impact, hazardous substance, collision and supply protection

(4) Location requirements: building plot, arrangement, links, connections and spacings, other location factors

(5) Economic requirements: multi-purpose use, minimal building, operation and maintenance costs; ability to fight fires, optimal service life, extendibility, low energy costs, flexibility, short transport routes for personnel, materials and operating supply materials.

(6) General requirements: structural integration, optimal infrastructure (traffic, supply and disposal), compact construction, architecture

The *specific use requirements* include the parameters that reside on the production level, which are derived from products, processes, machinery, plants and logistics facilities, plus the supply and disposal technology. They must be fixed for the main parts of the production, storage, administration and ancillary areas. The production and storage area including its combinations is the main area of focus. The parameters resulting from the user technology are to be derived for this.

In addition to the general information, such as industry sectors and production volumes, above all these are the product and process-oriented parameter influences. They can be determined for three categories:

- the product (product profile), including product parameters (e.g. measurements, weight, form, sensitivity), determines the manufacturing technology and can, for example, be decisive for the logistics and assembly process parameters
- the production technologies (e.g. production procedures according to DIN 8580) with the required machines and plants
- production-related units as integrated technology in the form of a grouping of machines and plants for processing and finishing including the THS facilities. These are depicted in the form of production areas, competence cells, production islands, segments, lines, systems, integrated production sections and fractals.

Fig. 5.2 clarifies the main parameters that correspond to the use requirements. They are geometric, load, supply and disposal, interference and other parameters (cf. chapter 6.2). The parameter properties differ for every production process and every production system and should be configured according to technical, organizational and economic aspects. Here the number of units per unit of time and the degree of automation play a particular role.

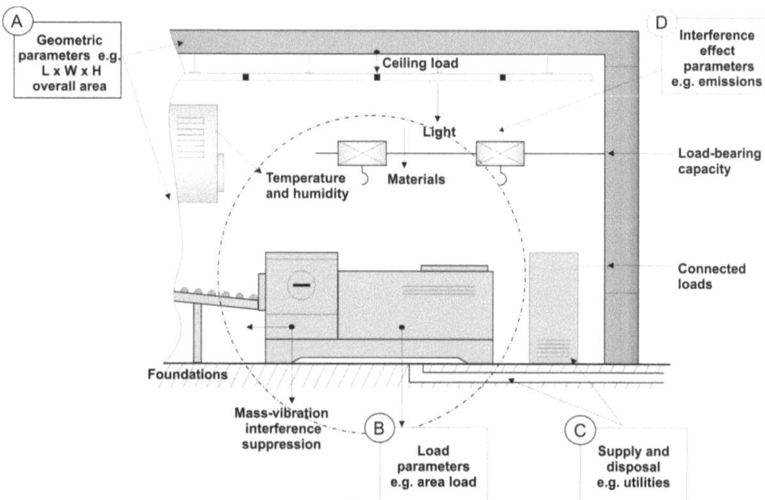

Fig. 5.2 Building requirements and main parameters (based on Rockstroh)

Adaptable structures presuppose that the building's structure is also capable of allowing modifications to be made. The key factor is the ability to change parameters that govern the building with - to a large extent - retention of the building structures that have already been laid out and need to be remodeled.

Ideally, adaptable structures should be laid out in such a way that they enable a strategic realignment of the production for novel products and services. The greater the degree of congruence between the requirements and the structural conditions, the greater the likelihood of being able to use existing buildings for new products and technologies. A degree of suitability can be defined for this.

Mobility potential is the effort required for relocation (mobility and accessibility) of the objects that need to be moved, which causes the development of an economically viable change in their spatial layout and structure. Mobility potential is defined by the difference between existing (ACTUAL) capabilities and those that can be achieved cost-effectively through a change of location. (Erfurth 2003)

The main problem consists in the filtering of the parameters that shape and are related to the structure and those that determine flexibility and mobility.

General indicators such as those contained in the "Checklist for recording use requirements" should be collected.

Example: checklist for recording use requirements

Company: .... **Date:** ....

Sector: ....

Contact person: .... **Telephone number:** ....

Number of employees: Production: ______ **Storage:** ______ **Administration:** ______

1. Geometric parameters

The current actual building requirements and those that may be necessary in the future should be recorded.

Number of spatially-separate production areas: ………..

Requirements		Production area			Storage	Administration	Secondary areas
		1	2	3			
Overall floor space	m^2						
Effective height	m						
Support grid	m x m						
Gate dimensions (width x height)	m x m						

2. Floor load

The maximum required floor load of the floor slab should be determined and recorded. Please place a check in the applicable column(s).

Floor load	Production area			Storage	Administration
	1	2	3		
up to 5 kN/m^2 (500 kg/m^2)	□	□	□	□	□
up to 10 kN/m^2 (1000 kg/m^2)	□	□	□	□	□
up to 15 kN/m^2 (1500 kg/m^2)	□	□	□	□	□
up to 20 kN/m^2 (2000 kg/m^2)	□	□	□	□	□
up to 30 kN/m^2 (3000 kg/m^2)	□	□	□	□	□
up to	□	□	□	□	□
uncoupled foundations required?	□	□	□		

3. Logistics / transport organization

The target situation should be indicated for all requirements.

A. Floor-based transport system	Production area 1 Max. carrying capacity (t)	Production area 2 Max. carrying capacity (t)	Production area 3 Max. carrying capacity (t)
1. Manual elevating platform truck			
2. Mechanical elevating platform truck			
3. Fork lift			
4. Transport trolley			
5. Floor-based conveying system (roller/chain conveyors, etc.)			

B Overhead transport system	Production area 1 Max. carrying cap. [t]	1 Height under hooks [m]	2 Max. carrying cap. [t]	2 Height under hooks [m]	3 Max. carrying cap. [t]	3 Height under hooks [m]
1. Elevator						
2. Overhead crane						
3. Slewing crane						
4. Overhead conveying system (electric telpher system, circular conveyors)						

4. Supply and disposal (building systems)

Required supply and disposal provided via mains networks. Please check as applicable. Power, heat, and water utilities are assumed to be basic requirements.

Supply and disposal	Production area 1	2	3	Storage	Administration
Exhaust air	□	□	□	□	□
Compressed air	□	□	□	□	□
Air conditioning	□	□	□	□	□
Ultra-clean room	□	□	□	□	□
Technical gases	□	□	□	□	□
Disposal of turnings & chips	□	□	□	□	□
Process heat/cold	□	□	□	□	□
Electrical power/drive	□	□	□	□	□

5. Interference effects

Interference effects	Production area			Storage
	1	2	3	
Natural illumination required	□ yes □ no	□ yes □ no	□ yes □ no	□ yes □ no
Noise emissions (dB)[1]				
Risk of fire[2]	□ low □ medium □ high	□ low □ medium □ high	□ low □ medium □ high	□ low □ medium □ high

[1] Emission guideline values for the noise rating level within the building

[2] Classification based on BGR 133

6. Products and auxiliary materials

- Processed materials: ………………………………………………………………………………..............

In the case of the materials to be specified, the transport dimensions (binding materials, bars, barrels, coils, etc.) as well as the heaviest or most bulky materials/products are important.

Materials and products	Length [mm]	Width [mm]	Height [mm]	Weight [kg]	Number of units [shift]	Area allocation			Logistics allocation (A1 – B4)		
						1	2	3	1	2	3
Basic products 1.											
2.											
3.											
Auxiliary materials 1.											
2.											
3.											
Final products 1.											
2.											
3.											

a) general indicators

- industrial sector (general information),
- product profile (type and scope),
- manufacturing type and principle,
- degree of readiness to supply and
- degree of vertical integration.

b) technology and process-related indicators

- process chain (as a whole, but through selected parts) including production processes and their spatial layout,
- dominant production units (with respect to geometry, load parameters and interference effects),
- logistics structures (transport, handling and storage facilities).

For arrangement into groups, it makes sense to combine individual requirements, of which similar and complementary influences have an effect on the function, dimension and structure of production buildings. The performance requirements for the area of production technology including service/management and THS areas are:

Geometry (A)

- dimensions of the machines and plants
- connected installation areas

Load (B)

- mass of the machines and plants (including integrated THS facilities)
- machine foundations
- load-bearing capacity (e.g. floor, ceiling)

Supply and disposal (C)

- connections for machines and plants for the supply and disposal of utilities (e.g. power, gas, water, compressed air)

Interference effects (D)

- emissions (e.g. waste heat, radiation, humidity, vapors, dust, odor, noise)
- climate (e.g. temperature, air pressure, clean room climate)
- fire protection (e.g. fire hazards, explosion protection)
- vibrations (vertical, horizontal, rotational)

These are summarized in chapter 6.2 Main building / production parameters. The performance requirements identified for buildings can be specified in greater qualitative and quantitative detail. They have different effects on the elements of the building structure and are thus determining factors for production-related platforms.

5.1.2 Building characteristics

A multitude of factors, which are determined by the building characteristics, influence the building (see fig. 5.3).

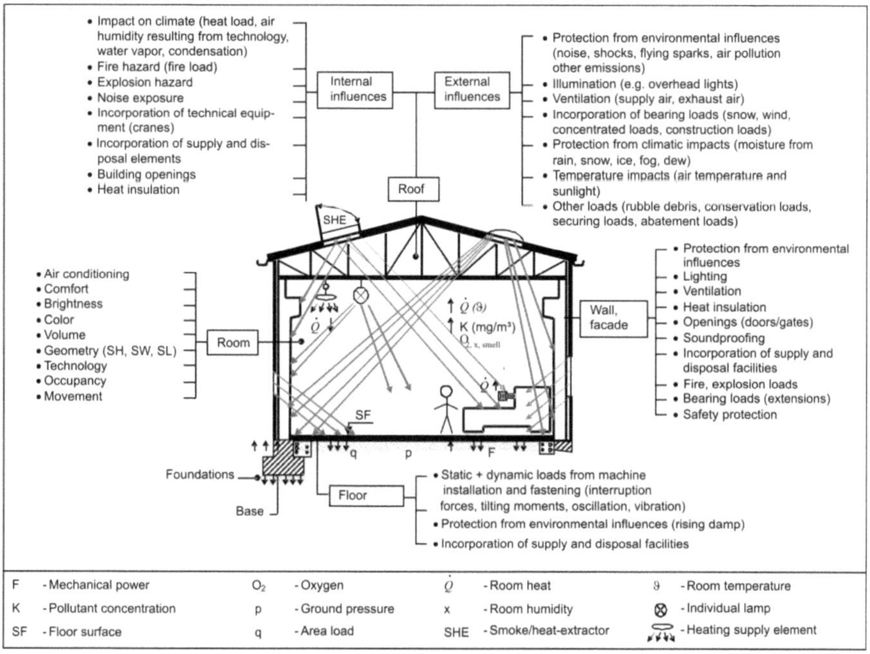

Fig. 5.3 Influencing factors on buildings (Helbing 2007, p. 391)

The use characteristics from a structural point of view are determined by the design element parameters of a building. (Erfurth et al 2003). These include:

- structural envelope,
- supporting structure including pillars,
- structural finishes,
- floors
- technical building systems
- operating costs and effort.

The foundations, loads, supporting structures, geometry, utility supply and disposal and transport arteries belong, among other things, to the primary structure.

Secondary structures include the roof, facade and building services and the interior fittings. As a general rule they can be configured in a flexible and mobile fashion. The number of stories (single and multi-story) in the building has a major influence on its use characteristics.

The smallest structural unit on which, on the one hand, requirements have an effect that is decisive for the building construction, and which, on the other hand, can be combined to form functionally and technologically necessary and economically-justifiable variants of building form, is a module (length x width x height) in the form of a "segment cell." From a structural point of view, it constitutes the basic unit that may be extended or reduced at will in terms of length, width and height over the modular dimensions (e.g. 6 x 6 x 6m). The main parameters of geometry, loads, supply and disposal, plus interference factors in accordance with chapter 6.2 have an effect on a room defined in such a way.

5.1.3 Building selection

Buildings have different useful lives. These depend on the enterprise objectives and range from permanent "fixed" to temporary "flexible" buildings. Decisions have to be made regarding the following aspects:

(1) Examine the concept of building-free production plant installations

A free-standing installation of production, supply and disposal facilities should be considered. Tents, lightweight construction buildings and production domes and similar structures offer viable alternatives for short-term product creation and production processes. They are cost-effective to use and recycle. The use of technical textiles as the envelope and functional elements is increasing steadily.

(2) Choose between general, special, multi- and single-purpose buildings

General purpose buildings are flexible in terms of use and are adaptable, but are associated with high costs. Special purpose buildings are dedicated to particular purposes and only designed for specific technologies (e.g. plumbing metalwork shop). Multi-purpose buildings are flexible in certain technology areas. The boundaries are fixed by the products and production processes. Single-purpose buildings are designed for one intended use. They dominate in chemical process technology (e.g. laboratories).

(3) Decide between low-rise, industrial hall-type and multi-story buildings

Low-rise buildings are single-story, "open" buildings (height $\leq$9.60 m, crane load $\leq$ 1.6 MN, width $\leq$ 42 m, floor loading high). Industrial hall-type buildings are single-story, heavy-duty buildings (height $\leq$78 m, crane load $\leq$ 10 MN, width $\leq$ 168 m). They are supplemented by a steel structure and plant construction. Multi-story buildings are multi-story, light-duty buildings with limited floor and ceiling load bearing capacity (story height 4 – 6 m, floor and ceiling loading 0.4 – 1.0 MN, width 24-42 m, modular dimensions 6 x 6 x 6; 12 x 12 x 12, (24 x 24 x 24) m. Suitable for manufacturing and assembly of small and light products.

(4) Determine crane provision and technical building systems for supply and disposal

Cranes ensure flexibility and mobility in the space. Disadvantages: capacity utilization, larger room heights and costs. Without cranes, flexibility and mobility can, for example, be achieved with fork lift trucks. Conduct functional and profitability assessment. Supply and disposal equipment can be installed in technical basements at low cost.

(5) Specify width, length and height plus extension and reinstatement possibilities
It should be possible to extend the building and technical plants in two directions.

(6) Strive for optimal coordination and incorporation of the building into the infrastructure of the entire factory site.

(7) Evaluation through establishment of use requirements (cf. table 5.1), secure adaptability, minimize interference effects and classify requirements (chapter 6.2)

Note: secure resource and energy efficiency and adaptability, minimize interference effects and evaluate requirements according to table 5.1.

Table 5.1 Checklist – Evaluation of requirements

☐	**System-related requirements:**					
	Maximum					
	☐	Safeguarding of throughput	☐	System-securing supply		
	☐	Incorporation of disposal volumes	☐	Circulation systems		
	☐	Continuity of supply and disposal				
	☐	Simultaneity of supply and disposal				
	☐	Integrity of supply and disposal				
	Should be zero					
	☐	Disruptions due to supply and disposal systems				
	Minimal					
	☐	Supply system volume	☐	Disposal system volume		
☐	**Factory environmental demands (should be zero)**					
	☐	Environmental load	☐	Aerosols	☐	Expansion
	☐	Air pollution	☐	Ground pollution	☐	Noise pollution
	☐	Visual pollution	☐	Heat loss	☐	Groundwater pollution
☐	**Ecological requirements**					
	Maximum					
	☐	Sanitation	☐	Comfort	☐	Elimination of impacts
	☐	Sterility				
	Should be zero					
	☐	Concentration of pollutants	☐	Toxic loads	☐	Noise pollution
	☐	Vibration	☐	Vapors/gases	☐	Carcinogenic pollutants
	☐	Air oil content	☐	Oil dispersal	☐	Hazardous substances
	☐	Protection from impact	☐	Protective measures	☐	Heat cone
	☐	Radiation				

	Minimum					
	□	Impact mitigation	□	Air movement	□	Physical contact
□	**Efficiency requirements**					
	Maximum					
	□	Heat energy use	□	Material reuse	□	Natural disposal
	Should be zero					
	□	Filtering costs	□	Technical compulsory disposal	□	Maintenance costs
	Minimum					
	□	Supply volume	□	Disposal volume	□	Operating costs
	□	Treatment costs				
□	**Overall factory requirements**					
	Maximum					
	□	Centralization	□	Communal disposal	□	State of the art technology
	□	Statutory compliance				
	O	Laws/regulations	O	Technical instructions	O	Technical guidelines/rules
	Should be zero					
	□	System peculiarities	□	Frequency of change		
□	**Spatial and technical requirements**					
	Maximum					
	□	Room condition (cleanliness)	□	Consistency of requirements	□	On-site coverage
	Should be zero					
	□	Spread of harmful substances in the room				
	□	Room condition				
	O	Mist/dust	O	BAT/TRK	O	Fumes/vapors
	O	MAK	O	MIK	O	MEK
	Minimum					
	□	Room humidity	□	Room utilization		
	□	Supply and disposal routes	□	Room temperature/heating		

5.1.4 Building selection – Case study

Task definition: Due to the situation, information for the planning of a production building is to be provided to the architect/structural engineer at a time when the technological project is not yet complete. For this purpose, prompt decisions and rough calculations are necessary for the performance requirements placed upon the building.

Required information: building area, room dimensions including building grid based on building dimensions according to appendix 6.2, fig. 6.4.

Initial data: mechanical manufacture of individual parts on 6 machines (lathes and drilling machines) or production areas for the workshop.

Basis: building area classification (cf. fig. 5.4)

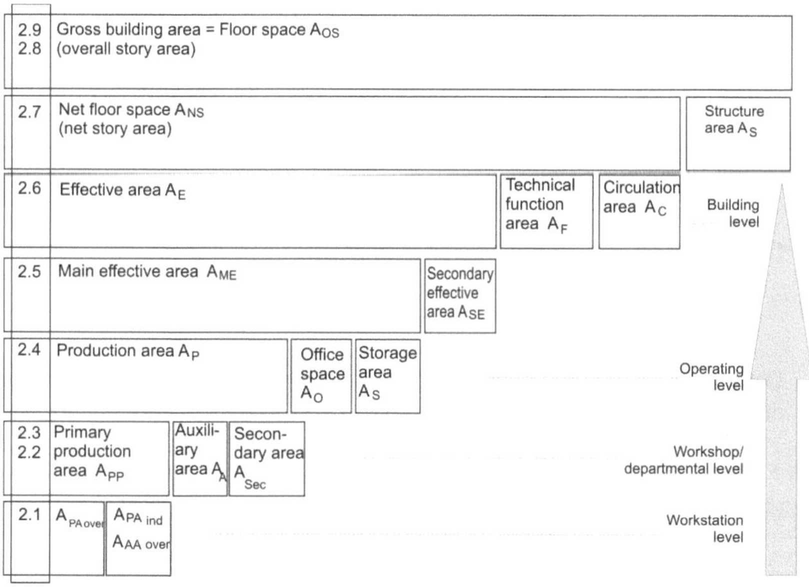

Fig. 5.4 Building area classification

Approach:

1 Building selection: steps 1-7 (implement as per point 5.1.3 and define specifications)

2 Carry out floor space determination (as per floor space dimensioning point 3.2.3.3)

2.1 Determination of the overall floor space for the workstations

$$(A_{FDover} = \sum A_{PA}, FO_j) \tag{5.1}$$

(individual workstations on the basis of the substitution area method (see table 5.2))

Table 5.2 Individual workstations based on the substitution area method

Production area floor space	Amount of working materials N_{WM}	Production area floor space A_{PA} [m²]	Production area $A_{PA,FO}$ [m²]
DRT 50	1	12.6	12.6
DRT 63	2	14.5	26.1*
DRT 80	1	16.3	16.3
BFT 63	2	12.5	25.0
Overall A_{PAover}	-	-	80.0

* Overlap due to number of machines (production area group)

2.2 Determination of the main workshop production area (AMP)

$$A_{MP} = 2 \bullet \sum A_{PA\ over} = 2 \bullet 80\ m^2 = 160\ m^2 \tag{5.2}$$

2.3 Determination of the auxiliary area (A_A) in percentage terms and main production area (A_{MP})

Table 5.3 Areas for auxiliary processes

Auxiliary production	% of A_P	% selected:
Maintenance	1.0 - 2.5	2.0
Electrical repairs	0.2 - 1.0	1.0
Tool and equipment making	1.7 - 3.8	3.0
Equipment storage	0.4 - 1.8	1.0
Tool grinding shop	0.7 - 1.3	1.0
Storage for technical gases	0.1 - 0.5	0
Storage for oils, paints, lubricants	0.1 - 1.0	0
Precision measuring and laboratory rooms	0.4 - 1.7	0
Total		**8.0**

$$A_{MP} = (1 - 0.08) \bullet A_P = 0.92 \bullet A_P \tag{5.3}$$

2.4 Determination of the production area (A_P)

$$A_P = A_{MP} / 0.92 = 160\ m^2 / 0.92 = 174\ m^2 \tag{5.4}$$

2.5 Determination of the main effective area (A_{ME})

A_{MU} = Production area (A_P) + Storage area (A_S) + Office area (A_O)

$$A_{MU} = A_P + A_S + A_O = 174\ m^2 + 41\ m^2 + 30.5\ m^2 = 245.5\ m^2 \tag{5.5}$$

A_S = central storage areas in buildings (23.5% of A_P)

$$A_S = 0.235 \cdot A_P = 0.235 \cdot 174\ m^2 = 41\ m^2 \tag{5.6}$$

A_O = Office and administrative areas (17.5 % of A_P)

$$A_O = 0.175 \cdot A_P = 0.175 \cdot 174\ m^2 = 30.5\ m^2 \tag{5.7}$$

2.6 Determination of the overall effective area (A_E)

Table 5.4 Floor space indicators

Type of sub-area	Reference basis and dimensions
Administrative area: - manufacturing technology - project design engineers, draftsmen, development engineers	5-6 (10) m² per person 6-7 m² per person 7-8 m² per person 8-11 m² per person
Desk	3-4 m² per person
Floor space for social rooms	1.8 m² per person
Floor space for dining area, kitchen and adjoining rooms	2.2 m² per person
Floor space for changing rooms	0.3 m² per person
Area for assembly and break rooms	1.5 m² per person
Toilet room	1 place for every 15-20 people
Wash room	1 place for every 5 people
Parking area	25 m² per car

2.7 Determination of the net floor space (A_{NF})

Effective area + technical function areas + transport area

$$A_{NF} = A_E + A_F + A_C = 260\ m^2 + 13\ m^2 = 273\ m^2 \tag{5.8}$$

A_C – Transport areas in buildings (only the areas that cannot be used in any other way due to the building construction, such as staircases, corridors, elevators and similar)

$$A_C = 0.03 \ldots 0.05\ A_N = 0.05 \cdot 260\ m^2 = 13\ m^2 \tag{5.9}$$

2.8 Determination of the required overall story area (A_{OS} = building area)

$$A_{OS} = A_{NF} + A_{Con} = 273\ m^2 + 14\ m^2 = 287\ m^2 \tag{5.10}$$

A_{Con} - construction area of the building (walls, pillars etc.)

$$A_{Con} = 0.03 - 0.05\ A_{NF} = 0.05 \times 273\ m^2 = 14\ m^2 \tag{5.11}$$

2.9 Determination of the building dimensions via the system dimensions (building grid)

Calculation for system width 12m:

calculated system length = $287\ m^2 / 12\ m = 23.9\ m$ selected system length (round up to a whole grid) = $4\ m \cdot 6\ m = 24\ m$ resulting building area:

$$A_{OS} = W_B \cdot L_B = 12\ m \cdot 24\ m = 288\ m^2 \tag{5.12}$$

3. Determination of the room dimensions (effective height) (as per Appendix 6.1-1)

- building system height, oriented to machine height (+ crane)
- crane rail height (only if crane)
- crane working area height (only if crane)
- pillar grid (12 x 12 or depending on supplier)

4. Estimation of the load parameters (cf. Appendix 6.2-2)

5. Estimation of the supply and disposal parameters (cf. Appendix 6.2-3)

6. Estimation of the interference parameters (cf. Appendix 6.2-4)
7. Combining of the decisions, information on choice of building and evaluation (building project with recommendations for structural design and execution), evaluation according to checklist: evaluation of requirements

5.2 Production-related solutions

Production-related systems are a part of the production facilities that are to be designed as independent parts manufacturing and assembly areas as well as a combination of the two. The production design combinations are shown in figure 5.5.

The characteristics of the products being manufactured, the production technology (processes, machinery, plants) and THS facilities are determining factors for the structure of the workstations and areas in the building. For each level of consideration, production, assembly, transport, storage and handling facilities, spatial structures and interference effects all have an individual and cumulative effect on the production facility (cf. chapter 6.1).

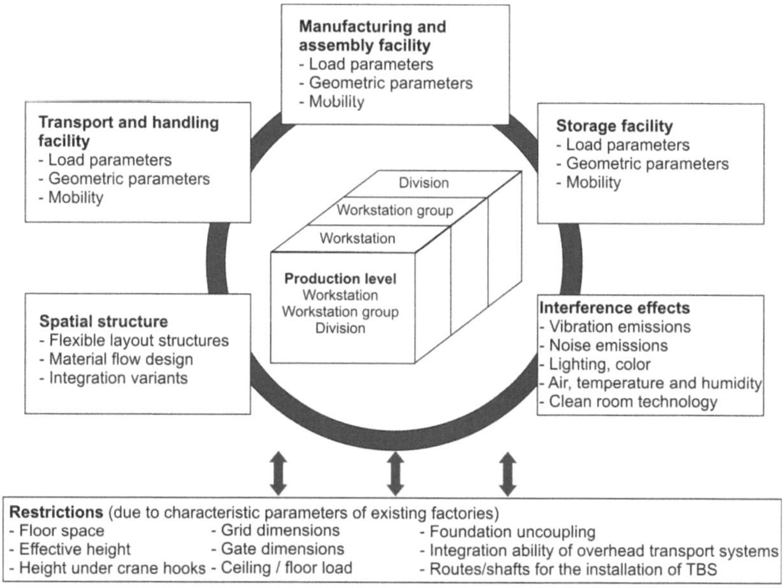

Fig 5.5 Production design components (IREGIA 2004)

The following should all be designed for parts manufacture and assembly areas from the workstation upwards taking the existing restrictions into consideration:

- the workstation including working area with the equipment/facilities and interfaces to supply and disposal both inwards and outwards
- the clearance measurements for the non-occupancy of floor space
- the room measurements for working and operating heights including operating distance, maintenance, transport route and safety measurements.
- Flow equipment for material, information, energy and workflows

Note: production-related parts manufacture and assembly systems have transferable design components and their workstations are configured in a similar way. They should be designed to be adaptable and modular in terms of the basic equipment layout as well as their flow equipment and interfaces.

5.2.1 Parts manufacture

The development from the individual machine to flexible and fixed multiple machine systems is shown in fig. 5.6 as a function of productivity and flexibility.

Workstation-related design solutions are based on the individual machines, e.g. NC machines, processing centers, production cells.

Area-related design solutions always include multiple-machine plants (e.g. integrated production sections, fractals, flexible production systems, transfer lines).

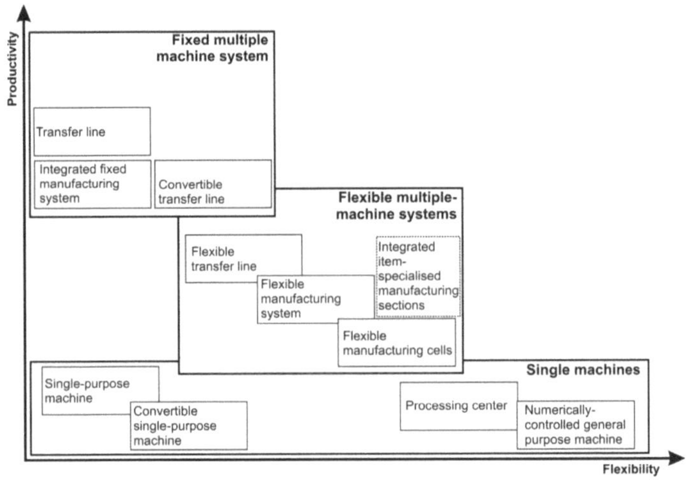

Fig. 5.6 Relationship between productivity and flexibility of different machines and production systems (Eversheim & Schuh 1999, p. 10)

5.2.1.1 Workstation-related design examples

Example: production area (equipment) with storage, transport and handling facilities as design variants for interface reduction (fig. 5.7)

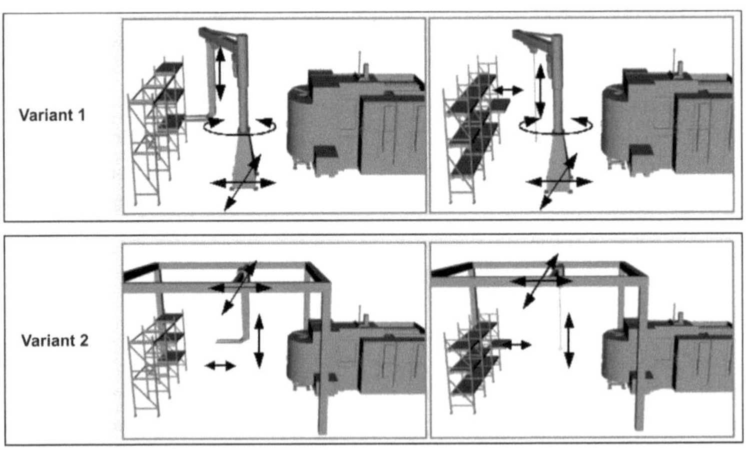

Fig. 5.7 Design variants for interface reduction within networkable production areas (Enderlein et al 2002)

From this it is clear that through expedient design and arrangement of the facilities in a production area, the characteristic parameters of existing building structures can be satisfied in different ways.

The mechanization and automation of individual items of equipment/machines are being further developed and in addition to the established conventionally-controlled machine tools (MT), in small and medium batch production NC-controlled machine tools, NC processing centers and flexible manufacturing and assembly cells are becoming accepted.

Special purpose machines, robots and machine systems in the form of transfer lines (flow shops) are suitable for large series and mass production.

They should be planned as complex units, primarily according to the building block or plug + produce principle.

Example: solution concept for modules with low and high flexibility (fig. 5.8)

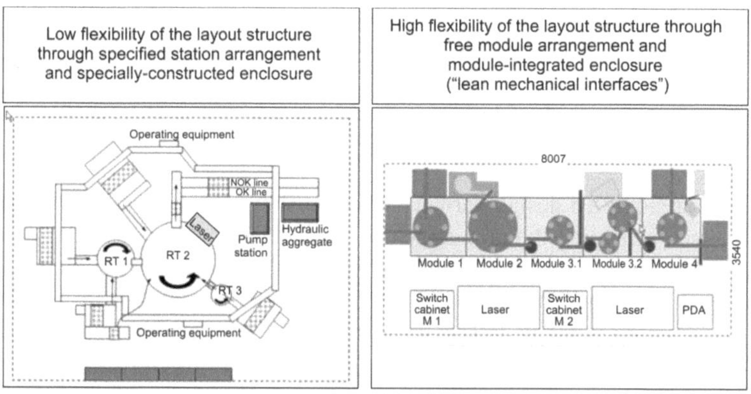

Fig. 5.8 Layout structure with different degrees of flexibility (source: SITEC)

Example: configuration of a production area according to the building block principle as stand-alone and integrated solutions (fig. 5.9)

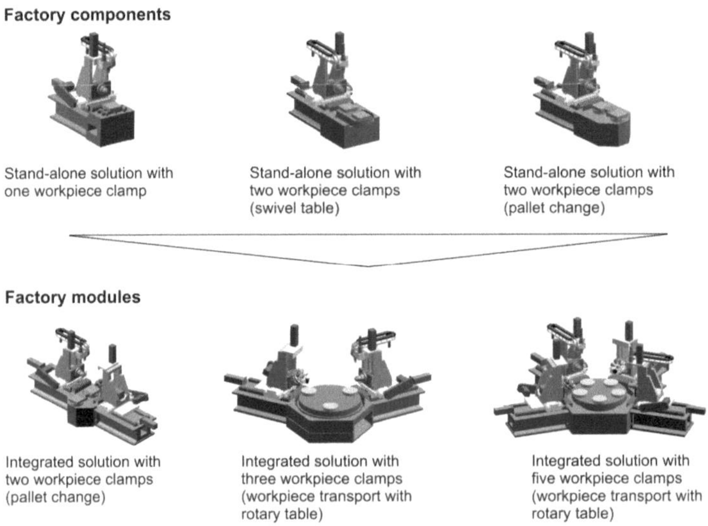

Fig. 5.9 "HPC Flexcell" configuration options (source: WEMA Vogtland)

5.2.1.2 Area-related design examples

In the case of area-related design variants attention should be paid to the fact that as early as when arranging the area structure, changes to machinery and production equipment, the supply and disposal technology, the information and communication technology and the building structure (including technical building equipment) should be integrated. The structuring principles of the organizational and spatial structure plus the manufacturing forms and the design of the processing facilities including the flow facilities themselves are decisive for this.

Example: plug + produce principle and case study for type workshops/factories (based on Hildebrand 2005, p. 196 ff)

(1) The plug + produce principle

The plug + produce philosophy reduces planning and implementation effort by approx. 60%, as it builds on standard processes and type solutions. Two forms stand out:

a) Franchising

Replication of products, personnel, plants (resources) and buildings, with the manufacturing technology being protected. The purchaser or operator takes over the ready-laid-out design solution with low investment risk. Application e.g.: differentiation of fast-food restaurants, department stores, discount chains, gas stations

b) Type workshops/factories

Successive setting-up of production facilities on the basis of type solutions, which can be used multiple times (building block principle). Advantages of type workshops are:

- quick planning and implementation combined with high quality through reuse of documents, approvals and design solutions
- low-requirements, gradual change (expansion, reduction) without disrupting the ongoing operations
- quick opening up of new markets with minimum-risk investments through markct-dependent potentials
- series products, standard products and product variants only result in low set-up costs
- suitable for plant-intensive processes with high technology utilization and little manual activity
- highly-responsive, situation-driven planning and implementation

Application e.g.: (automobile) supply industry, components manufacturers, factory outfitters, consumer goods industry

The functional schematic for the plug + produce principle is shown in fig. 5.10. It shows the approach in the case of an existing production facility that must gradually deal with change.

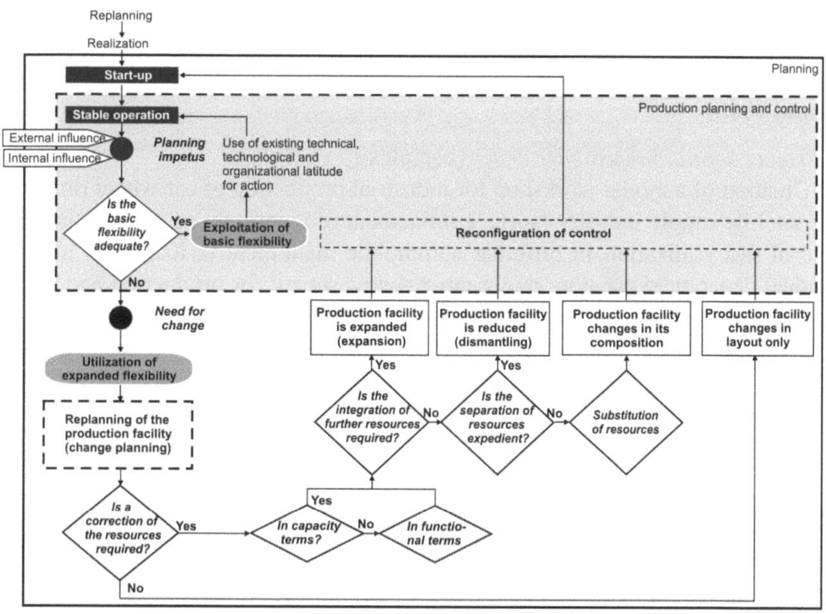

Fig. 5.10 Plug + produce functional principle (based on Hildebrand 2005, p. 28)

This building block, module or plug + produce principle forms the basis for a flexible, temporary factory/production facility (Wirth 2000, Enderlein 2003) It requires the reconfigurability of the equipment and plants and their evaluation (Heile 2004, Daschenko 2006, Heger 2007).

The planning concepts for type and adaptable production facilities are shown in figure 5.11.

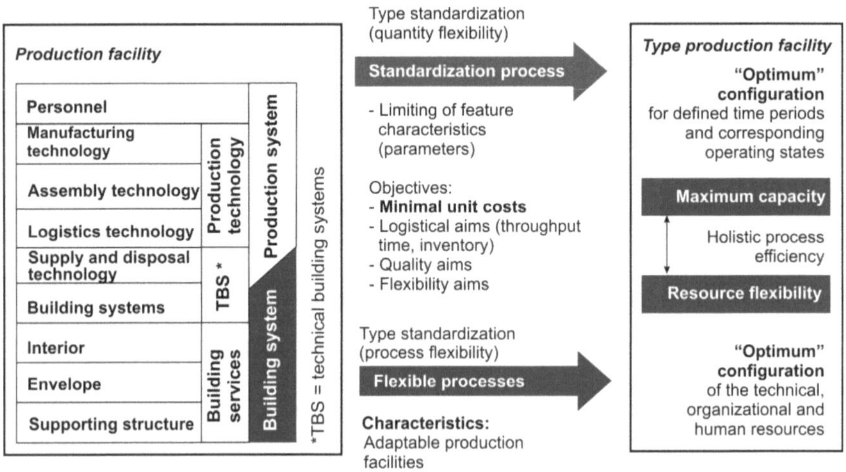

Fig. 5.11 Planning concepts for type and adaptable production facilities (based on Hildebrand 2005, p. 188)

(2) Case study: Type workshop for manufacture of car wheel rims (Hildebrand 2005)

0 *Project definition/customer order* (complex I - checklist)

Creation of a model workshop for industrial production of car wheel rims in the automobile supply industry (surface refinement of aluminum rims) with the objective of fast realization in different automobile manufacturer locations; low cost, gradual capacity expansion; minimum-risk investment for production and marketing and minimal unit costs. Investment costs, timespan of the workflows and expansion stages and resources are predefined. The product is shown in fig. 5.12.

Size:
Ø 14" – 22"
corresponds to approx. 350 mm - 550 mm

Weight:
10 kg – 25 kg

Fig. 5.12 Product specification

The approach is iterative and integrated situation-driven planning for technological, organizational and equipment-related aspects.

5 Project development (complex II – checklist)

The checklist of different planning steps (5/1-5/5) between conventional and modified planning is shown in fig. 5.13.

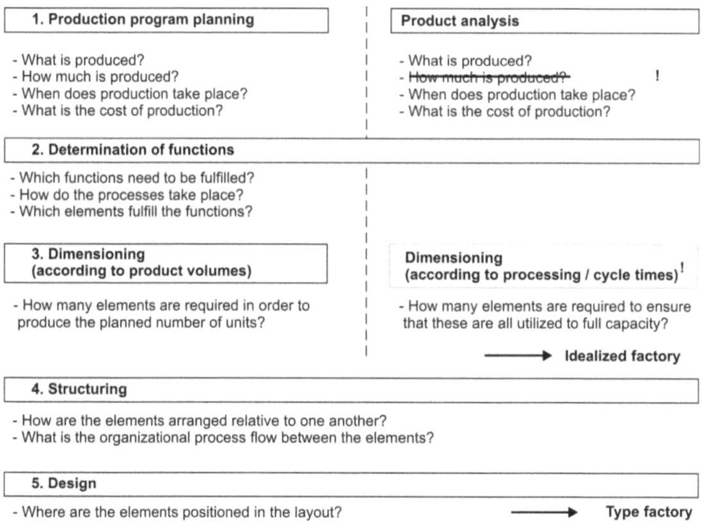

Fig. 5.13 Differences in the planning process flow

The technology graph was created for the *determination of functions* step (5/2). It shows the specific questions for production and logistics technology. The technology graph for production technology is apparent from fig. 5.14.

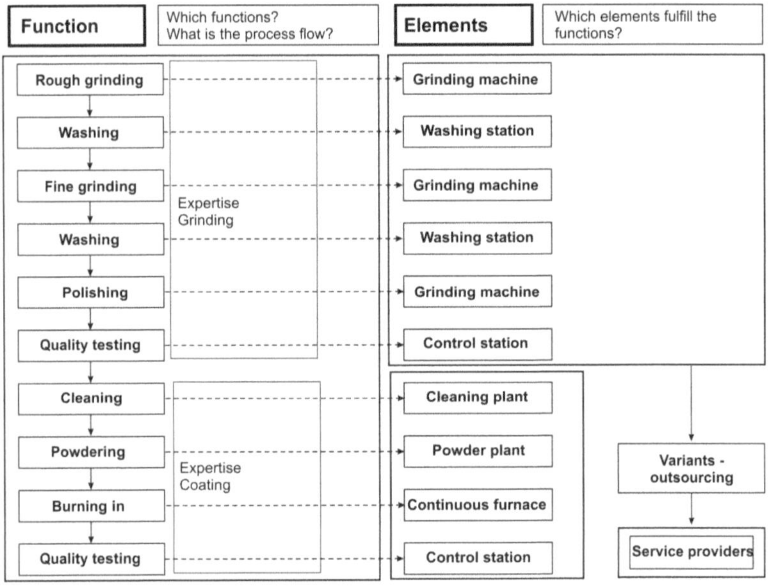

Fig. 5.14 Technology graph for production technology

The technology graph for logistics technology can be seen in fig. 5.15.

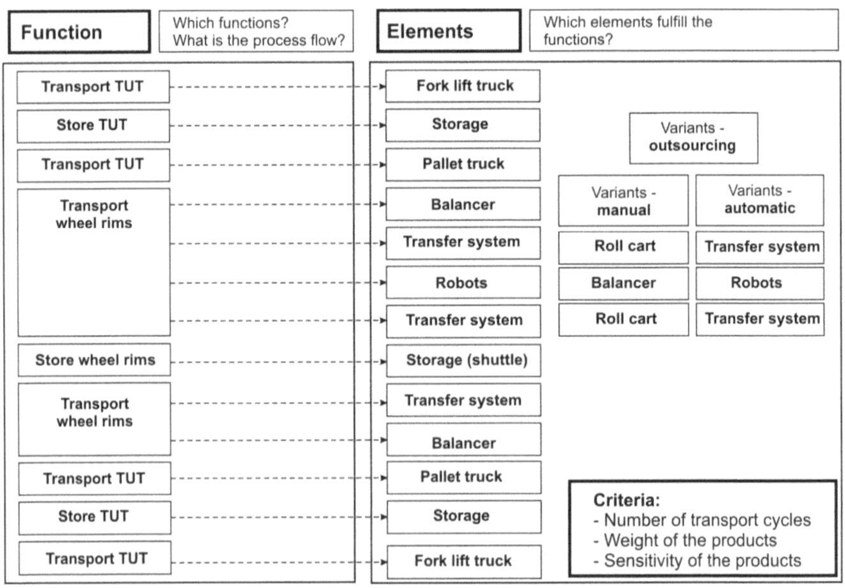

Fig. 5.15 Technology graph for logistics technology (TUT = Transport utilities)

The *dimensioning (5/3)* was done on the basis of the determination of functions (5/2). The outcome was the determination of the type and quantity of equipment: 46 grinding machines, 1 cleaning plant, 1 continuous furnace and 12 robots plus 11 workers. This configuration allowed a capacity of 1,300 wheel rims per day.

The *structuring (5/4)* and *design (5/5)* steps resulted in three layout variants as shown in fig. 5.16.

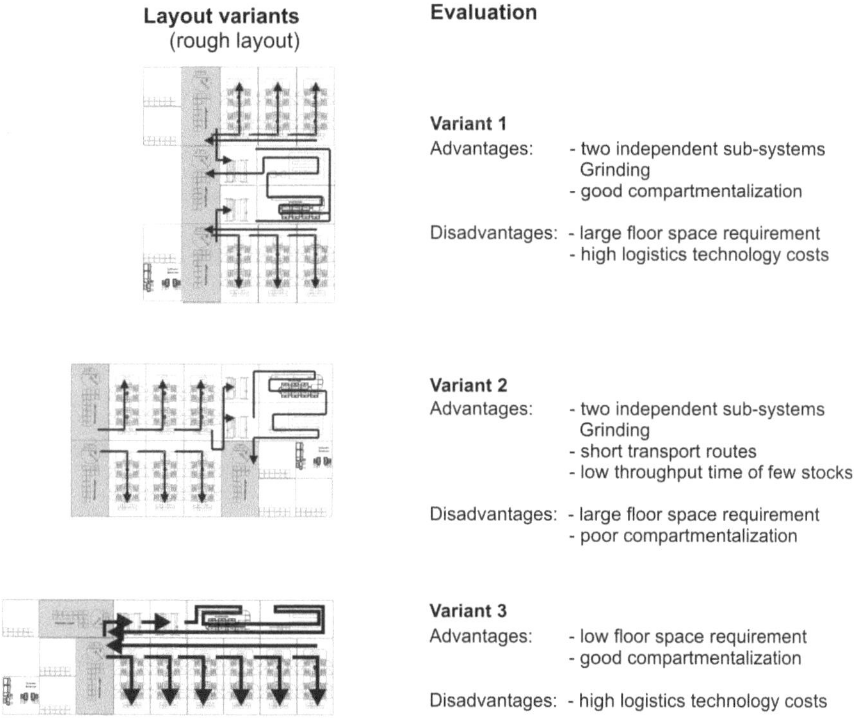

Fig. 5.16 Layout variants with evaluation (TPT = throughput time)

As a result of the evaluation, and taking into account the determined number of workers, the type factory shown in fig. 5.17 was selected for development.

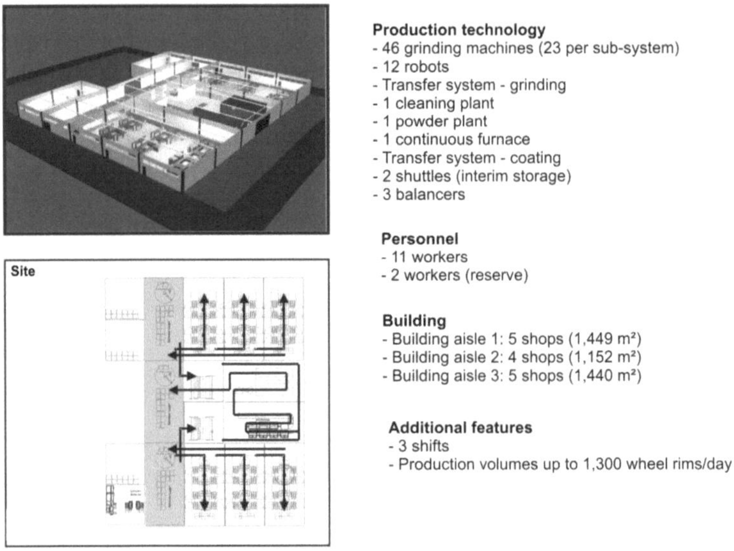

Fig. 5.17 Type workshop

The planning outcome clarifies the expansion stages as a function of costs and number of units, as shown in fig. 5.18.

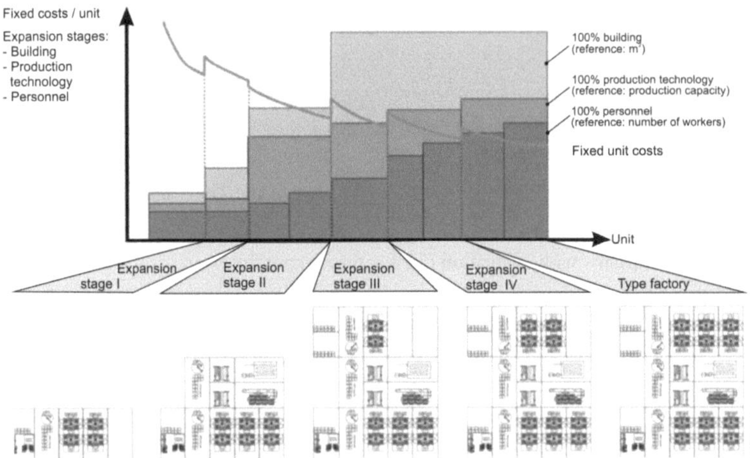

Fig. 5.18 Expansion stages of a workshop

5.2.2 Assembly

For assembly processes, similar manufacturing forms exist, which are distinguished by their sequential nature (pre-, parts, interim / component, system and final assembly). The assembly process constitutes the final phase of production in almost all sectors. The proportion of time taken by the assembly process amounts to between 25 and 40% of product creation. The assembly and parts manufacturing processes are separated based on a division of labor across the supply industry. While in machinery construction, the combination of parts manufacture and assembly still predominates (although admittedly this trend is decreasing), in automobile construction, division of labor is so far advanced that the customer-supplier chain has developed into pre-assembly/final assembly. Since assembly processes are becoming ever more automated (especially in the automobile industry), robots, and their steel mountings within the building, predominate.

It is possible to differentiate here between *stationary* (assembly object/unit is stationary) and *mobile* assembly (assembly object moves).

Construction site and group assembly (individual location) count as stationary assembly. They have a high degree of mobility, as here it is predominantly a matter of the movement of workstations.

Bank and flow forms of production line assembly, including clocked, are categorized as mobile assembly. Here the mobility is limited, since the object movement is fixed spatially by technical equipment.

The development of manual assembly workstations for automatic assembly includes hybrid assembly (automated and manual assembly). Fig. 5.19 demonstrates the *ranges of application of the assembly concepts* that are used for the assembly of components and/or products.

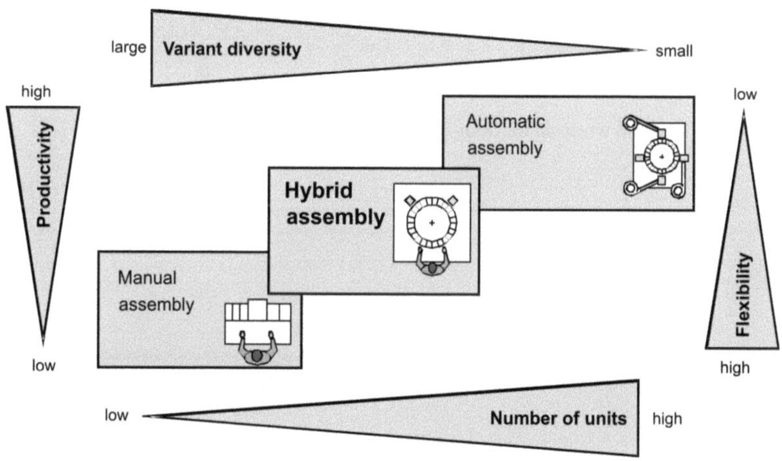

Fig. 5.19 Ranges of application of manual, hybrid and automatic assembly concepts (Lotter & Wiendahl 2006, p. 193)

Workstation-related design solutions are based on the individual manual workstations, flexible single-station machines and inflexible special purpose machines.

In addition, area-related design variants include multiple-station machines with different time synchronizations (e.g. revolving and rectilinear transfer machines, assembly lines).

The prerequisite for assembly workstation design is compliance with the ergonomic design principles of the individual (manual) workstation. They are linked by transport, handling and storage facilities according to organizational (production form), ergonomic, technical and spatial principles.

Ergonomic design of assembly workstations – examples in figures 5.20 and 5.21.

The height of the workstation area must be suitable for workers of differing body heights.

- Make sure that there is adequate room for maneuver under the workbench.

- Note that the size of the workpiece and the jigs and fixtures on the workbench have an influence on the worker's body posture.

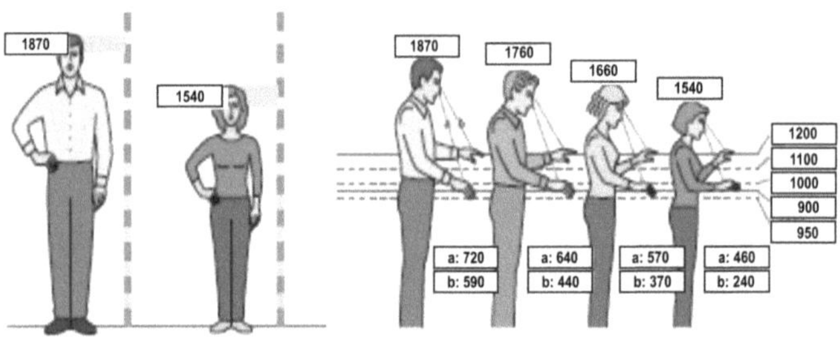

Fig. 5.20 Body size and workstation floor space

Avoid working positions above heart height, as this reduces the blood circulation. This results in the performance of your workers dropping off rapidly.

- Introduce seated/standing or standing/walking concepts in order to enable a change of load on the body.

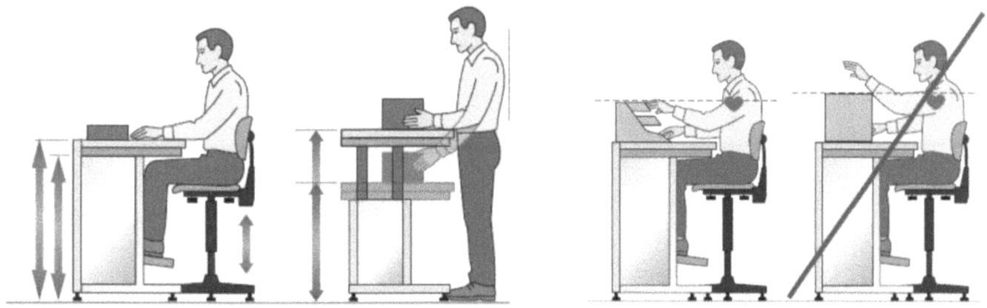

Fig. 5.21 Heart height and working position

Correctly positioned working materials reduce strain, decrease levels of absence and simultaneously increase productivity and performance.

- Arrange your production tools in such a way that they are tailored to the worker and his or her task.

- Use the adjustment options of the equipment to enable your workers to adopt a body posture that does not cause exhaustion and is thus more efficient.

Fig. 5.22 Work equipment placement

In principle all parts and tools and all accessories should be located within the worker's reach in order to avoid placing heavy stresses on the body.

It is possible to differentiate between three different categories of reaching distance in the workstation:

A = Maximum reaching distance
B = Optimal reaching distance
C = Two hands area

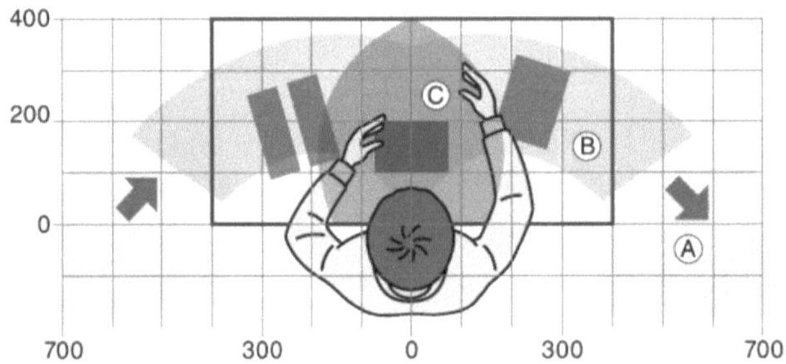

Fig. 5.23 Reaching distance optimization

Bins holding the most frequently-used parts should always be easy to reach with a minimum of movement (short reaching distance).

Heavy parts should always be immediately above the workbench.

- Position as many containers as possible within the optimal reaching distance B.
- Use the possibility of working with both hands to save time.
- Select the bin sizes in accordance with the geometry of the parts and the number of parts required.

Fig. 5.24 Bin arrangement optimization

Constant refocusing and frequent changes of angle of vision overtax and tire the eyes.

- Adapt lighting to your workers' tasks.

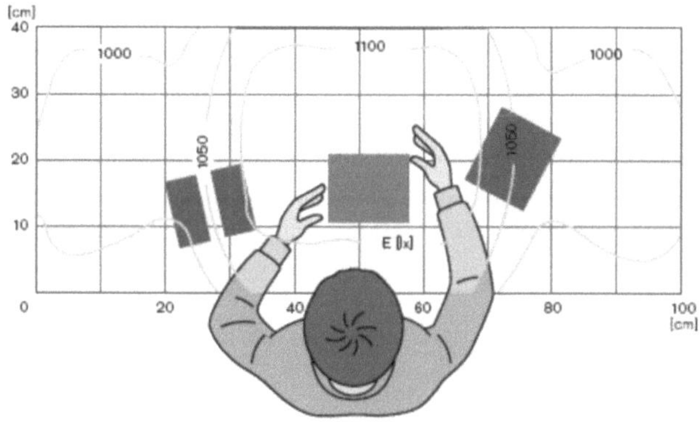

Fig. 5.25 Tailor light intensity to task

Constant refocusing and frequent changes of angle of vision overtax and tire the eyes.

- Arrange frequently-used materials in the optimal field of vision A, in order to avoid unnecessary head and eye movements.
- Position the picking bins at equal distances.

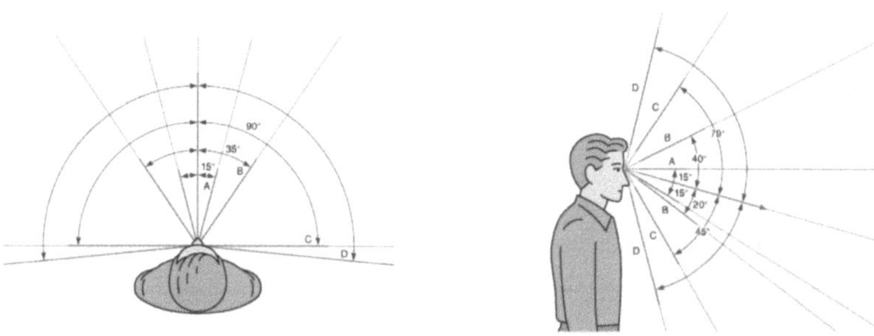

Fig. 5.26 Visual range

5.2.2.1 Workstation-related design examples

The requirements of the design of manual assembly workstations relate to:

- the workers: ergonomics, labor protection, operator inspection
- fittings: basic structure (product-neutral), parts supply (order-related), mechanization level (hybrid), adaptability
- profitability: assembly principle (piecemeal, routine), cost

Piecemeal assembly for products with a footprint over 1,500 cm^2 (machines, large household appliances).

Routine assembly is characterized by the repetition of individual assembly processes. For products with a footprint under 250 cm^2 (small parts).

Example: manual assembly workstations with typical basic equipment (fig. 5.27)

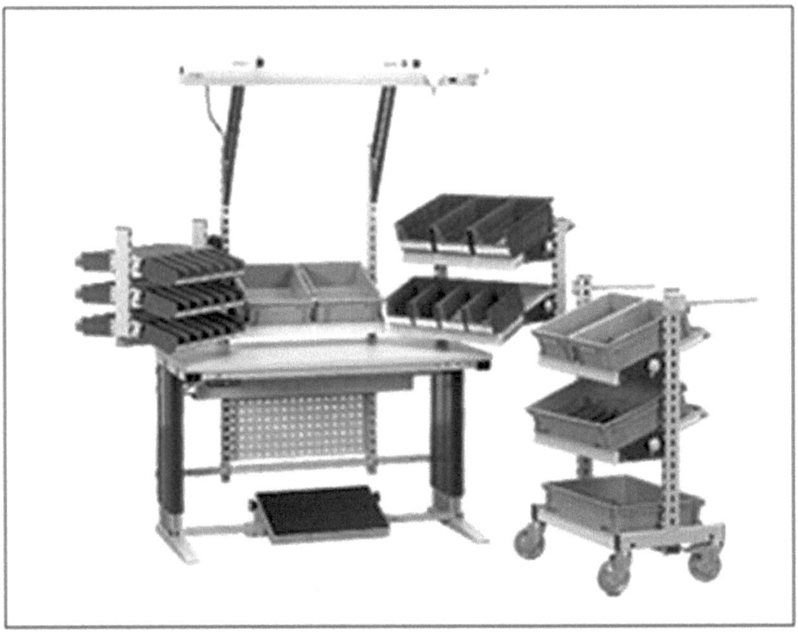

Fig. 5.27 Example of an assembly workstation with typical basic equipment (Lotter & Wiendahl 2006, p. 137)

Example: piecemeal process flow: provision of the parts by continuous parts elevator (precision engineering, electrical components). Here the reach path is almost always the same (fig. 5.28).

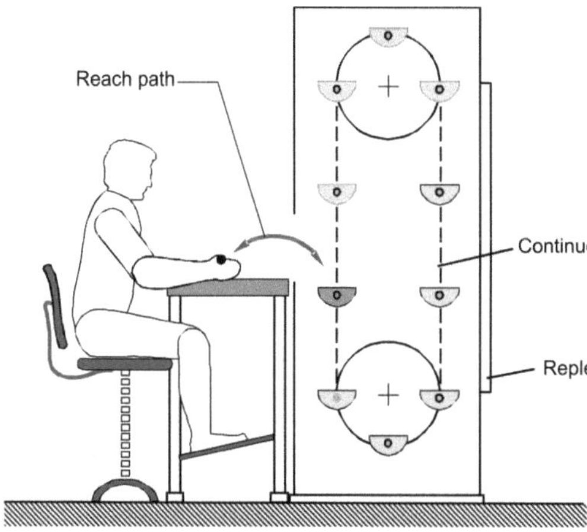

Fig. 5.28 Provision of parts in a workstation with a continuous parts elevator (Lotter & Wiendahl 2006, p. 138)

Example: stationary assembly objects are characteristic of on-site assembly. Here: individual parts assembly (fig. 5.29)

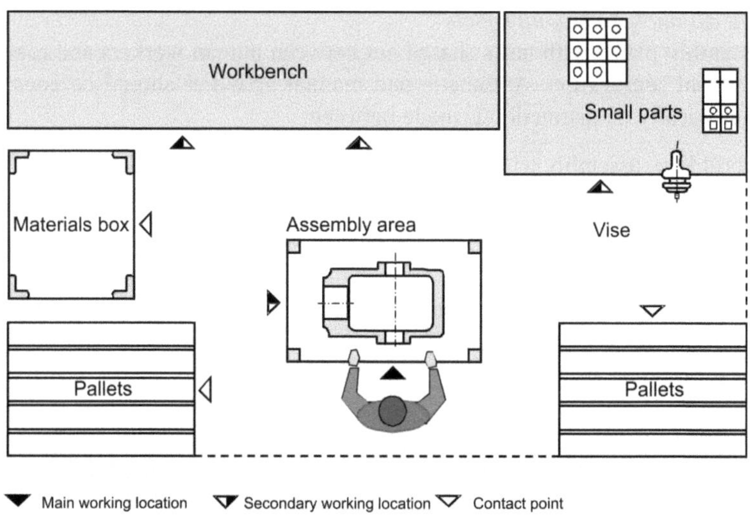

Fig. 5.29 Stationary assembly station for a gearbox (after Schilling)

Note: building block systems exist for different assembly workstations, which can be integrated either as a single workstation or flow line. Over and above this, assembly robots with automatic assembly stations with and without robots in different installation variants, plus assembly cells as flexibly automated assembly stations with workpiece and tool supply for complete product assembly are also used.

5.2.2.2 Area-related design solutions

General assembly lines with different specifications and layouts in the assembly workstation are:

Manual assembly lines

The assembly effort is divided across several connected (interlinked) workstations. The assembly/flow line is designed according to material flow-related ways of connecting individual stations (manual, mechanical, synchronized conveyance). The layout principles are similar to the conventional and integrated forms of manufacturing.

Field of application: series production with large product range

Assembly lines with robots

The equipment modules are controlled automatically. There are automatic, fixed and flexible assembly systems.

Field of application: series production, small and medium-sized products

Hybrid assembly lines with robots

Assembly plants with tasks shared out between human workers and robots and with spatial segregation. Automatic and manual activities should be coordinated chronologically. A distinction is made between:

- hybrid flow assembly with routine process flow and use of robots
- robot assembly machine with routine process flow

Field of application: medium and large series for small and medium-sized products

The building block / plug + produce principles underlie all design solutions.

Assembly lines are planned and implemented according to the building block principle based on modules.

Example: assembly system modules (fig. 5.30) whose configuration is specific to the task according to the plug + produce principle. It consists of assembly stations, basic, robot, palletizing and manual workstation modules.

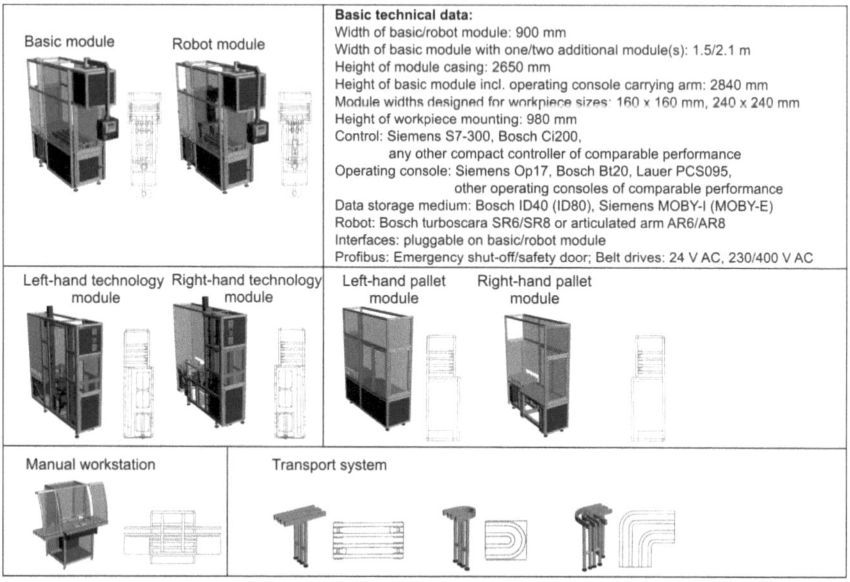

Fig. 5.30 Modular assembly building block system (source: System 21)

Example: assembly plants for mechatronic precision parts, planning according to the 0 + 5 + X model (based on building block VARIOMODUL, SITEC Company).

0 Project definition/customer order (complex I - checklist)
Product (specified in the form of drawings and prototype); technical availability 95%, shift operation, approx. 3,000 assembled workpieces (assembly units; 0.5 workers for provision, removal of the parts and plant control)

5 Project development (complex II – checklist)

(0) Fixing of the planning and implementation tasks (together with the client), specification of the chronological process flows, initial presentation of solution concept based on the building block principle

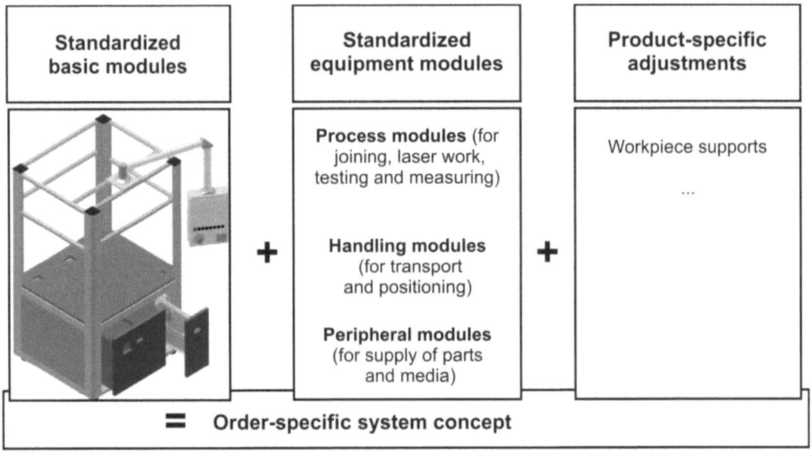

Fig. 5.31 Plant concept – VARIOMDUL (source: SITEC)

(5/1) Determination of the defined production program – requirements list, product analysis (design engineering, technology)

(5/2) Determination of functions – function analysis, assembly process flow schematic, assembly technology (joining stations, joining sequence of the individual processes) – functional distribution among the modules (fig. 5.32), predetermination of the modules from building blocks (catalogue)

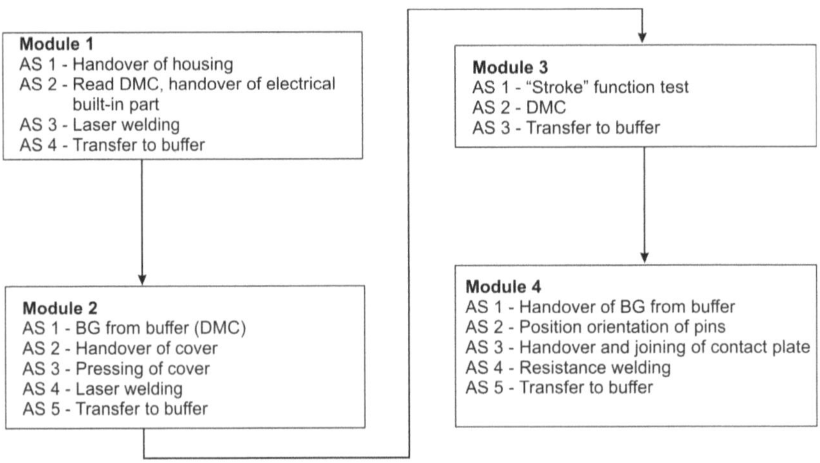

Fig. 5.32 Functional distribution of the assembly tasks among the modules

(5/3) Dimensioning – determination of the amount of assembly equipment and the cycle times, predetermination of assembly and production areas plus workers

(5/4) Structuring – spatial and chronological arrangement of assembly and THS facilities in variants. Fix variant evaluation with respect to parts quality, number of stations, availability, arrangement concept (fig. 5.33)

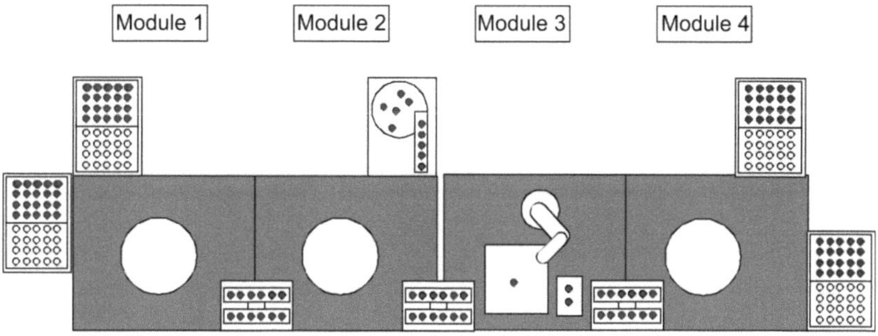

Fig. 5.33 Arrangement concept in the VARIOMODUL system

(5/5) Design – create and evaluate layout for the assembly plant (fig. 5.34)

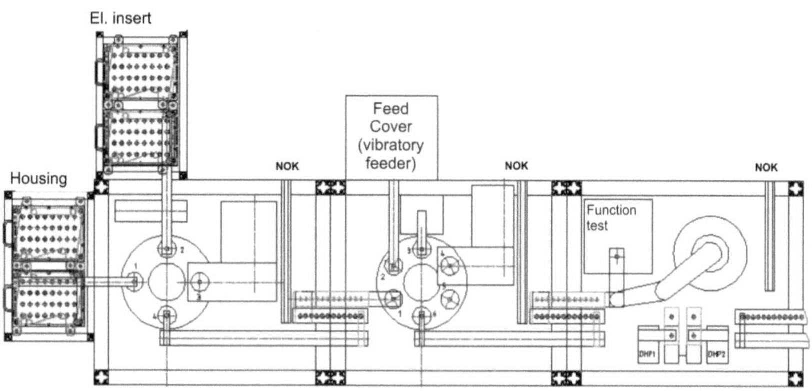

Fig. 5.34 Layout for assembly plants for mechatronic precision parts (source: SITEC Company)

X Project implementation (complex III – checklist)
Develop implementation project on the basis of

- binding requirement/technical specifications and contracts for all objects (with delivery deadline and procurement price/work center costs)

- check evaluation (functional and competitive) of the optimized overall solution and project implementation using checklists

(X/6) Profitability assessment – carry out investment appraisal.

(X/7) Implementation (fig. 5.35) – plant set-up, testing, functional handover, acceptance

Fig. 5.35 Installed assembly plant (source: SITEC Company)

(X/8) Operation – production according to start-up and process flow plan (normal operation), operational use, secure production planning and control (PPC) in terms of personnel and materials, have service and maintenance available.

Example: hybrid assembly plant for pumps (solution concept)

Project definition

Assembly is mainly executed using pressing in and bolting joining procedures (6 main parts [including an externally assembled, supplied component assembly] plus approx. 40 small parts), cycle time of 25 s, corresponds to a shift output of approx. 950 pumps where technical availability $A_T = 0.92$. Depending on capacity utilization, up to 6 operators and a coach (line manager) will be operating in the hybrid system.

Project development (design of solutions) – fig. 5.36

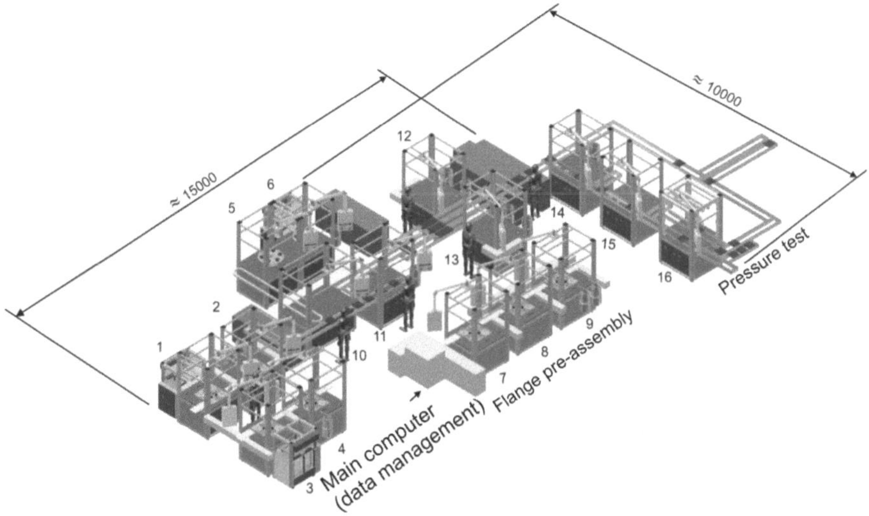

Fig. 5.36 3D model of the plant (hybrid system, source: SITEC)

The plant for partially-automated pump assembly consists, in addition to the main assembly line (7 modules, nos. 10-16), of 3 complexes (separated by storage areas) for pre-assembly

- modules 1-4 housing preassembly
- modules 5-6 control element preassembly
- modules 7-9 flange preassembly

whereby modules 12 and 13 are identical in terms of construction in order to increase throughput (reduction of the average cycle time).

In module 16 at the end of the line, the assembly is checked for leaks. In addition, in several modules, completeness and parameter monitoring is undertaken in order to determine whether parts are quality-compliant or not. The results are collected by an integrated MDA/PDA software program and evaluated as a function of time or event.

Example: basic variants: parallel and sequential configuration of assembly systems. They open up alternative solutions for flexible assembly flow as well as capacity-related changes. (fig. 5.37)

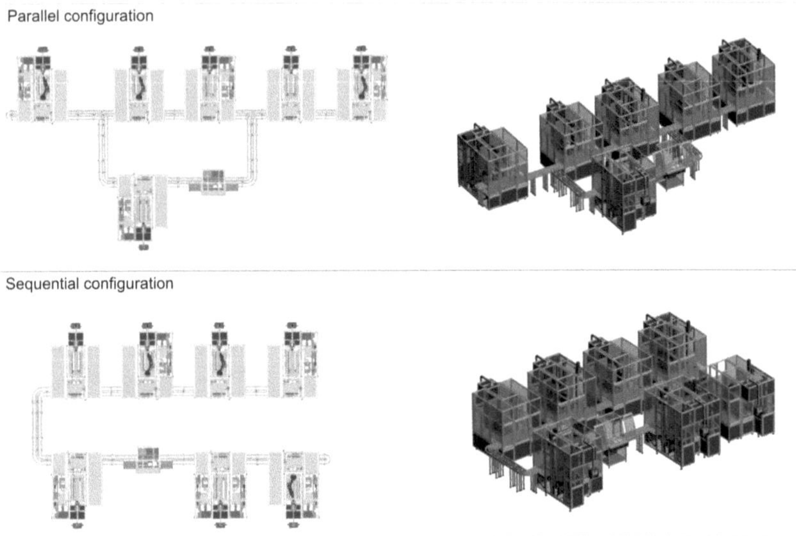

Fig. 5.37 Configuration of assembly systems (source: USK)

5.2.3 Integrated design solutions

Parts manufacture and assembly production facilities can be designed to be both separate and integrated. The "worm gear production" planning project shows an integrated solution.

In most cases there are separate parts manufacture and assembly areas. In the case of integrated areas, the final individual parts production stages merge into the assembly plant. Here, parts in finished condition are processed in full in the assembly plant and assembled directly. The trend is towards integration of the final production stages for individual parts in to the assembly plant.

In the pre-production stage variant-neutral and variant-specific parts and components are introduced into the final assembly process.

Example: integrated solution for complete processing (parts manufacture and assembly for cylinder head covers) – (fig. 5.38)

Plant capacity: 2000 parts per day, with a cycle time of 32 seconds

Process flow: (1) unfinished part processing on rotary transfer machine, (2) pre-washing of cylinder head cover, (3) seal test, (4) assembly, (5) mechanical processing of the assembled cylinder head cover, (6) mechanical processing and measuring, (7) deburring, (8) cleaning/washing

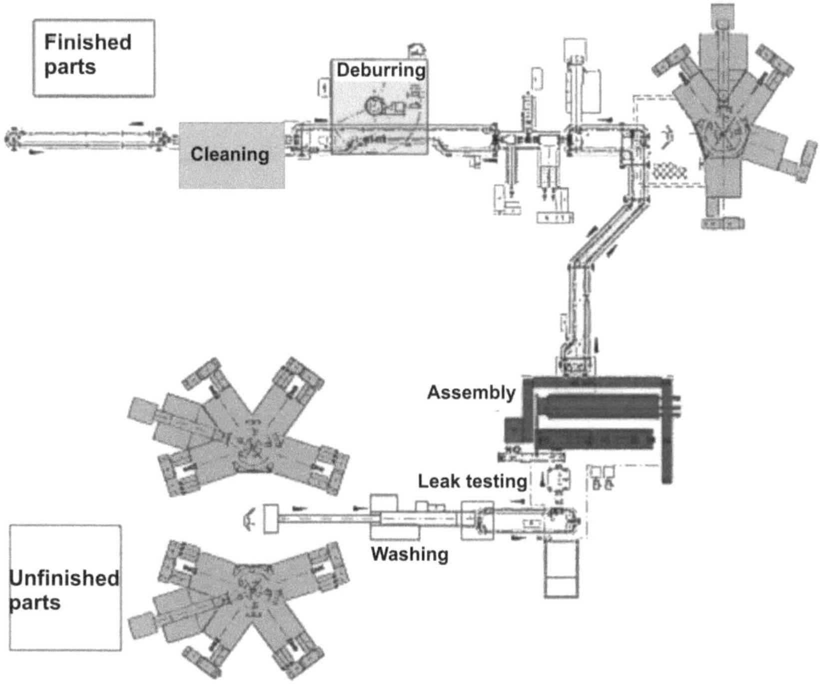

Fig. 5.38 Integrated solution for complete processing

Additional functions include rotary transfer, processing units, cooling lubricant equipment, disposal of turnings and chips, control units.

Example: layout of an overall assembly from parts supply via assembly to the finished goods store (fig. 5.39)

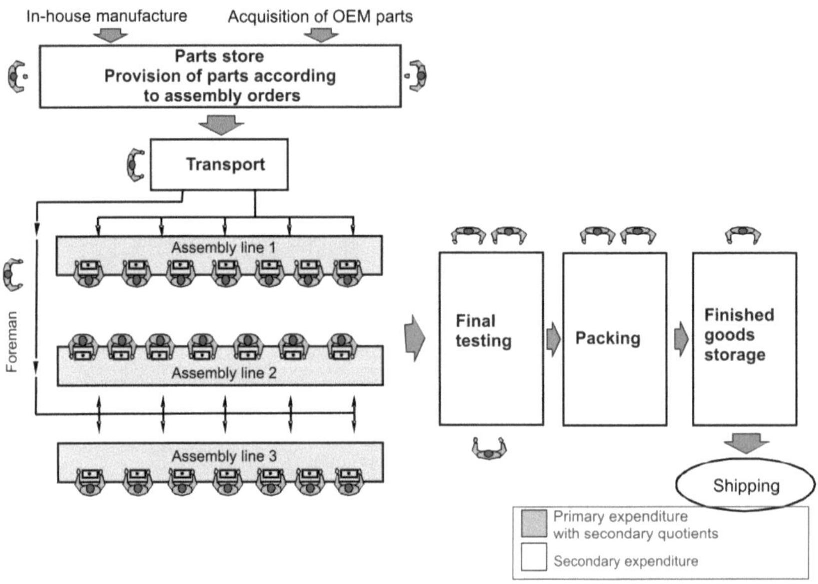

Fig. 5.39 Layout of an entire assembly process (Example Lotter & Wiendahl 2006, p. 80)

References

Dashchenko AI (2001) Reconfigurable Manufacturing Systems and Transformable Factories, Springer, Berlin

Enderlein H, Hildebrand T, Müller E (2003) Plug + Produce. Die Fabrik der Zukunft aus dem Bauskasten, wt Werkstatttechnik, online 93/4

Erfurth R (2003) Strategischer Handlungsrahmen, lecture notes, Projektmanagement, IBF, TU, Chemnitz

Erfurth R, Olschweski T, Wirth S (2003) Mobilitätsstufenabhängige Fabrikplattformen, wt Werkstofftechnik, online H 4

Eversheim W, Schuh G (1999) (Hrsg) Produktion und Management 4 – Betrieb von Produktionssystemen, Springer, Berlin

Heger C (2007) Bewertung der Wandlungsfähigkeit von Fabrikobjekten (Evaluation of the transformability of factory objects), PZH-Verlag, Hannover

Heisel U, Michaelis M (2004) Progress in Reconfigurable Manufacturing System, 2nd International Conference on Reconfigurable Manufacturing, Ann Arbor, USA, 20./21.08.04

Helbing K (2007) Fabrikprojektierung. Book manuscript, Rostock

Hildebrand T, Mäding K, Günther U (2005) Plug + Produce – Gestaltungsstrategien für die wandlungsfähige Fabrik, IBF, TU Chemnitz

Lotter B, Wiendahl HP (2006) Montage in der industriellen Produktion, Springer, Berlin

IREGIA (2004) Leitfaden Flexibilisierung bestehender Fabrikstrukturen, IREGIA eV, Chemnitz

Wirth S (2000) (Hrsg) Die flexible, temporäre Fabrik – Arbeitsschritte auf dem Weg zu wandlungsfähigen Fabrikstrukturen, VA 16, Ergebnisbericht, Karlsruhe

Wirth S, Hildebrand T (2001) Neue Anforderungen an den Industriebau. IndustrieBAU, H 5, S 54 - 57

Zäh MF, Bayerer P (2004) Gestaltung und Betrieb mobiler Produktionssysteme – Abschlussbericht des Forschungsprojektes ProMotion, Herbert Utz, München

6 Appendix

6.1 Investment Appraisal

6.1.1 Static methods

Static methods are very popular in practice and since above all, they are easy to use and involve low costs and little time for implementation, they are used very frequently. Furthermore, they also only need a small amount of straightforward initial data. They are known as static methods because they do not or only partially take into account the different times when payments are received and made, and they only consider one planning period.

They are recommended in the case of:

- investments of low value and/or little significance,
- uncertain initial data,
- profitability assessments that need to be carried out quickly and easily.

6.1.1.1 Cost comparison method

This is the simplest investment appraisal method. It is used to compare the costs accrued in a planning or usage period for two or more alternative investment objects. It is used to compare new objects (expansion investment) or old objects with new ones (replacement investment). In the case of expansion investment, the question is: "Which of the assessed alternatives results in the lowest costs over a usage period?" The problem of replacement investment is, by contrast: "When should the old investment object be replaced by a new one?"

If the capacities (quantitative output) of the investment objects are not the same, then instead of the cost for a period being compared, the unit costs must be compared. The cost comparison method therefore works on the premise that the investment objects have equal capacities. All relevant costs must be taken into account when using this method. If the investments in question are used over multiple periods, the average costs per period are applied. Revenues are not accounted for with this method, since it is assumed that each alternative earns equal revenue as they have equal capacities. The relevant costs are:

Operating costs

- personnel costs (wages, salaries, social benefit contributions)
- material costs (production and operating materials and ancillary supplies)

M. Schenk et al., *Factory Planning Manual*, DOI 10.1007/978-3-642-03635-4_6,

- maintenance and repair costs
- occupancy costs
- energy costs, etc.

Capital costs

- imputed depreciation

Advantages:

- ease of use

Disadvantages:

- in practice, itemizing the cost functions of the systems under comparison frequently involves difficulties
- the method can only be used for short-term investment projects since it is static (a comparison of two states) and therefore cannot take future developments into account
- revenues are not taken into account, so no information can be provided about returns on the capital
- differences in the qualitative output of investment objects are not taken into account
- possible effects on product sales prices due to higher production and sales volumes are not taken into account

6.1.1.2 Accounting rate of return method

The cost comparison method is not adequate for many investment projects since earnings may also vary subject to varying qualitative outputs (especially in the case of expansion investments). In order to assess the investment alternatives, therefore, earnings must also be taken into account as well as the costs. In this respect the accounting rate of return method represents an extension to the cost comparison method.

Where the earnings per quantity unit are the same, both methods produce the same results. The accounting rate of return method can be used to assess individual investments or to compare several alternatives. In this case, for individual investments, any that return a profit greater than zero are favorable. When comparing alternatives, the investment opportunity for which the average annual anticipated profit is greatest is the one that should be selected.

Advantages:

- the investment projects that have the greatest influence on earnings can be assessed more effectively, as is the case for new and expansion investments

Disadvantages:

- the accounting rate of return method is essentially subject to the weaknesses of the short-term static method
- no information can be provided about the rate of return on the capital invested

6.1.1.3 Rate of profit method

The static rate of profit method establishes the relationship between the profit and the capital invested, since the investment profits are often earned with differing levels of capital investment and unlimited capital is generally not available. This investment calculation, which is often used in American business practice, is also known as "Return on Investment" (ROI) or "Return on Capital Employed". The profitability for the period is calculated, and where alternatives are compared, the alternative with the maximum rate of profit is selected:

$$ROI = Profit/Turnover \bullet Turnover/invested\ capital \tag{6.1}$$

$$Rate\ of\ return = \frac{average\ \mathrm{Pr}\,ofit}{average\ capital\ employed} \tag{6.2}$$

Profit/Turnover = Turnover Success
Turnover/invested capital = Capital turnover

Advantages:

- ease of use

Disadvantages:

- short-term approach that does not take future changes into consideration
- it is difficult to assign turnover and revenue to the individual investment alternatives
- the difference in the useful life of individual objects is not a concern, i.e. the capital of the shorter-lived investment must earn the same return despite the difference in useful life

So that the rate of profit can be assessed, it is often compared with the minimum rate of profit desired by the investor. If it is greater, the investment is favorable; if it is smaller then it will not be made.

6.1.1.4 Static payback period method

The payback period method represents an extension of the cost and accounting rate of return methods. This method determines the period of time that it takes for the original capital investment to be returned completely in the form of revenues, thus making it possible to roughly evaluate the risk of various investment objects. An individual investment object is favorable if its payback period is shorter than the investor's target specification. When comparing alternatives, the alternative that has the shortest payback period is the one that should be selected.

$$Payback\ period = \frac{Capital\ investment}{Average\ return} \tag{6.3}$$

Advantages:

- as a complement to the rate of profit method, this method provides valuable information with respect to the risk assessment of investment projects; the longer the capital is tied up for (or payback period) the more uncertain is the likelihood of capital recovery; with different useful lives of investment alternatives, however, it makes less sense to make the investment decision exclusively on the basis of the payback period method, as the annual depreciation depends on the useful life and thus has a significant influence on the payback period.

Disadvantages:

- the method is based on the assumption that annual payments in and out will remain the same,
- the target payback period is based on estimates

In practice, for risk reasons the target payback period is usually set at not longer than 3 – 5 years, even if the intended usage time is 10 years and more.

6.1.2 Dynamic methods

The profitability of investments can be assessed much more effectively using dynamic investment appraisal methods than it can with static methods. However, these are much more complex and difficult to use.

- by using dynamic (or financial theory-based) methods the known weaknesses of the static methods can be eliminated:
- the single-period approach that is based on average values is replaced by the accurate recording of payments in and out during the entire useful life of the investment object (multi-period approach).
- payments in and out are evaluated according to the time when they accrue.

The following dynamic methods will be explained in more detail:

- net present value method,
- internal rate of return method,
- capital recovery method.

They are based on:

Year of usage

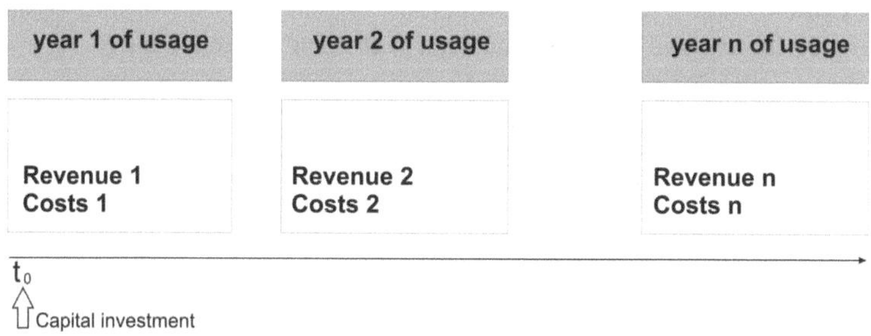

Fig. 6.1 Year of usage

Interest rate

Is generally related to a year and specified in percent

Compounding

If the annual interest is added to the amount of capital after a year has elapsed, the following capital development results:

$$1^{st}\ year:\ CI_1 = CI_0 + CI_0\,p = CI_0\,(1 + p) \tag{6.4}$$

$$2^{nd}\ year:\ CI_2 = CI_1 + CI_1\,p = cI_0\,(1 + p)^2 \tag{6.5}$$

$$n^{th}\ year:\ CI_n = CI_0\,(1 + p)^n \tag{6.6}$$

$$q_n = (1 + p)^n \tag{6.7}$$

is known as the compounding factor. It increases as the interest rate and length of the underlying period increase.

Discounting

This method is used to determine the value of an amount at a point in the past. The discounting factor is the reciprocal value of the compounding factor.

$$\frac{1}{q^n} = \frac{1}{(1+p)^n} \tag{6.8}$$

It decreases as the interest rate and length of the underlying period increase.

Capital investment (CI)

Investment sum and costs for the procurement or manufacture of the required fixed and liquid assets.

Net present value NPV_0

Difference between the discounted costs and discounted revenue. At t_0, C_0 corresponds to the present value of an investment.

Present value

The value of cost and revenue dates at a particular reference date.

Example: The following cost stream exists:

1st year €1000, 2nd year €2000, 3rd year €3000, 4th year €3000; interest rate 8 %

What is the present value?

Present value = 1,000 x 0.9259 + 2,000 x 0.8573 + 3,000 x 0.7938 + 3,000 x 0.7350
= € 7,226.90

(using static methods the result would be € 9,000!)

Discounted cash flow

Describes discounted and compounded returns (return as the difference between income and expenditure)

Recovery factor

Is the factor by which an amount of capital (CI) must be multiplied in order to calculate the annual constant annuity amount (A) that is to be paid at the end of each year (in arrears) with a term of n years and an interest rate (p) for

- amortization (depreciation) and
- interest

6.1.2.1 Net present value method

Question: What is the total surplus of an investment?

The net present value method fulfills the requirement for the evaluation of payments made and received at different times by determining the present value of all payments in and out (revenue and costs). For this, all payments are discounted at the time $t = 0$ (time of the first payment that is made in connection with the investment).

The net present value of the investment is calculated from the difference in the present values of all incoming payments (including the proceeds of asset liquidation) and the present values of all outgoing payments.

Discounting is done on the basis of an imputed rate of interest i, which represents the desired minimum rate of return on the capital.

$$NPV_0 = -I_0 + \sum_{t=0}^{n} (E_t - A_t) * (1+i)^{-t} + \frac{L_n}{(1+i)^n} \tag{6.9}$$

NPV_0 = Net present value I_0 = Investment expenditure where t=0
E_t = Earnings in period t $(E_t - A_t) = R_t$ = Returns in period t
A_t = Costs in period t i = Imputed rate of interest
t = Period (0, 1, 2 - n) L_n = Liquidity proceeds

3 cases:

a) NPV_0 = net present value = 0
The investment repays its cost from the returns and achieves a yield in the amount of the imputed rate of interest applied.

b) NPV_0 = net present value > 0
In addition to the recovery of the capital employed, the aim of the investment is to achieve a return that is above the imputed rate of interest applied and thus to generate a profit.

c) NPV_0 = net present value < 0
The investment does not achieve the required imputed return on the capital employed or it does not even achieve the payback of the capital employed.

The net present value method is suited to assessing the profitability of investments in comparison to investment at the imputed rate of interest.

In the event of a shortage of capital, the net present value method is not suited to generating rankings since the net surplus achieved is not related to the capital investment.

6.1.2.2 Internal rate of return method

Question: How high is the actual profitability of an investment?

The effective yield of an investment is known as the internal rate of return. The internal rate of return method does not calculate the net present value with a given rate of return on capital as the net present value method does; instead, it calculates the internal rate of return for a net present value of zero. For this calculation, therefore, the same approach is used as for the net present value method and the net present value is set as $NPV_0 = 0$. The imputed rate of interest (i) thus becomes the internal rate of return (r):

$$NPV_0 = -I_0 + \sum_{t=0}^{n} (E_t - A_t) * (1+r)^{-t} + \frac{L_n}{(1+r)^n} = 0 \qquad (6.10)$$

NPV_0 = Net present value E_t-A_t = Returns in period t

r = Internal rate of return L_n = Asset liquidation proceeds in period n

Resolving the equation for r is laborious in the case of multiple periods, so therefore an approximation procedure is frequently used, which delivers sufficiently precise results according to the following method:

- an imputed rate of interest is selected for which a net present value > 0 is calculated;
- then an imputed rate of interest is selected at which the net present value < 0;
- using linear interpolation, the rate of interest is determined at which the net present value is precisely 0.

The profitability of an investment can be determined with additional knowledge of the imputed rate of interest if the internal rate of return is greater than or equal to the imputed rate of interest. It is clear that the imputed rate of interest must be specified as a benchmark for comparison.

If the aim is to compare alternatives, then assuming that the alternatives are complete, the one that demonstrates the highest internal rate of return must be selected. However, this must also be greater than the imputed rate of interest. If it cannot be assumed that the alternatives are complete, then they must be examined using fictitious investments.

6.1.2.3 Capital recovery method

The capital recovery method can be regarded as a variant of the net present value method. The payment streams related to the investment are converted into equivalent, equidistant, and uniform amounts. The annuity thus corresponds to a notional payment series, i.e.:

- the present values of the capital value series are equivalent to the present value of the annuity,
- the payment times are equidistant from one another,
- the payments are uniform in size.

The annuity of an investment (i.e. the average annual payment surplus) is calculated with the help of recovery factors. The following applies:

$$A = NPV_0 * \frac{i(1+i)^n}{(1+i)^n - 1} = NPV_0 * RF_n^i \qquad (6.11)$$

NPV_0 = Net present value A = Annuity
i = Imputed rate of interest RF_i^n = Recovery factor for i and n

An investment is favorable if its annuity ≥ 0. This means that the tied up capital yields at the imputed rate of interest and a period surplus is also generated. If investment alternatives are compared, then the one whose positive annuity is highest is considered to be the most favorable. Like the net present value method, the capital recovery method also rests on the assumption of a perfect capital market and assumes that the imputed rate of interest is known. Therefore, the points of criticism of the net present value method also apply to the capital recovery method.

6.1.3 Lease-buy methods

In addition to conventional external financing, additional financing instruments also come into consideration, such as leasing and operator concepts like pay on production (the operator generally receives the investment made by his customer paid per unit produced). Generally, it is assumed that all investments serve to immediately increase the efficiency of production. Facilities on the premises and immovable assets such as production buildings and general operating installations such as compressed air or crane systems are generally necessary investments that are required to guarantee the environment for the processing machines and, in particular, for production expansion. The strategic decision-making for a custom financing model requires the sequence of steps described below to be applied (carried out in hybrid form).

Value added chain → classification of the company within the value added chain

Essentially, an investor must undertake a classification in the value added chain.

OEM – End manufacturer of products (vehicles, machines, household appliances)

Is able to determine the product change strategy and the investment need for this independently, depending on market conditions.

SS – System supplier for OEM, e.g. engine manufacturer, manufacturer of fluid systems.

Is only partially able to determine the product change and investment need for new products independently.

CS – Component and parts supplier

Determines the investment needed largely on the basis of own rationalization requirements. If necessary, makes investments to increase skills base. Fig. 6.2 shows the cooperation between groups of companies.

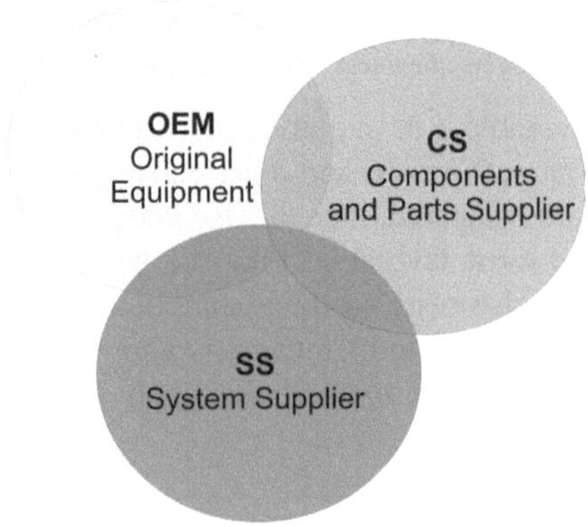

Fig. 6.2 Types of companies

In practice, these types of companies occur largely in hybrid form. They determine the financing strategy of an investment to a considerable extent.

Reason for investment → disclosure of the reason for investment
Reasons for investment can be:

- primarily transition to a new product (OEM, SS)
- sudden production expansion (OEM, SS)
- rationalization (OEM, SS, CS)
- creation and acquisition of new skills (CS)

In business, several reasons for investment may overlap. They are significant for the effectiveness potential and also for participation in financing. The transition to new products in production in the case of an OEM can, for example, lead to new investments in a CS or SS, which are financed from the OEM's budget. For rationalization investments, the cost savings can generally be shown clearly; the potential, however, becomes ever smaller with the increase of the useful life of the hitherto existing systems.

Operator concepts → Selection of operator models

Operator models are production and financing models for which the end producer (OEM or SS) has transferred a part of the production including maintenance to an operator (usually CS) for a limited time.

Here the investment for the capital goods is made by the operating company. The operator generally receives the investment made by the client by unit produced. Frequently the system manufacturer is also the operator. The "pay on production" operator model is especially well-suited to leasing.

Useful life → assessment of the useful life

The financing model largely depends on the intended useful life of the capital goods. A long useful life frequently results in complete self-financing. A useful life shorter than the depreciation for wear and tear (DWT) values is recommended for leasing. Classifying intended investments in groups of capital goods is therefore beneficial for decision-making. Table 6.1 shows a grouping of capital goods by depreciation for wear and tear/useful life, which is significant for the refinancing of an investment.

Table 6.1 Official depreciation for wear and tear tables, Verlag Neue Wirtschafts-Briefe Herne, Berlin, loose-leaf collection

Group	Useful life in years	Example
1	< 5	Personal computer, printer, scanner, vending machines
2	$> 5 \leq 8$	Automobiles, mobile machine tools, special purpose machines, test benches
3	$> 8 \leq 13$	Cut-off machines, welding machines, packaging machines, office furniture
4	$>13 \leq 20$	Presses, milling machines, steam generators, guide rails, work platforms, high-rack warehouses
5	>20	Crane systems, production buildings

Ranking → Ranking of the financial situation

The financing of the investments must go together with a ranking of the financial situation of the whole company. Generally it must be decided what the proportions of self and third-party financing should be. This usually depends on the capital resources and the company objectives. Primary objectives can be: unburdening of the balance sheet, tax optimization, protection of credit lines and equity capital.

The rating that determines the creditworthiness results from an optimization of these objectives according to the following optimization criteria: achievement of a minimum creditworthiness with a reduction in the risk costs or an extension of the term.

Here the company's own creditworthiness, security of results, third party guarantees, distribution and joint liability must be variable.

This ranking should be obtained either from the company's own bank, from a leasing company or from a consulting company. The outcome should be a rough structuring of the investment in terms of the proportion of equity capital, subsidies, and third-party financing (credit, lease finance).

Financing models → selection of financing models

The financing forms are now assigned to the respective investment objects. Ultimately, the following financing forms produce the financing model: cash purchase, finance purchase (purchase with credit), rental, rental-purchase and leasing.

These forms have a multitude of special structuring possibilities that are influenced by both international and many domestic laws. For instance, using the "Improving the regional economic structure" subsidy program in the Federal Republic of Germany excludes leasing.

Table 6.2 shows the influences on the financing model in global terms only, since as a result of legal regulations, there is much creative leeway but also some restrictions, which require optimization by or advice from a bank or financial auditor.

Table 6.2 Orientations for financing models

		Cash purchase	Finance purchase	Rental	Rental-purchase	Leasing
Position in the value added process	OEM	+	+	0	0	+
	SS	0	0	+	+	0
	CS	0	0	-	+	+
Reason for investment	New products	+	+	0	0	+
	Sudden production increase	+	+	0	0	+
	Rationalization	+	+	0	0	+
	Acquisition of new skills	+	+	0	0	-
Operator model	Final producer	+	+	-	0	+
	Pay-on-production	-	0	0	-	+
Useful life according to depreciation for wear and tear	Group 1	+	0	0	+	+
	Group 2	+	0	0	+	0
	Group 3	0	+	0	+	-
	Group 4	0	+	0	+	-
	Group 5	0	+	-	+	-

+ Recommended; 0 Possible, optimization with several factors; - Not recommended

Leasing → selection of the leasing variant

Leasing is the relinquishment of an item for a specified duration by the owner (lessor) to the user (lessee). An agreed usage fee (leasing rate) is paid for its use. Since the asset does not transfer to the user's ownership and thus balance sheets and taxes are influenced for the lessee, there are all kinds of domestic legal regulations.

Generally, it is assumed that the asset is easily identifiable (can be removed from a production system without being destroyed) and fungible (reusable and can continue to be used) (Weise 2006). Another condition is that the first usage phase should be much shorter than the useful life according to depreciation for wear and tear.

Both terminable leasing and operate leasing are suitable for the leasing of moveable goods. The two forms are distinguished by the fact that in the case of terminable leasing, the contract may be terminated early as a result of unforeseeable order changes, while in the case of operate leasing, a fixed leasing duration, leasing rate, and a fixed residual value availability are agreed upon. Operate leasing is especially well-suited to OEMs, while terminable leasing is suited above all to CS and SS. Fig. 6.3 shows the sub-steps for the initiation of a terminable leasing contract.

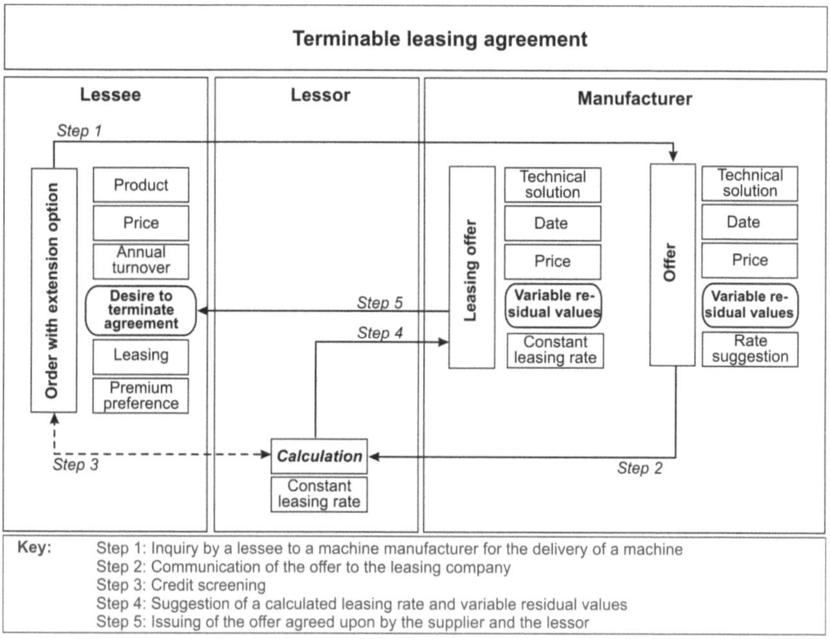

Fig. 6.3 Terminable leasing sub-steps

This procedure assumes that the investor (subsequently the lessee) has decided in the "financing model" step that leasing is advisable for the asset. This means:

- the product to be produced using the supplied equipment has already been determined.
- the price and annual turnover are known for the product.
- the production duration is regulated contractually for at least an initial usage phase. Extension options are possible.
- an ideal leasing rate has been calculated.

Leasing also exists in the following forms:

- full amortization leasing
- partial amortization leasing

The many tax aspects that must be heeded with these leasing forms make an optimization taking all conditions into account necessary.

Example: Financing of investments

A machine construction company intends to increase turnover by developing new products and applying effective operator models. It plans to expand the production area including the required auxiliary systems as well as to procure a number of special machine tools and special purpose machines. Some of the special purpose machines are needed for the pay-on-production operator model. Four-year supply contracts have been agreed for this. The following sequence of steps is recommended:

Value added chain:

- OEM as machine manufacturer
- CS as component manufacturer for an OEM in another industry

Reason for investment:

- sudden production increase as a result of good market conditions
- expansion of pay-on-production

Operator model:

- 20% of the area expansion and 40% of the equipment will be required for pay-on-production

Table 6.3 Assessment of useful life

Asset	Usage group in € 1000					DWT Useful life
	1	2	3	4	5	
Production building					800	25
Street and parking spaces				50		19
Supply and disposal system			400			12
Storage				20		15
Workshop equipment				80		14
Automatic lathes, grinding machines		150				8
Special purpose machines		400				8
Other			100			10
Total: T€ 2,000	0	550	500	150	800	

Ranking: The following ranking is the result of an assessment of the financial situation with the company's bank:

Financing from own funds:	T€ 600
Financing from subsidy funds:	T€ 300
Third-party financing:	T€ 1,100
of which credit:	T€ 600
Leasing:	T€ 500

Financing model: The following financing model results from the ranking and legal options:

Cash purchase:	€ 700,000
Finance purchase:	€ 800,000
Leasing:	€ 500,000

Table 6.4 Leasing

Type	Operate leasing	Terminable leasing
Acquisition costs	€ 500,000	€ 500,000
Term	36 months	84 months (termination possible from 36 months)
Monthly payment	€ 7,750	€ 7,300
Residual value	€ 300,000	36 months € 311,000 48 months € 240,000 60 months € 164,480 72 months € 84,870 84 months € 0

6.2 Main Building/Production Parameters

6.2.1 Geometric parameters

Figure 6.4 shows building areas, space and dimensions.

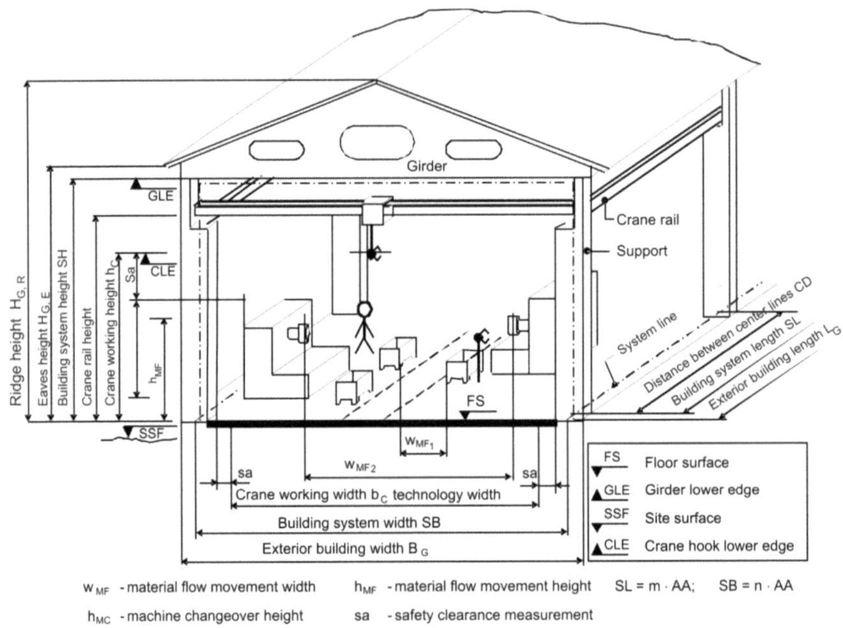

Fig. 6.4 Building dimensions (after Helbing 2007)

Building's effective height

The required height of a production building (vertical geometry) is generally determined by (cf. fig. 6.5):

- the roof level → (1)
- the upper installation level → (2) (installation zone TBS on the upper space enclosure)
- the over-floor transport level → (3)
- the production level → (4)
- the lower installation level → (5)
- the under-floor level → (6)

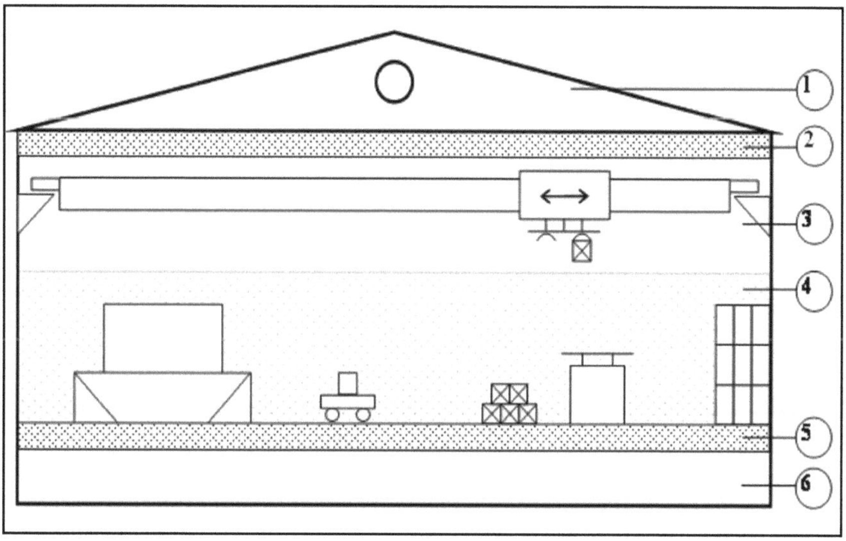

Fig. 6.5 Geometry of production buildings (IREGIA 2004)

From the point of view of the production-related requirements, the necessary effective height must be defined, i.e. the height of the production level (4) and if applicable the over-floor transport level (3). This is calculated from:

machine height
+ height of items being transported (product, machine, auxiliary materials)
+ height of means of suspension
+ height of installation height of the over-floor means of transport
+ safety clearances

= effective height

When using the formula it should be noted that the height of the item being transported that is applied here depends on the transport path. The machines, products, and auxiliary materials are not transported directly over the equipment in every case. Transport can also be done between the machines, e.g. above the transport routes using ground-based means of transport.

When determining the effective height, the installation heights of overhead cranes, for example, are important where over-floor transport systems are used. The following table shows an excerpt for overhead cranes for loads between 3,200 and 10,000 kg, using the "ABUS" company as an example.

The additional details are based on statistical investigations and are included in IREGIA 2004.

Table 6.5 Average crane installation heights

Bearing load [kg]	Span/ projection [mm]	Installation height [mm]			
		Single girder overhead traveling crane	Ceiling traveling crane	Single girder overhead traveling crane with side crane trolley	Double girder overheard traveling crane
3200	4000 - 17000	780 - 1040	978 - 1135		
3200	6000 - 34000			910 - 1660	1370 - 2070
5000	4000 - 17000	910 - 1150	1225 - 1305		
5000	6000 - 34000			960 - 1710	1370 - 2320
8000	4000 - 17000	1050 - 1330	1375 - 1445		
8000	6000 - 34000			1120 - 1770	1560 - 2410
10000	4000 - 17000	1080 - 1380	---		
10000	6000 - 34000			1120 - 1780	1560 - 2420

If the use of over-floor means of transport is not necessary, the effective height can be derived from the dimensioning of the machines and systems. The following table shows that the average height for a majority of machines, except presses, does not exceed 3.30 m.

Table 6.6 Typical machine heights

Machine heights in mm	Ø Height	Percentage coverage of all machines			Percentage of the max. height			
Manufacturing process		50%	75%	90%	50%	75%	90%	100%
Lathes	2,477.6	2,500.0	2,800.0	3,000.0	1,700.0	2,550.0	3,060.0	3,400.0
Drills	2,081.0	1,904.0	2,408.0	2,600.0	1,655.0	2,482.5	2,979.0	3,310.0
Milling machines	3,274.5	2,853.0	3,913.0	4,750.0	3,375.0	5,062.5	6,075.0	6,750.0
Grinding machines	2,752.2	2,700.0	3,200.0	3,500.0	1,900.0	2,850.0	3,420.0	3,800.0
Sawing and cutting-off machines	2,017.0	2,100.0	2,230.0	2,570.0	1,355.0	2,032.5	2,439.0	2,710.0
Bending machines	1,794.8	1,700.0	1,955.0	2,165.0	1,162.5	1,743.8	2,092.5	2,325.0
Presses	4,636.3	4,450.0	5,600.0	7,000.0	4,100.0	6,150.0	7,380.0	8,200.0
Average overall	**2,818.4**	**2,705.3**	**3,250.8**	**3,721.8**	**2,233.8**	**3,350.6**	**4,020.8**	**4,467.5**

Pillar grid

The grid measurements are made up of the grid width and grid length. This forms the skeletal structure for the production facilities that is of great significance for the production, assembly and storage processes, as well as for the integration of internal logistics.

When specifying the optimal measurements for the pillar spacing, there are two diametrically-opposed requirements:

- pillar spacings that are as wide apart as possible in order to achieve a high level of flexibility in terms of floor space utilization and to enable changes to be made at a later date
- limited pillar spacings in order to be able to use cost-effective supporting structures (especially when installing over-floor transport systems)

Specifying the grid measurements is thus a difficult planning task. It is recommended that you map initial structural variants in different rough layouts and test the use of typical pillar grids. Attempts to standardize pillar spacings in manufacturing plants are still largely left open, despite various attempts at regulation. Recommended grid lengths and depths range through 3m, 6m, 9m, 12m, etc.

Height/width of thoroughfares and gates

The height and width of thoroughfares and gates is determined by the measurements of the largest separable transport unit plus the means of transport. This means that the dimensions of the units to be transported (e.g. products, machines) as well as the height and width measurements of the ground-based means of transport (e.g. front loaders) must be determined and used as the basis for the dimensioning of the openings. Measurement guidelines for ground-based means of transport can be taken from the following table.

Table 6.7 Typical dimensions for ground-based means of transport

Means of transport	Bearing load [kg]	Width [mm]	Minimum height [mm]	Maximum height [mm]
Manual elevating platform truck	Up to 5000	520 - 770	800 - 1990	1200 - 1990
Electric elevating platform truck	Up to 2500	700 - 1150	1670 - 2645	1790 - 4540
Electric front loader	Up to 3000	1006 - 1530	2010 - 3120	3450 - 7550
Electric front loader	Up to 8000	1190 - 2100	2075 - 3800	3805 - 7150
Diesel front loader	Up to 3000	1098 - 1740	1975 - 3120	3450 - 7745
Diesel front loader	Up to 8000	1250 - 2100	2250 - 2950	3805 - 6950
High-rack fork lift	Up to 1500	1160 - 1600	2400 - 5900	4250 - 15210
Reach truck	Up to 2500	900 - 1670	1740 - 4930	2920 - 12240
Low lift picker	Up to 3000	760 - 980	1080 - 1340	1080 - 1340
High-rack picker	Up to 1500	970 - 1475	1460 - 6530	2940 - 11655

When dimensioning the free openings, the flow elements both from and into the factory must be taken into account. For example, if a material delivery is required by driving through a part of the building with a truck, gate dimensions must be planned according to the largest relevant truck dimensions.

6.2.2 Load parameters

Upper floor(ceiling) and floor load

This describes the load in the form of the forces that industrial floors and ceilings must be able to accommodate. In production, assembly, and storage areas, these are frequently determined by the production and storage equipment used as well as the means of transport. In addition, the weight of raw materials and intermediate and finished products can determine the floor load requirements depending on their dimensions.

Purely static load values can be derived from the relationship of the weight to the relevant effective area (e.g. shelf storage systems). The floor load required for the installation and operation of machines and systems must be determined by asking the equipment manufacturers. Here the concern is to determine the static and possibly dynamic load values as a result of the functional effect. Similarly, this must be applied with respect to the means of transport. In addition to the means of production and transport, the products or product units to be processed/transported must be taken into account.

Typical load values in the metal-processing industry fall into the 10 - 50 kN/m² range. An analysis of the required floor load in companies in this industry resulted in the following values:

Table 6.8 Typical values for floor load in manufacturing plants

Floor load kN/m²	Frequency in %	Frequency in % (cumulative)	Flexibility in % [30 kN/m² = 100 %]
5.00	6.78	6.78	16.70
10.00	13.56	20.34	33.30
15.00	3.39	23.73	50.00
20.00	28.81	52.54	66.70
30.00	27.12	79.66	100.00
50.00	20.34	100.00	166.70

Transport loads

Transport organization is of great significance as a building block of internal logistics. On the one hand, transport processes are essential for a structured, timely and quantity-compliant material flow as the condition for efficient manufacturing processes. On the other hand, the transport organization (especially the means of transport) significantly influences the demands placed upon building structures. The transport tasks in the operation can be divided into three categories:

- transport between operational areas,
- transport within the operational areas (e.g. means of production - storage space),
- transport between the workstations.

These must be taken into consideration when applying the following criteria plan for the assessment of transport tasks and the material-flow compliant organization of transport systems.

Table 6.9 Criteria plan for selection of transport systems

Category	Criteria	
Commodity to be transported	- Packaging (type, size (volume), weight) - Transport quantity per unit of time - Handling of the items to be transported (transportability, stackability) - Special properties (hazard to the surroundings, hazard to the items being transported)	
Functional units to be served	- No spatial limitations - Limited space	- Limited distance - Particular location
Transport route	- Freely-selectable transport route - Partially-specified route	- Fixed course
Transport intensity (frequency)	- Occasional - Interrupted	- Intensive - Continuous
Required installation	- Under-floor - In and on the floor	- At working height - In the upper building space
Direction	- One-level (horizontal) - Sloping down - Sloping up	- Sloping up - Multi-level - Vertical
Possible combinations	- Only in one direction - In both directions	- Linking of several addresses (turnouts)
Mechanization and automation	- Manually-operated and manually-controlled - With automatic control	- With automatic unloading - With automatic loading

In order to identify the appropriate means of transport (or systems), it is important to define the transport tasks, especially the primary criteria of items to be transported, transport quantity and intensity plus the transport type. In order to integrate transport systems into existing structural envelopes, the circumstances and modification possibilities must be taken into account with respect to the transport route (inclines, height differences, etc.) and especially the realizable effective heights. The following overview can be used to preselect the appropriate means of transport based on the transport tasks to be performed.

Table 6.10 Selection of suitable means of transport – assignment to the conveyance task

		Means of conveyance										
		Belt conveyor systems	Chain conveyor systems	Roller conveyor systems	Single-track vehicle	Fork lift	Tractors, electric carts	Hand truck	Lift truck	Overhead crane	Mobile crane	Elevator
Area served	Unlimited space					x	x				x	
	Limited distance	x	x	x	x							
	Limited space							x		x		
	Point								x			x
Installation	Under-floor		x								x	
	On the floor		x	x		x	x	x				
	At working height	x	x	x								
	In the air	x		x	x				x	x		
Route	Route not specified					x	x	x			x	
	Fixed, transportable distance	x	x	x					x			
	Fixed route	x	x	x	x					x		x
Frequency	Occasional							x	x			x
	Interrupted			x		x	x	x	x	x	x	x
	Continuous	x	x	x	x							
Direction	Horizontal	x	x	x	x	x	x	x		x	x	
	Sloping down	x	x	x	x							
	Sloping up	x	x		x							
	Vertical downwards					x			x	x	x	x
	Vertical upwards					x			x	x	x	x

After selecting the appropriate means of transport or systems for performing the required transport tasks, the requirements for building and infrastructure can be specified.

Table 6.11 Requirements profile for means of transport or transport system

Parameter type	Requirements for means of transport or transport systems
General	– required installation (under-floor, in and on the floor, in the upper building space) – design of driving surface, track, support structures (evenness, surface)
Geometric parameters	– dimensions: width, height and length of the means of transport – necessary safety clearances – required areas for turning/maneuvering – loading and unloading (type: manual, mechanized, automated; loading height; operating space)
Load parameters	– carrying capacity of the transport routes (wheel load, empty weight, total weight, consider also elevators in multi-story buildings)
Interference parameters	– mode (drive type, type of energy supply) – operating type/control: driver on-board or accompanying, remote control, automatic control)

6.2.3 Supply and disposal parameters, building systems

Numerous supply and disposal systems are required in industrial buildings. In principle, a distinction can be made between systems that are integrated in the building (building systems) and plant-related supply and disposal systems. Plant-related supply and disposal systems are influenced by the type of use and the connection needs of the equipment. (IREGIA 2004)

A part of these systems serve both the building-related technical facilities and plant-related supply and disposal (e.g. power and water supply). What needs to be determined is which utilities should be supplied or disposed of via distribution networks.

The following overview (cf. fig. 6.6) includes important supply and disposal systems for industrial buildings organized according to building systems, general systems and production plant-related systems.

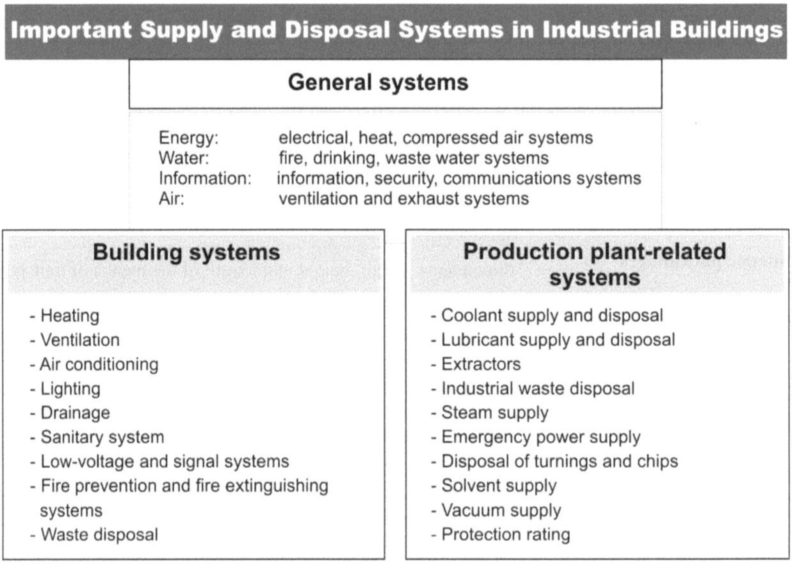

Fig. 6.6 Important supply and disposal parameters (building systems)

The distribution networks connect the production resources with the central/decentralized supply and disposal systems. The way in which they are arranged conforms to the locations of the corresponding production resources and systems (cf. fig. 6.7).

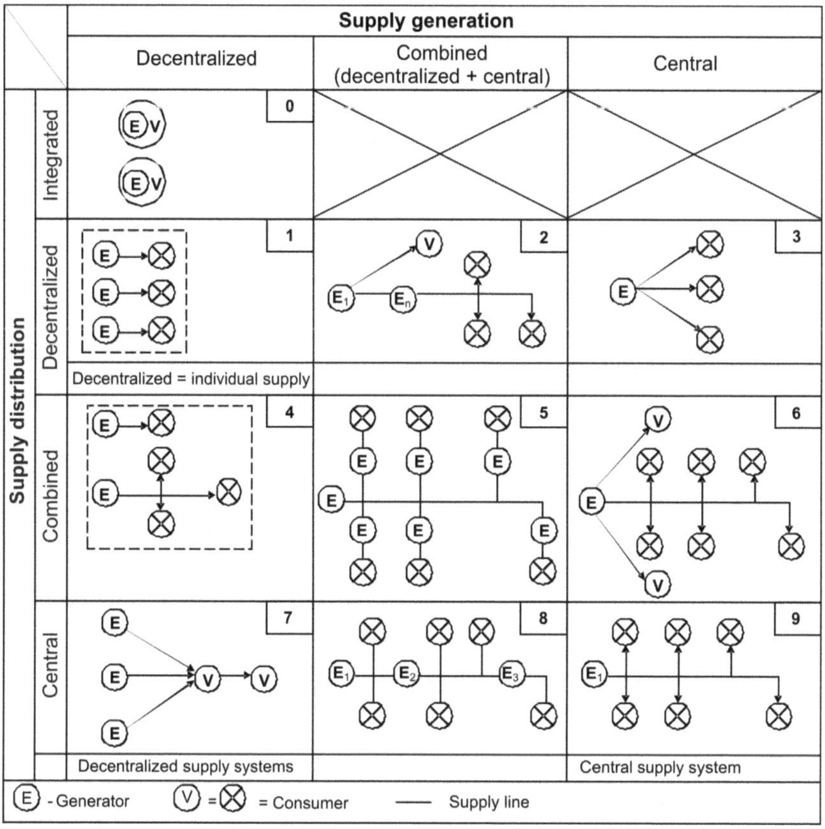

Fig. 6.7 Supply principles

From the point of view of mobility it is recommended to develop main distribution networks and connecting lines combined with creating supply sections. One the one hand, this enables a partial shut-down in case of breakdowns or the execution of repair work. On the other hand, it is possible to expand the distribution network according to requirements and to connect additional production resources or to bolster their supply by laying only short connection lines. In principle, there are the following possibilities for installing distribution networks (cf. fig. 6.8):

- underneath the working level (connection from below),
- on the working level (as base or threshold line),
- above the working level (connection from above).

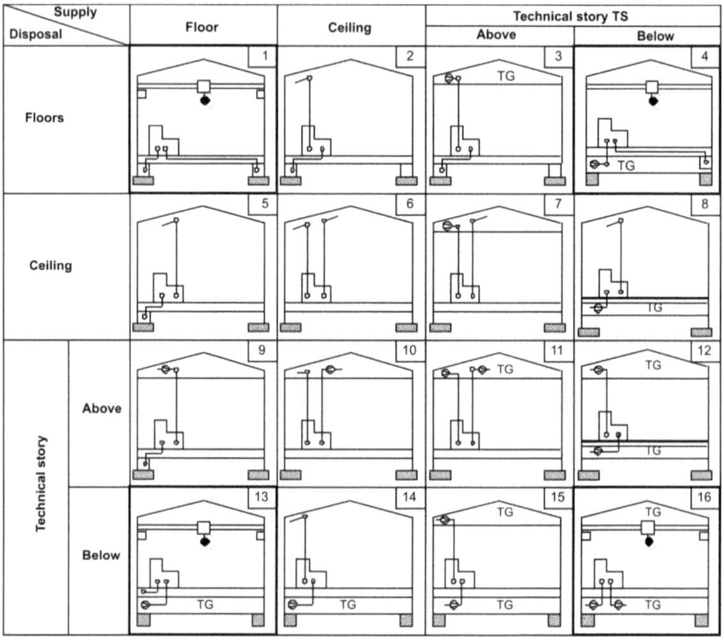

Fig. 6.8 Installation of supply and disposal lines (Helbing 2007)

The essential advantages and disadvantages of the different types of installation are depicted in fig. 6.9 and table 6.12

Installation / Criterion	Relationship to the working level: Under-floor (Useful for basement and multi-story buildings)	Floor	Over-floor (Only of limited use with overhead cranes/slewing cranes)
Flexibility with respect to changes	□	◩	■
Freedom from encumbrance of transport area	■	□	◩
Accessibility (possibility for repair and inspection)	□	■	◩
Space requirements (potential effective area)	□	■	□
Construction effort	■	◩	◩
Usability for the various utility types	Limited	Not limited	Limited

Key: □ Low ◩ Medium ■ High

Fig. 6.9 Approach for assessing installation methods

Table 6.12 Advantages and disadvantages of the distribution of power connections from above and below

Supply and Disposal Systems	
Connection from above	**Connection from below**
Advantages – Good inspection and repair possibilities – Easy adaptation in case of retrospective changes – Possibility for use of flexible connection systems (electricity, compressed air, etc.) – Possibility for use of combined, half-height distribution systems	*Advantages* – No disruptive influence on the operation – Advantageous in case of basement and multi-story buildings – Natural incline possible – For single supply (only electricity, compressed air, water, vacuum), floor ducts must be used
Disadvantages – Not always possible with overhead cranes – Only possible to a limited extent with slewing cranes – Can be disruptive for particular operating functions – Can be disruptive in the case of intensive fork lift operation – Visually disruptive in the case of thick conduits – Risk of damage	*Disadvantages* – Numerous openings required in the ceiling – Difficult in case of retrospective changes (less flexible) – Requires costly, accessible under-floor channels, if no basement – Costly assembly type – Precise coordination when planning

6.2.4 Interference parameters

When designing variants for maintaining flexibility on the production level, it is necessary to consider the interference parameters that arise during the manufacture of products in connection with the technologies used, and which have a disruptive effect on people and goods.

Interference parameters can be roughly grouped into:

- emissions (e.g. heat, radiation, moisture, vapors, dust, odors, noise),
- climate (e.g. temperature, air pressure, clean room climate),
- fire protection (e.g. fire hazards, explosion protection) and
- vibrations (vertical, horizontal, rotational).

There are extensive technical instructions, regulations, ordinances and laws that dictate the handling of interference effects that primarily affect people and secondarily goods. These are subject to constant change and (re)specification.

Here the focus is on the effects of noise, air, climate (temperature), lighting, color and clean room technology. Interference effects arise in the workplace and have an impact on their surroundings. They influence the working environment.

Interference avoidance strategies – There are two strategies for reducing and eliminating interference.

1st approach: elimination of interference through substitution of technological methods, processes and machines (e.g. laser welding instead of gas or electric welding).

2nd approach: reduction of interference through spatial encapsulation of technological methods, processes and machines (e.g. noise during grinding process by encapsulating the cause of the noise or all of the equipment).

The following principle applies: if the root of the interference cannot be eliminated through substitution, then spatial encapsulation is necessary. Avoidance strategies for interference effects: are to modify, avoid, encapsulate or neutralize.

The reduction of exposure to interference can be achieved through suitable layout design by combining workstations (means of production) with the same or similar interference effect parameters in the same place (e.g. machines that generate dust during grinding, noise during pressing, vibrations when striking).

Fig. 6.10 shows the structuring principles for various cases of interference elimination or prevention using avoidance strategies.

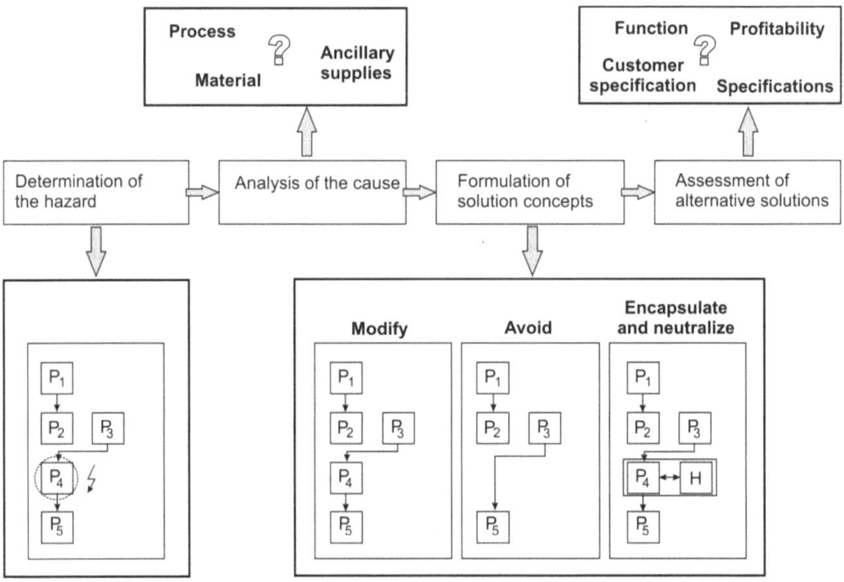

Fig. 6.10 Avoidance strategies (manufacturing processes)

These avoidance strategies must be taken into account as early as when designing the layout, since they affect flexibility and mobility.

NB: General design principles

(1) Design layout of the structural units (workstations – areas) and their connecting elements (logistics facilities) so that the production process is more economical and less prone to breakdown (VDI 2325). Here the spatial circumstances must be considered and selected taking the avoidance strategies into account.

(2) Place machines and systems together that have the same and similar interference factors with respect to load parameters, noise parameters, radiation, dust and vapors, odor, temperature and climate, air pressure, noise, fire hazards and explosion protection. Combined solutions can also be targeted here.

(3) Direct elimination of interference factors by changing technologies and materials (process and material substitution) as well as through organizational flows

(4) Indirect elimination of interference factors through encapsulation or separate housing of sources of interference or whole workstations and areas in connection with measures for protecting people and things (e.g. hearing, dust, gas protection)

(5) Clean rooms must be provided for products and processes with a particularly strong need to be kept free from dust, and to have constant temperature and climate conditions. The guideline values of the European "Clean room technology" standard ENV 1631 ff must be adhered to for this.

(6) For the incorporation of technologies and equipment with high quality requirements, containers work well; these are installed as black boxes with the corresponding connection points (interfaces to the outside) as closed systems in existing rooms. Installing container solutions offers new possibilities for reducing interference and, at the same time functions as a production-related platform.

The following specifications for the working environment are based on Unger, 2008. Significant interference parameters such as noise, air, climate, lighting, color design and clean rooms are described below in outline, together with references to the current specific literature.

6.2.4.1 Noise

The impact on a person of noise depends on:

- noise factors (VDI 2058, Bl. 1),
- the person's current activity,
- the person's state of health.

The health consequences due to noise exposure:

- aural impact (affecting the ear) = hearing impairment,
- extra-aural impact (affecting the human organism).

The noise level must be kept as low as possible in work rooms. The level that exists and must be assessed in the workplace, also called the "noise rating level," may not exceed the maximum values.

There are three possibilities for design solutions to avoid noise:

Encapsulation of the source of the noise

On the one hand it is possible to encapsulate individual sources of noise (machines or plant); on the other hand, under certain circumstances it is possible to reduce the specific effort required for encapsulation through a spatial concentration of individual sources of noise (e.g. special rooms that have been soundproofed accordingly). When encapsulating sources of noise using appropriate structural measures, attention must be paid to ensuring that functionality is not impeded (loading, checking, maintenance, heat removal, etc.). If it is not possible to encapsulate the plant or equipment completely, then noise screens, dividing walls, etc. can be used.

Noise protection cabins

These cabins are special rooms, e.g. for the personnel in control rooms, control centers and measuring stations. They provide suitable protection for specific requirements.

Noise-absorbing room lining

Noise reflection can be reduced with the help of noise-absorbing linings.

Legal principles

1) Directive 2003/10/EC (2/6/2003)
Daily noise exposure level LEX, 8h

Exposure limit:	LEX,8h = 87 dB(A) npeak = 140 dB(C) or ppeak = 200 Pa →
Upper trigger level:	LEX,8h = 85 dB(A) npeak = 137 dB(C) or ppeak = 140 Pa
Lower trigger level:	LEX,8h = 80 dB(A) npeak = 135 dB(C) or ppeak = 112 Pa →

2) German federal immission protection law
3) Technical instruction on noise (TA Lärm)
4) German workplace ordinance (BGBl I 1983, p. 1057)
5) German accident prevention regulation (UVV) "Noise" (BGV B3)
6) German accident prevention regulation (UVV) "Occupational health precautions" (BGV A4)
7) Administrative guidelines of the German federal states

Table 6.13 shows the permissible immission noise rating level in workplaces. The following guideline values apply for construction sites.

Table 6.13 Immission guideline values for various construction areas according to TA Lärm and VDI 2058

Construction site (building use ordinance)	Immission guide value N_g in dB(A)	
	Day (6am-10pm)	Night (10pm-6am)[1)]
Industrial area	70	70
Commercial area	65	50
Business zone, mixed area, village	60	45
General residential, suburban area	55	40
Purely residential area	50	35
Spa area, hospital, nursing home	45	35
Noise transmission within buildings (**measurement in room** - closed windows and doors)	40 (TA Lärm) 35 (VDI 2058)	30 (TA Lärm) 25 (VDI 2058)

N_g: Rating level (equal to average level L_{eq}) across the assessment period T_A (e.g. 16 hours for the day and 8 hours for the night or the loudest night hour).
Brief noise peaks day ≤ 30 dB(A), night ≤ 20 dB(A) above guideline value are permissible.
Measurement generally 0.5 m in front of the open window (see VDI 2058 Bl.1)

Three measures are available for noise minimization (fig. 6.11).

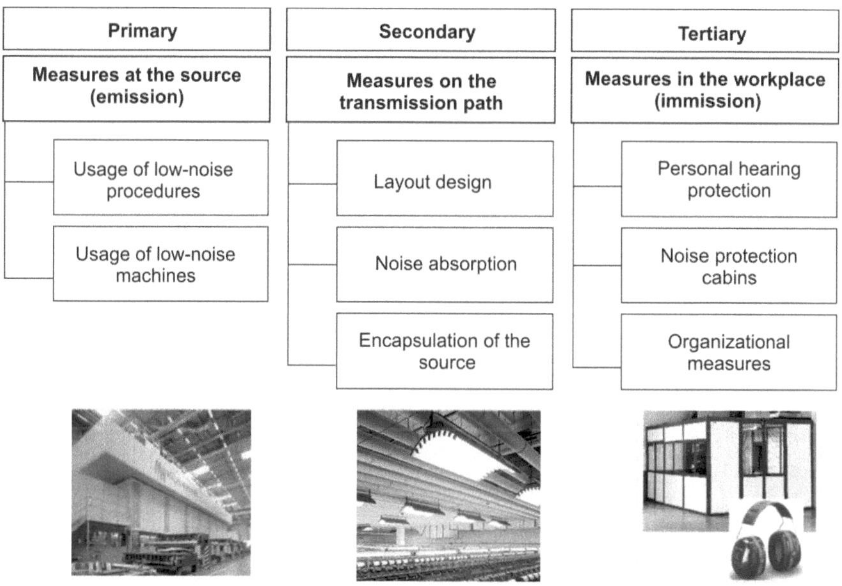

Fig. 6.11 Overview of the design levels for noise minimization measures

6.2.4.2 Air pollution

Fig. 6.12 depicts the classification of air pollution .

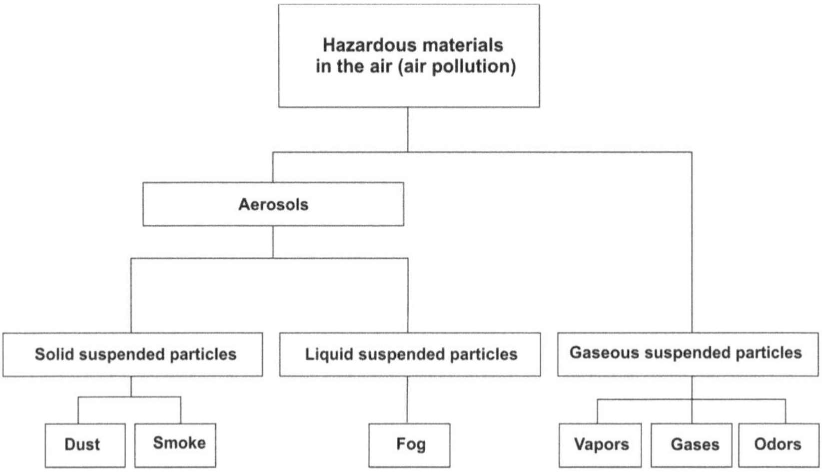

Fig. 6.12 Classification of types of air pollution

Dust is hazardous to human health. It is classified according to its respirability. Fig. 6.13 shows these relationships.

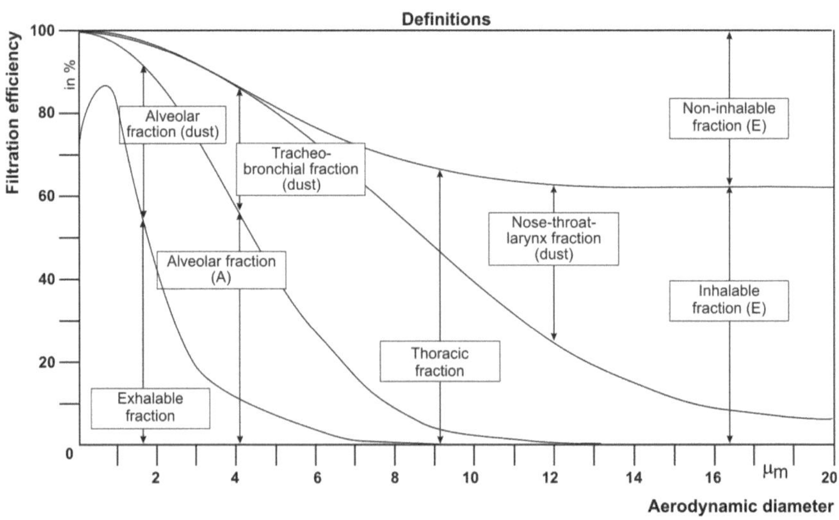

Fig. 6.13 Dust contents and dust-related specifications depending on aerodynamic diameter according to the filter model (Deutsche Forschungsgemeinschaft MAK and BAT values list example)

Air pollution is assessed using different values (MAK, TRK, BAT, EKA, MIK and MEK). Fig. 6.14 explains one system.

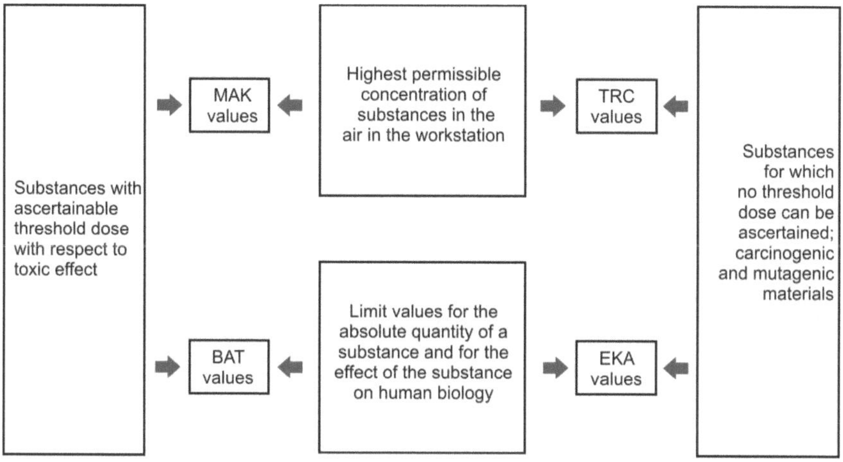

Fig. 6.14 System of limit values

MAK value (maximum allowable concentration value):

This is the highest permissible concentration of a substance in the air in the workplace, which according to the current state of knowledge does not generally endanger the health of employees and does not disproportionately trouble them even with repeated and long-term exposure, of generally 8 hours a day within the limits of a 40-hour working week.

TRC value (technical reference concentration):

This is the concentration in the air in the form of gas, vapors or suspended particles that, according to the state of the art, can be reached and must be used to give an indication for the protective measures to be taken and for measurement-related monitoring in the workplace.

BAT value (biological tolerance value):

This is the highest permissible quantity of a material or metabolized product or the deviation triggered by it of a biological indicator from its standard which, according to the current state of scientific knowledge generally does not endanger the health of the employee even if it is reached regularly due to influences in the workplace (in the case of exposure of 8 hours a day or 40 hours a week).

EKA value (exposure equivalent for carcinogenic substances):

The EKA value specifies how high the material or metabolite concentration is in biological material (e.g. blood, urine) given a particular concentration in the air in the workplace. From this it is possible to determine what the internal impact would be solely through absorption of the material by inhalation.

It is only specified for carcinogenic substances. Carcinogenic substances are not documented with BAT values since at present it is not possible to specify a permissible limit value whereby exposure can be considered harmless.

MIK value (maximum immission concentration):

Limit value given an environmental exposure time of 24 hours that according to current knowledge is generally considered harmless.

MEK value (maximum emission concentration):

Limit value for emissions, e.g. from chimneys and exhaust systems of combustion engines.

Trigger level:

This is the concentration of a substance in the air in the workplace or in the body, which if exceeded requires additional health protection measures to be taken.

6.2.4.3 Climate

Figure 6.15 shows the influence of the climate on human beings.

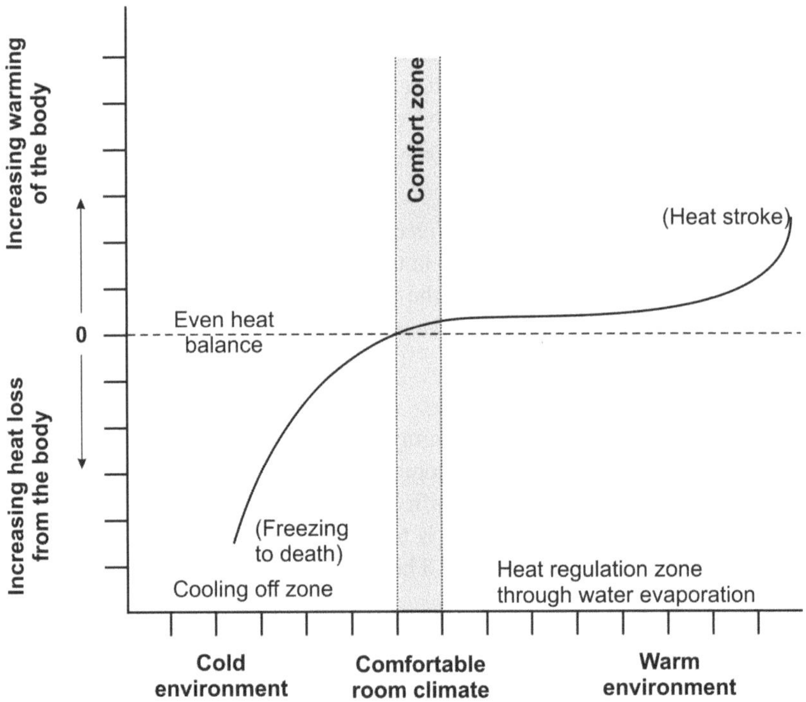

Fig. 6.15 Heat balance of the body under various climatic conditions according to GRANDJEAN

An excerpt from the German workplace regulations (ASR 6) for room temperatures provides information about normal interference conditions for temperature.

Table 6.14 Excerpt from the German workplace guidelines (ASR 6) for room temperatures and ASR 5 - Ventilation

2.1	In work rooms, the minimum temperature must be	
	a) for activities that primarily involve sitting	19°C
	b) for activities that primarily involve standing	17°C
	c) for heavy physical activity	12°C
	d) in office spaces	20°C
	e) in salesrooms	19°C
2.2	The minimum temperatures should have been reached when work begins.	
2.3	The room temperatures may be below those specified according to b, c and e if lower temperatures are necessary for operational reasons. For seated activities in salesrooms, e.g. at cash registers, it may be necessary to set the minimum temperature according to no. 2.1 e at a higher level.	
2.4	The room temperature in workrooms should not exceed 26°C; workrooms with heated workstations are excepted from this.	

Excerpt from ASR 5, Ventilation

4.2.2 Room air speeds

The ventilation systems must be designed so that there are no unreasonable drafts in the workstations. Drafts depend primarily on the temperature of the air, the air speed and the type of activity (e.g. heat generation due to physical work). Up to a temperature of 20°C, with an air speed of less than 0.2 m/s, there are normally no drafts.

4.2.3 Humidity

The relative humidity should not exceed the following values:

Air temperature °C	Relative humidity %
20	80
22	70
24	62
26	55

6.2.4.4 Lighting

Lighting must be in accordance with the type of visual task. The strength of the general lighting must be at least 15 Lux [lx] (§ 7 ArbStättV). In constantly occupied workstations in interior rooms, a nominal illumination level of at least 200 lx must be provided (DIN 5035).

Depending on the type of the interior room or the type of the outdoor workplace, there are minimum guideline values on which the planning and assessment of a lighting system should be based. These are listed in DIN 5035, Part 2 (for excerpt, see table 6.15).

Table 6.15 Guideline values for nominal illumination levels for interior rooms according to DIN 5035 (selection)

Type of interior room or activity	Nominal illumination level [lx]
Storage room where it is necessary to search for non-identical items	100
Welding	300
Precision assembly in metalworking and metal processing	500
Inspection, automobile construction	750
Assembly of precision parts, electronic components	1500

In the course of layout planning, the following must be taken into account:

- It is advantageous to align the workplaces according to the daylight.
- Artificial light sources must be set out so that with equal amounts of daylight and artificial light, no areas of the workplace can be under-lit ("twilight").

Fig. 6.16 shows the effect of light on human beings.

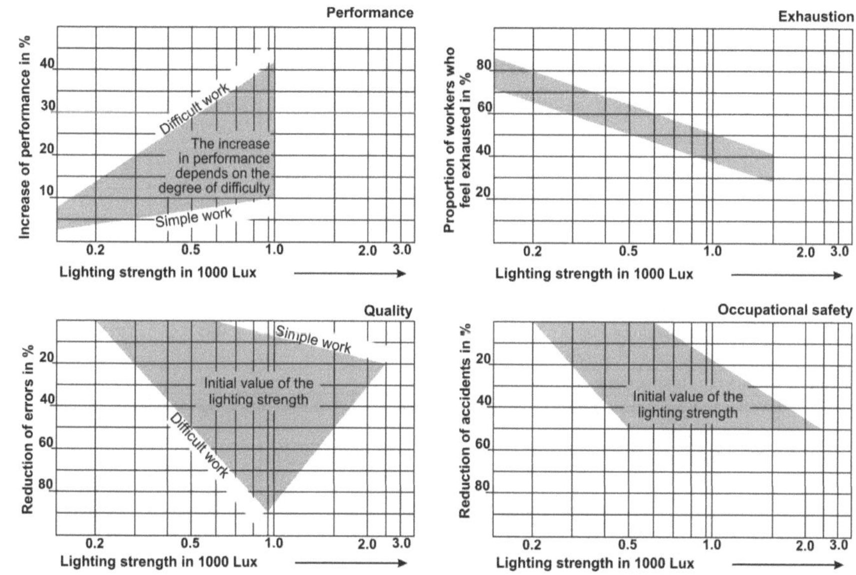

Fig. 6.16 The positive effect of light (Fördergemeinschaft Gutes Licht (Association for the Promotion of Good Lighting): Good lighting for commerce, trade and industry)

The standard lighting distribution as shown in fig. 6.17 clarifies the lighting design options.

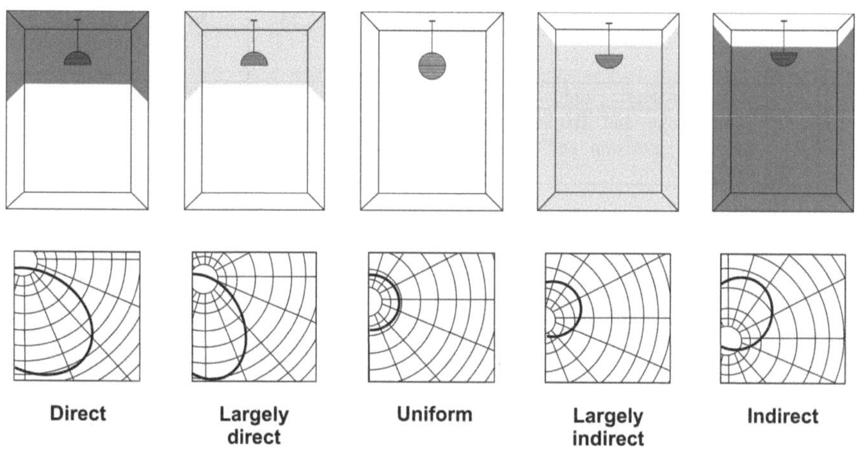

Fig. 6.17 Standard light distribution curves

The maintenance value of the lighting strength $\bar{E}_m$ is decisive for interior rooms. It is depicted in table 6.16 for metalworking (excerpt from DIN EN 12 464 – 1).

Table 6.16 Lighting strength maintenance value

1	2	3	4	5	6
2.13 Metalworking and processing					
Ref. no.	Type of room, task or activity	Maintenance value lighting strength $\bar{E}_m$ [lx]	UGR limit value UGR_L	Color rendering index R_a	Comments
2.13.1	Open die forging	200	25	60	
2.13.2	Drop forging	300	25	60	
2.13.3	Welding				
2.13.4	Rough and medium machine work: *tolerance* $\geq$*0.1 mm*	300	22	60	
2.13.5	Precision machine work, grinding: *tolerance* < *0.1 mm*	500	19	60	
2.13.6	Marking, inspection	750	19	60	
2.13.7	Wire and pipe drawing, cold forming	300	25	60	
2.13.8	Working of heavy metal sheets: *thickness* $\geq$ *5 mm*	200	25	60	
2.13.9	Working of light metal sheets: *thickness* < *5 mm*	300	22	60	
2.13.10	Manufacture of tools and cutting tools	750	19	60	
2.13.11	Assembly work: rough medium-fine fine very fine	 200 300 500 750	 25 25 22 19	 80 80 80 80	Assembly work: For tall production buildings, see 4.6.2
2.13.12	Galvanizing	300	25	80	For tall production buildings, see 4.6.2
2.13.13	Surface processing and painting	750	25	80	
2.13.14	Tool, gauge and fixture construction, precision and micromechanics	1000	19	80	

NB: Use energy-saving bulbs!

6.2.4.5 Color design

Color design is not just an important component when constructing new production facilities; it is also very important for the design of existing production facilities. Here the psychological effects (according to table 6.17 and table 6.18), color effects in the space and the ergonomic color design (table 6.19) are important.

Table 6.17 Psychological effect of colors

Color range	Effects
Yellow	sunny, bright, light, fresh, warm
Orange	sunny, stands out, dry, warm
Orange-red	fiery, bright, strong, loud
Purple	powerful, compact, crushing, solemn, strong, dignified
Violet	sultry, heavy, dark, dim, dull, oppressive
Blue	cool, fresh, shady, reticent, reinforcing, solidifying
Turquoise	languid, distant, watery, cold, reticent, icy
Green	quiet, natural, limiting, safe, balanced, simple
White	cool, bright, light, hygienic, pure
Black	dark, heavy, crushing
Beige	bright, warming, dry, light
Pink	fragrant, tender, weak, sweet
Brown	crude, heavy, earthy, aging, solidifying
Dark blue	dark, heavy, threatening
Dark green	restrained, quiet, oppressive
Light blue	airy, watery, wide, cool
Light green	neutral, sophisticated, distinguished
Dark gray	grim, threatening, dreary

Table 6.18 Psychological effect of safety and order colors

Red (RAL 3000)	Requires (instinctive) action. It has the greatest stimulus value, creates a mood of alarm and is especially emotionally loaded.
Yellow (RAL 1004)	Directs the attention as towards a light source. Together with black, it warns (cf. warning colors in the animal kingdom!). Primarily emotional.
Green (RAL 6001)	Offers the idea of refuge and security, indicates - unlike cyan – an absence of tension. Characterized as emotional and rational. Attention-grabbing when used on an illuminated sign.
Blue (RAL 5010)	Makes aware, encourages consideration and decision-making. Rationalizes our actions (permission, but also a lack of permission) and is especially rational. Used to raise the alarm when in the form of a flashing illuminated sign!
With this in mind:	Red: Alarm color - prohibition Yellow: Warning color - warning Green: Safety colors - information Blue: Order colors - command

The information reproduced in table 6.19 applies correspondingly according to DIN 4818 for ergonomic color design.

Table 6.19 Use and meaning of safety and order colors

Red (RAL 3000)	for immediate danger, stop, prohibitions, emergency switches (cf. traffic signs and pictograms with red slash across for no smoking, no entrance, no parking, not for drinking, end of the driving area, etc.). *Contrast color:* white *Shapes:* circles and slashes or striped patterns as for barrier tape if simultaneously warning, also triangles (e.g. in the case of traffic signs).
Yellow (RAL 1004)	for caution, warning about hidden dangers, collisions, falls and crushing Illustrated pictographically by, for example: caution danger of slipping, machine in motion, caution fork lift traffic, caution danger of explosion, caution suspended load; radioactivity, etc. Warnings signs with printed explanations for accident prevention. *Contrast color:* black *Shapes:* triangle Cf. also safety markings in radiation protection according to DIN 25 430, safety markings for electrotechnology according to DIN 40 003, Parts 1-6.
Green (RAL 6001)	for safety and rescue. Identification of emergency exits (also with green light over the door). Pictographic indications of escape routes, emergency ladders, exits. Rescue and assistance if the white cross or staff of Asclepius for medical doctors is added. (cf. also DIN 4844) *Contrast color:* white *Shapes:* square
Blue (RAL 5010)	for safety-related commands, operational orders, permits, driving direction indications. Also with pictograms for wheelchair users (white on blue). Recommendations: wash hands, wear head protection, wear mask, wear hearing protection, etc. permissions to smoke, park. *Contrast color:* white *Shapes:* rectangle

6.2.4.6 Clean rooms and clean room technology

Basic principles

Clean room technology is necessary for creating spatial product creation conditions with low interference from dust and climate and the minimization of contamination.

Areas of application

The development of clean room technology can be traced to more demanding production technologies in industry as well as sharply-increased precision and cleanliness requirements in recent decades in the areas of pharmaceutical, medical and precision engineering, food production and especially semiconductor production technology. Particular manufacturing processes can no longer be utilized in the normal environment since the contamination from the environment reduces the production quality significantly or even makes production impossible. In microelectronics, with miniaturization we have entered an area in which particles with dimensions of 0.3 μm and smaller can result in products being rejected.

Based on tasks of this nature, production in clean rooms has prevailed in industry for the purpose of minimizing contamination due to external influences. The aim is to keep products in as clean an environment as possible throughout the entire production cycle. In order not to compromise the environmental conditions, the production systems to be installed in these clean rooms must also satisfy very high cleanliness requirements.

For products in the industries listed as examples below, environmental purity has an influence on product quality:

- electronics industry (microcomponents, mask production)
- pharmaceutical products (blood substitutes, blood filters)
- food industry (germ-sensitive foods)
- film industry (coating of films)
- plastics industry (films)
- precision engineering (miniature ball bearings, precision scales)
- etc.

The environmental cleanliness requirements of the production tasks listed vary in their extent. Therefore, for cost reasons, efforts should be made to select the clean room class (cf. fig. 6.18) that is suitable for manufacturing the specific products. The costs of investment and the operation of the facilities are significant. For example, a 1,000 m² class 10 clean room requires investment costs of approx. € 10 million and annual operating costs of € 1.5 million.

Today, the greatest demands are posed by semiconductor production technology. The installation of clean rooms in existing buildings demands that the specific aspects listed below be taken into consideration. They can be achieved through spatial encapsulation and clean room containers for various cleanliness classes.

Cleanliness classes

Clean room technology is concerned with the design of systems for the manufacture of products whose production poses special requirements for environmental air that is extremely low in particles. Even very small quantities of the smallest airborne particles cause high levels of loss in semiconductor production or precision assembly, for example, or they increase the risk of infection in medical products. Clean room technology combines the areas of air and climate technology as well as filtration, particle measurement and systems engineering.

The degree of cleanliness is quantified in the standards through particle distribution. The US Federal Standard 209 has prevailed as a standard worldwide. Currently, FS 209 D is valid. Other countries have developed their own standards based on this American standard. In Germany, VDI guideline 2083 applies. It is currently being reworked and is available as preliminary standard DIN V ENV 1631.

The diagram in fig. 6.18 shows the subdivision of clean room classes according to US Federal Standard 209 D and VDI guideline 2083. For specific applications in the pharmaceutical, medical and biotechnology industries, there are additional provisions with respect to the biological purity of the air.

Clean and gray room

A clean room is a confined area that fulfills the specification of a defined degree of air cleanliness; in the case of the strictest requirements, particles larger than 0.1 μm are significant.

In the production system, the gray room is a generally uncontrolled area from the point of view of the clean room; the supply systems for process systems are integrated here.

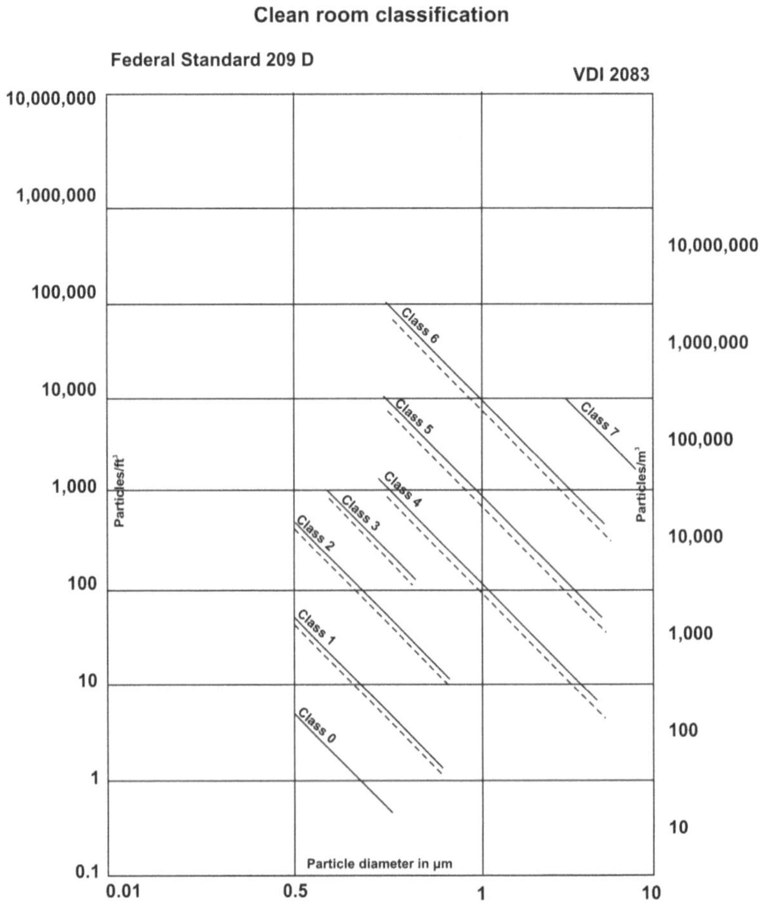

Fig. 6.18 Scope of cleanliness classes according to US Federal Standard 209 D and VDI guideline 2083

The design solutions comply with the requirements and development status of particle contamination minimization. Fig. 6.19 presents a summary of the relationship structure.

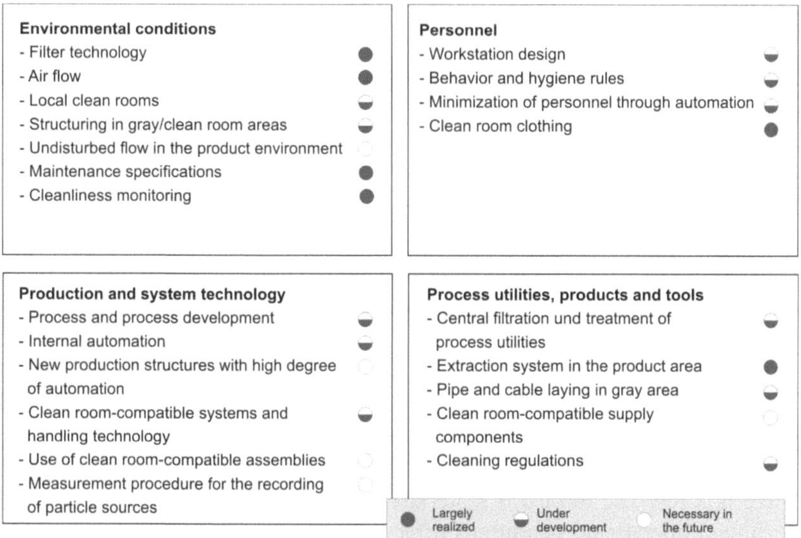

Fig. 6.19 Development status of particle contamination minimization (Geißinger 1989, p. 25)

The factors influencing particle contamination in the production process that cause a reduction in product quality can be seen in fig. 6.20.

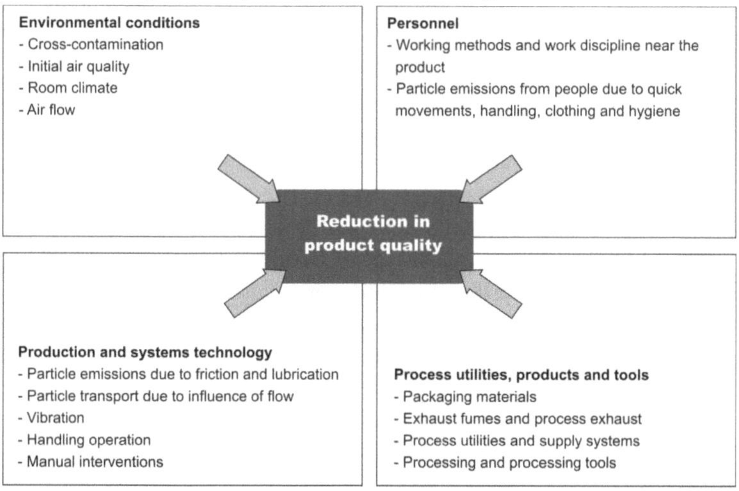

Fig. 6.20 Factors influencing particle contamination that cause a reduction in product quality (Geißinger 1989, p. 26)

It is estimated that the percentage share of contamination sources in clean room production is 25% for the production systems, 15% for the supply of utilities, 25% for the production process and 35% for people.

6.2.4.7 Fire and Health Protection, Occupational Safety, Environmental and Safety Protection

(1) Fire protection

Fire protection plays a significant role in the design of industrial buildings. However it is difficult to make general statements about fire protection in buildings since this is subject to countless influencing factors. From the point of view of usage requirements, the possible fire hazards must be classified according to the following hazard risks:

- low fire hazard
- medium fire hazard
- high fire hazard

There is a low fire hazard if materials with low flammability are present and the local and operational conditions offer only slim possibilities for a fire to take hold and if in case of a fire, little spread of the fire is anticipated.

There is a medium fire hazard if highly-flammable materials are present and the local and operational conditions are favorable to fire, however no great spread of a fire in the initial phase is anticipated.

There is a high fire hazard if, due to the presence of highly-flammable materials and the local and operational conditions, there is significant potential for fire and a fire would spread quickly in the initial phase. There is also a high fire hazard if it is not possible to classify the risk as medium or low.

The fire protection goals are directed primarily towards the protection of personnel and, increasingly, property. The German industrial guidelines (IndBauR) focus among other things on the danger of fire, fire detection and sealing off a fire. This is regulated by DIN 18230 "Smoke and heat extraction systems in industrial buildings" and DIN 4102 "Fire behavior of building materials and building components." In planning, fire protection simulation calculations are carried out using zone models (approximate) or field models (detailed). Fire safety and fire protection classes plus safety categories for fire compartments are regulated on the basis of fire resistance classes according to DIN 4102.

Fire risk categories apply for selected production areas as per table 6.20.

Table 6.20 Examples for the classification of fire hazards

Fire hazard		
low	medium	high
Brickyards, cement plants, manufacture of glass and ceramics, paper manufacture in wet areas, canning factories, manufacture of electronic items and devices, breweries, beverage manufacture and filling, steel manufacture, machine construction, nurseries, electro-plating shops, mechanical metal processing, turning shops, milling shops, drilling shops, punching shops,	Bakeries, leather and plastic processing, manufacture of rubber goods, plastic injection-molding shops, paperboard factories, assembly of vehicles/household goods, construction sites without work involving fire, metalworking shops, vulcanization, leather/artificial leather and textile processing, back operations, electrical workshops	Furniture manufacture, chipboard manufacture, weaving and spinning mills, manufacture and processing of paper in dry areas, grain and feed mills, construction sites with work involving fire, foam production, roofing material production, processing of paints and glues (flammable), paint and powder coating plants, refining, oil hardening shops, printing shops, petrochemical plants, processing of flammable chemicals, vehicle workshops, carpenters' and joiners' workshops, upholstery workshops

(2) Health protection

Precautions and radiation, breathing, infection and accidental contact protection

(3) Occupational safety

Accident, collision, contact, and motion protection, protective guards

(4) Environmental protection

Noise, aerosol, vibration, air, floor, water, emission and immission protection, conservation

(5) Safety protection

Natural disaster, break-in, radiation, explosion protection

6.3 Selected FTT, Supply and Disposal Facilities

6.3.1 Jigs & Fixtures, Tools and Testing Equipment flow facilities

Jigs and fixtures, tools and testing equipment (FTT) are resources that are required in order to carry out the manufacturing process. FTT are all individual flow objects with their own flow systems. The FTT flow systems have the task of securing the optional and timely availability of all required FTTs in the correct location in the production areas. The FTT flow is part of operational FTT management. (Wirth 2000)

The process of FTT supply consists of the following sub-tasks (technologies) that need to be carried out:

- preparation (set-up, pre-installation, adjustment, commissioning),
- storage, provision,
- transportation, transfer, handling,
- post-processing (dismantling, decommissioning, checking),
- maintenance (repair, monitoring, servicing).

Fig. 6.21 shows the chain of principle functions to be worked through.

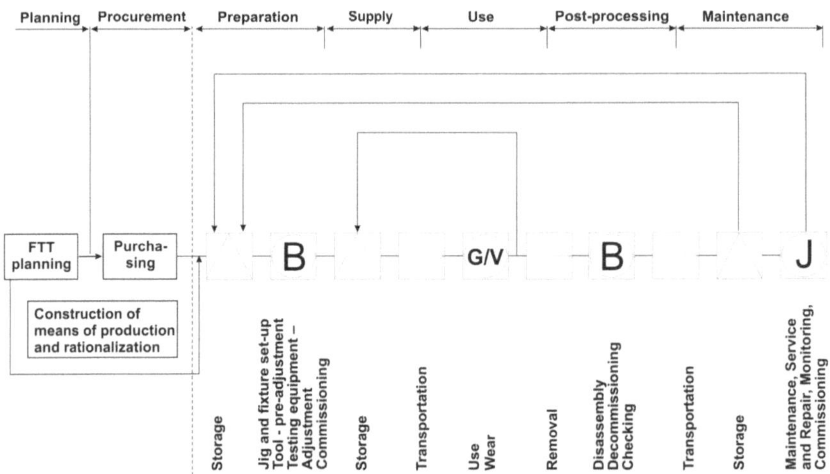

Fig. 6.21 FTT flow structure

The FTT provided for by planning via purchasing and in-house production arrive in a central or peripheral warehouse in which the FTT are prepared for production.

Jigs and fixtures are assembled for production according to order (modular systems) or produced and provided specially (special fixtures). They are used for fixing the position and clamping of workpieces for transport and machining. As a general rule, jigs and fixtures are required for one or more workpieces (cf. workpiece flow).

Generally, the FTTs are transported and stored using special pallets or working storage in conjunction with the workpiece flow (figure 6.22).

Fig. 6.22 Mode of operation for the flow of jigs and fixtures

The tools required for each work stage must be assembled (fig. 6.23) and preset to the parameters of the manufacturing equipment in relation to the task.

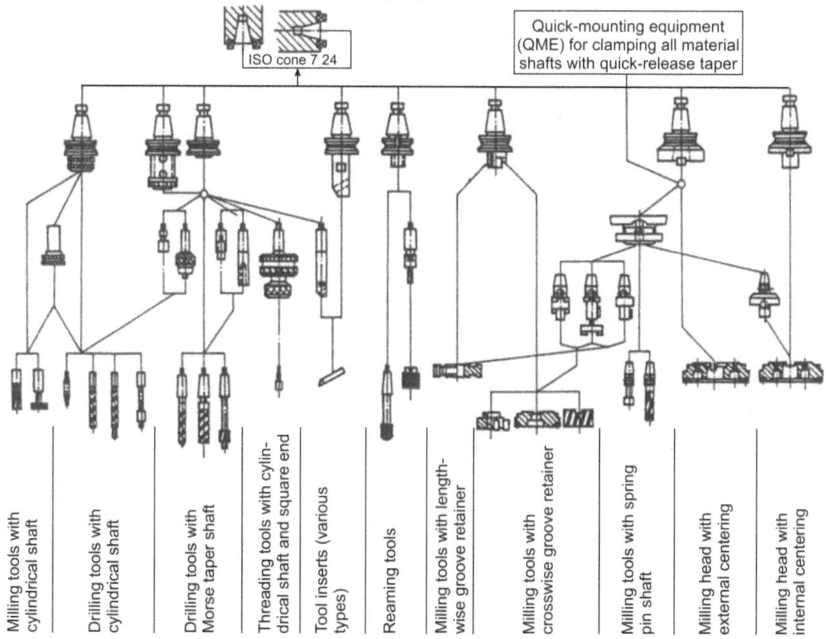

Fig. 6.23 Arrangement of tools for automatic tool change

There are three possibilities for setting up the tools:

- manual set-up according to settings lists,
- computer-aided set-up (offline operation) - the setting values are saved on the computer and transferred physically via data storage as a data set,
- computer-integrated set-up (on-line operation) – the setting values are transferred via direct data transfer to the NC machine (paperless variant).

NC presetters are a main component of a presetting station, which can be located either centrally or peripherally.

In the same way as the tools, the testing equipment must also be assembled and preset, i.e. adjusted. The adjusted testing equipment is made available in the same way as the jigs and fixtures via special or modular systems.

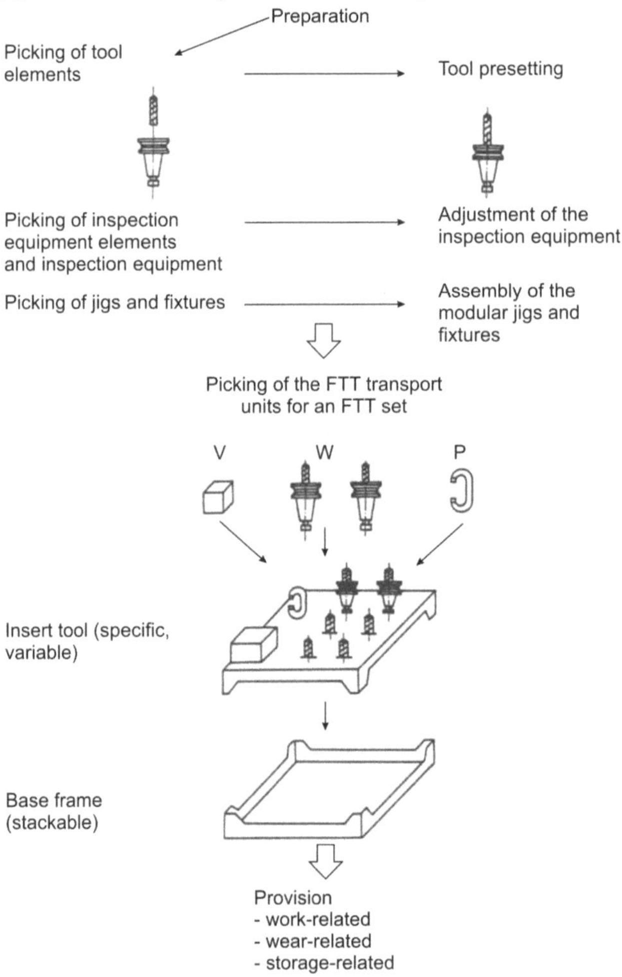

Fig. 6.24 Preparation, picking and supply steps

After prefabricating the jigs and fixtures and setting up the tools and inspection equipment, the FTT must be picked for the purposes of timely, order-compliant, transport and handling-ready, storage-appropriate provision for the actual manufacturing process. The picking of the FTT includes the task-oriented assembly of FTT sets into transport units (fig. 6.24). They are actually provided for manufacturing in transport and storage units.

Technical solutions

The FTT flow can then be arranged in a meaningful way in terms of technology, organization, control and finance, if

- there are strong relationships to the main process of the manufacturing systems,
- the technical and chronological utilization of cost-intensive workflow elements will be improved by the integration of FTT flow elements
- the automation of the processes of FTT provision, preparation, picking and scheduling including the FTT's transport, handling and storage process can only be realized through integration into the main process.

When determining the FTT flow element integration variants, it must be assumed that the specifics of the FTT (mass, dimensions, sensitivity) generally require a variety of flow systems and therefore offer heterogeneous conditions for integration. The four levels of integration depicted in figure 6.25 result from the significant technical and organizational integration conditions for the inclusion of the FTT operations in the main process.

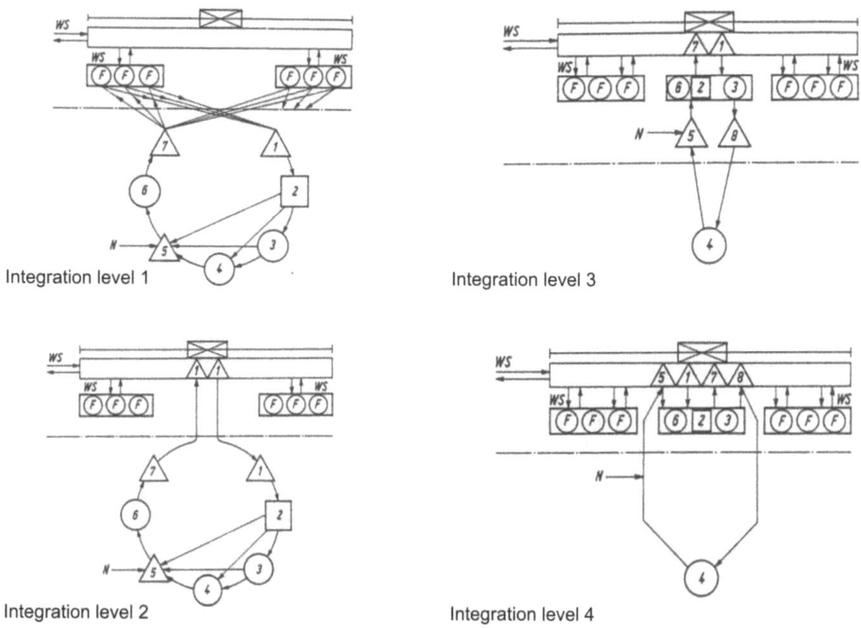

Fig. 6.25 FTT flow integration levels

Integration level 1	-	local connection of the FTT flow operations to the 3rd and 4th level manufacturing systems
Integration level 2	-	integration of FTT flow transport operations into the workpiece flow systems, especially the transport of the FTT to the production areas
Integration level 3	-	direct connection of the FTT flow operations to the internal central storage
Integration level 4	-	additional inclusion of the FTT storage in the central storage (complete merger of the peripheral "FTT" flow system with the workpiece flow systems)

Integration levels 3 and 4 are of particular significance for the design of manufacturing systems (manufacturing groups and area). In both cases, the FTT storage is located within the system limits ("on site"). Preparation and post-processing are integral parts of the system solutions. Maintenance is only carried out outside the system. For both integration levels, the principle of tool storage in the manufacturing workstation can be realized with the solutions depicted in fig. 6.26 via various storage and transport variants.

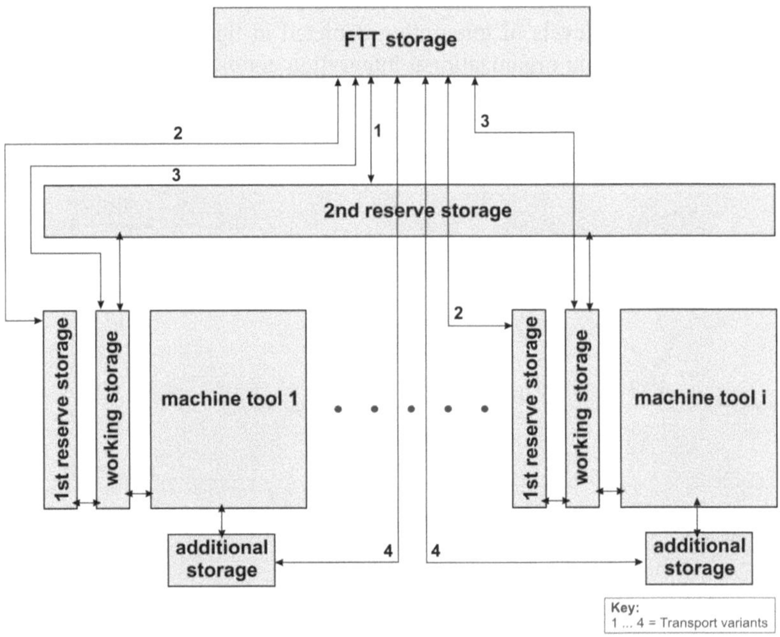

Fig. 6.26 Storage structures

Figure 6.27 shows practical possibilities for increasing tool storage capacity and the tool exchange principle for prismatic workpieces.

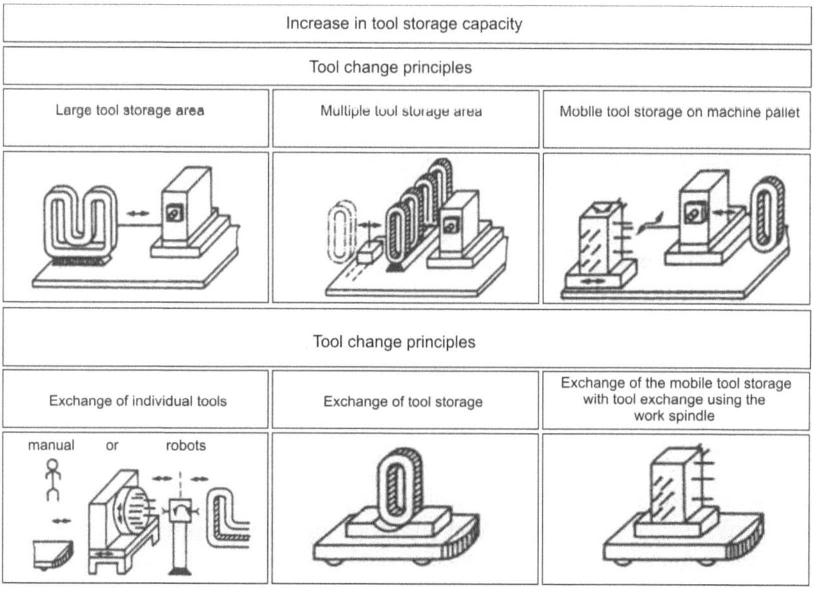

Fig. 6.27 Tool storage and tool flow in flexible machine systems (FMS) for prismatic workpieces

In summary, for the design of FTT flows, the following instructions should be taken into account:

- create self-contained FTT circuits;
- strive for integration of workpiece and FTT flows as well as of transport and handling. Generally for manufacturing systems for prismatic parts, the workpiece and tool flows are separate;
- implement product range-appropriate tool storage in the case of automatic tool flow and constant presence of tools needed for replacement including automatic monitoring and testing of tool wear on through to tool checking;
- guarantee independent tool exchange in case of order change or wear and tear;
- strive to maintain optimal stocks of FTT in peripheral storage to increase the ability to react in case of breakdowns in the production process;
- strive for sensible division of labor between peripheral and central preparation, post-processing and maintenance.

6.3.2 Supply and disposal systems

The supply and disposal systems required for product/material flow essentially include the flow systems (Wirth 2000). They are summarized in table 6.21:

Table 6.21 Supply and disposal systems

Electrical power (high-voltage, low-voltage)		Building systems
Compressed air		Building systems
Technical gases		Building systems
Heat energy (room heating)	Air conditioning	Building systems
Air (supply, exhaust air)	Air conditioning	Building systems
Water (drinking & process water – waste water)		Building systems
Ancillary manufacturing supplies (e.g. cooling lubricants)		
Waste - Production waste (e.g. waste turnings and chips) - "Municipal"-type solid waste		

As part of the detailed technical planning of the entire product/material flow, ready-to-use design and implementation of the supply and disposal technology are frequently commissioned from specialist suppliers based on the requirements that have been formulated (this applies to the building systems in particular).

By contrast, the detailed planning of the auxiliary manufacturing material supply and disposal and the removal of turnings and chips (for decentralized, machine-based systems) is usually done directly in connection with the manufacturing systems based on the performance program.

Particularly in the case of the integrated detailed planning of supply and disposal systems, a decisive contribution to environmental protection and preventative health protection can be made, among other things with respect to

- energy savings through optimization of the transmission equipment (minimization of line losses);
- avoidance and reduction of waste, among other things by optimizing the use of cooling lubricant (careful handling of the cooling lubricant) as well as waste recycling, especially for secondary raw materials such as waste turnings and chips (collection of a single type, drying of wet turnings/chips) and waste oils
- avoidance of emissions of substances that are hazardous to health and the environment (waste oils, acidic dips, alkaline solutions, dyes, paint, solvents, grinding dust etc.), gases and vapors (cooling lubricant, solvent vapors etc.) and similar, through avoidance of line and container leaks and incorporation of highly-effective separation technology (filters, settling tanks, etc.) in the supply and disposal systems.

Essentially:

preventative product and production-integrated environmental protection takes precedence over end of the pipe solutions. This way, environmental damage will not arise or will be kept to a minimum. Appropriate product developments and procedure and process selection and design are cheaper in the long run than remedial measures.

The operational supply and disposal systems are planned according to the following steps:

(1) formulation of the performance program,
(2) determination of functions,
(3) dimensioning,
(4) structuring and design.

Table 6.22 Operational supply and disposal flow systems

Flow system category	Flow groups	Flow objects	Supply/disposal processes
Material flow	Auxiliary manufacturing materials (AMM)	Emulsions, fluids, oils	AMM supply and disposal
	Waste products (solids, etc.)	Turnings / chips, core and sheet metal scrap	Turnings/chip disposal
	Industrial water	Drinking, process, waste water	Water supply and waste water treatment
	Exhaust products (gaseous)	Harmful gases, vapors, polluted exhaust air	Exhaust
Energy flow	Electrical energy	High voltage	High-voltage supply
	Heat energy	Hot water, water vapor, hot air Coolants, room air	Heating Cooling, ventilating, air conditioning
	Chemical energy	Technical gases	Technical gas supply
	Mechanical energy	Compressed air	Compressed air supply

A few main activities for formulating selected performance programs for supply and disposal technology are listed below.

6.3.2.1 Auxiliary manufacturing material flow, supply and disposal

Auxiliary materials are materials that are only used indirectly for the production of a part, an assembly or a product and can only be detected in insignificant quantities in the product (Fröhlich 2004).

Auxiliary manufacturing materials (AMM) are materials of all aggregate states whose function consists of and ends with supporting the manufacturing process.

In the production system, fluid AMMs are used primarily as emulsions and fluids (miscible with water) and grinding and cutting oils (not miscible with water). Their job is to cool, grease, protect against corrosion and to some extent to extract turnings/chips. They influence the manufacturing process technically and economically since the utilization of the manufacturing equipment, the useful life of tools, corrosion protection and occupational and health protection depend on their selection (composition of the AMM) and their care (purity, temperature, formation of bacteria, etc.). The process of supply and disposal of AMM for the manufacturing equipment must fulfill the sub-tasks shown in fig. 6.28.

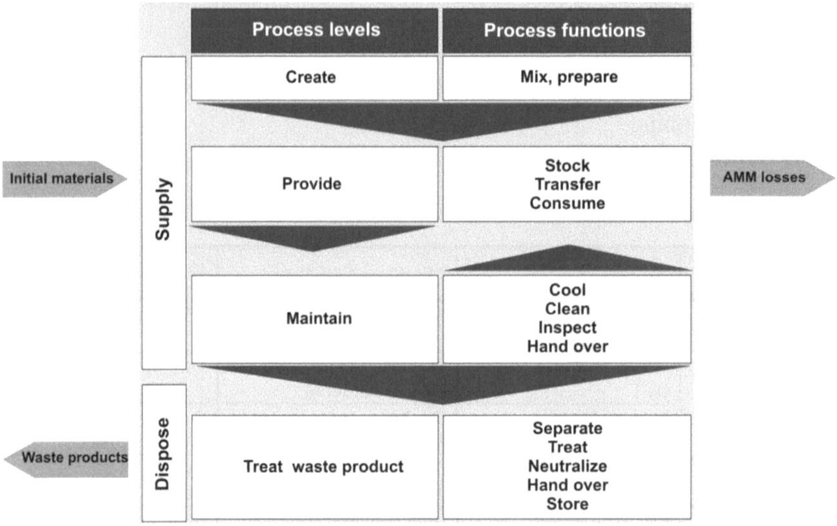

Fig. 6.28 General AMM supply and disposal process tasks

The AMM must fulfill such requirements as physiological safety (skin tolerance), the highest possible durability, harmlessness with respect to materials and color coatings, low price, low tendency to generate emissions.

The general process of AMM supply and disposal is expressed by a functional chain as depicted in fig. 6.29.

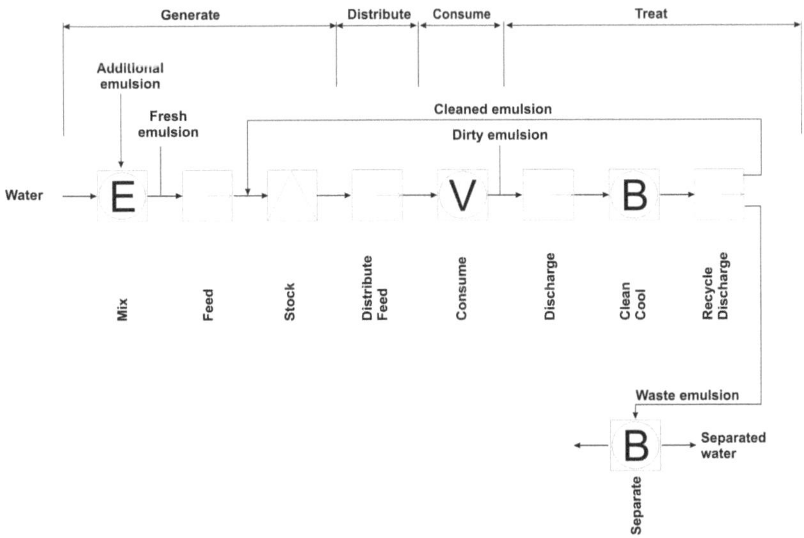

Fig. 6.29 General functional chain for the supply and disposal of AMMs

Depending on whether materials should be supplied and disposed of for an individual manufacturing machine or several machines simultaneously in one process, a decision must be made between peripheral and centralized AMM supply and disposal.

Examples of peripheral and central supply and disposal of emulsion and for cooling lubricant purification are shown in figures 6.30, 6.31 and 6.32 respectively.

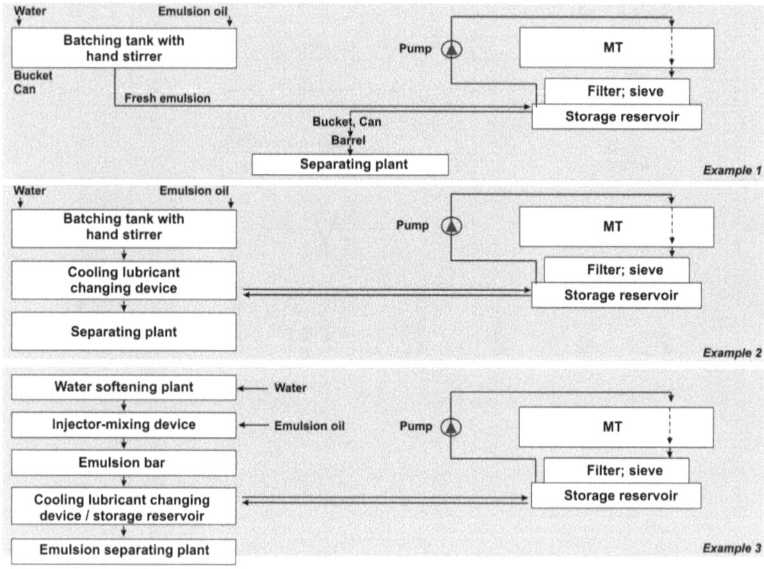

Fig. 6.30 Process variants of peripheral supply and disposal of emulsion

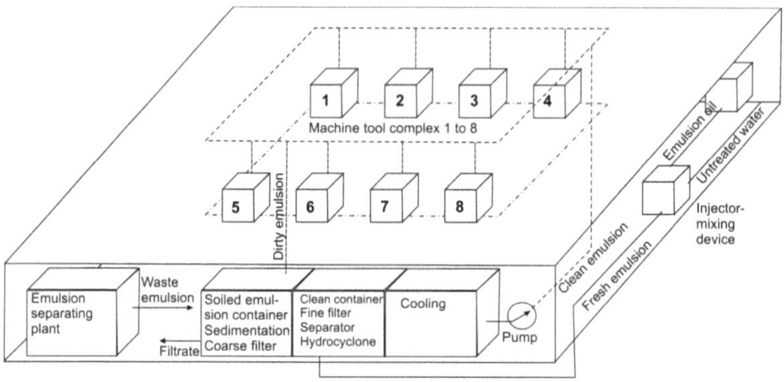

Fig. 6.31 Central supply and disposal system for emulsion

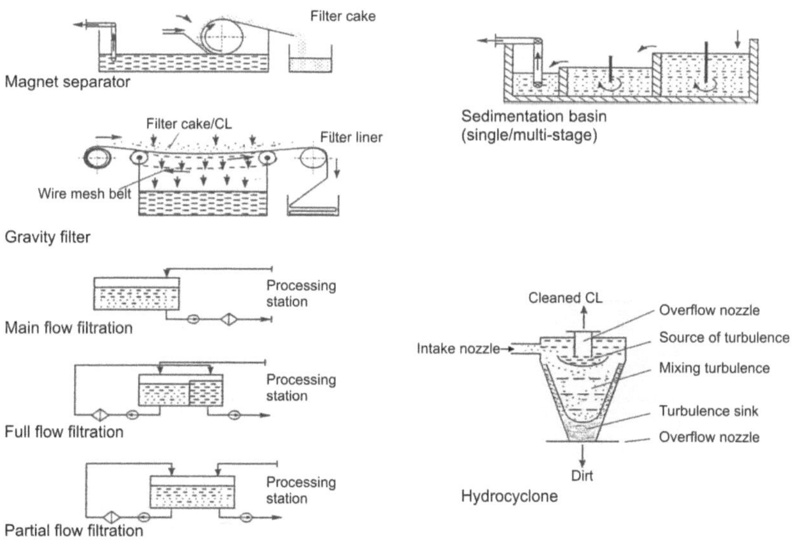

Fig. 6.32 Operating principles of cooling lubricant purification

6.3.2.2 Flow of chips and turnings

Chips and turnings are the most significant waste products in parts manufacturing systems. They are categorized as scrap (scrap turnings and chips) and are organized like all scrap types primarily according to material in types, groups and alloys.

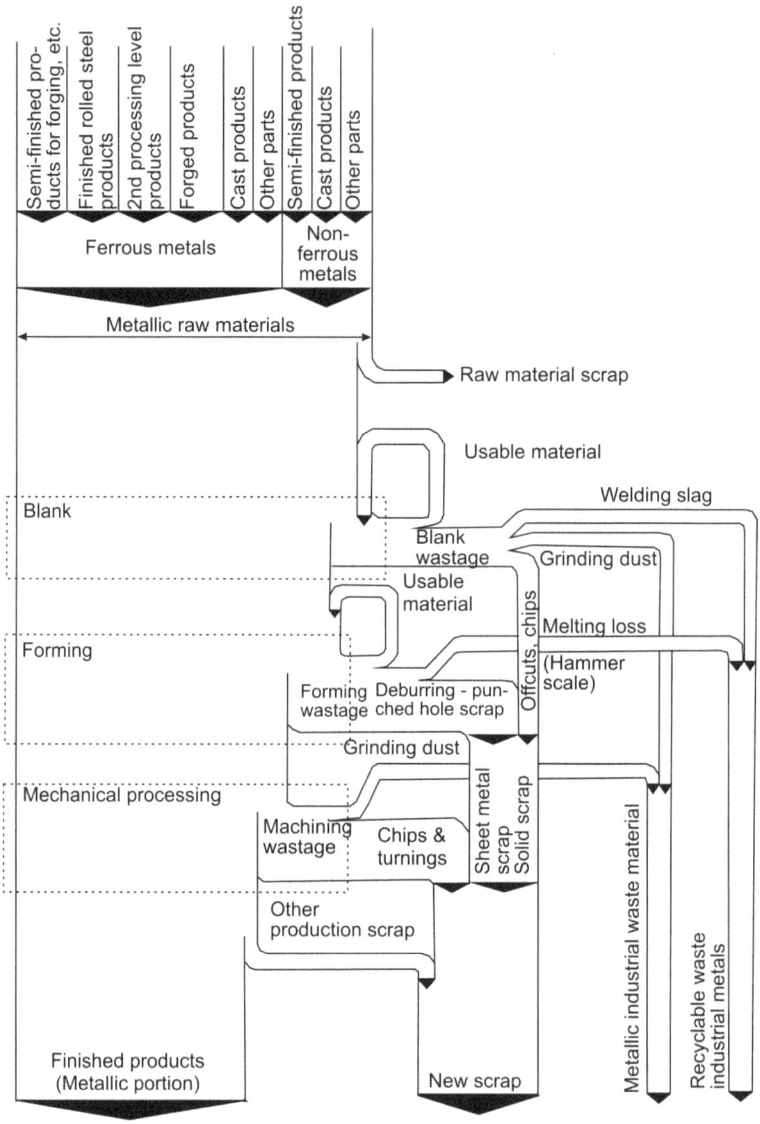

Fig. 6.33 General material flow balance for metallic raw materials

The scrap turnings and chips generated on the manufacturing equipment in the object being planned must be collected so that they can be recycled, treated if necessary and stored (by type and free of adhesions such as cooling lubricant). The general process and corresponding functional chain are depicted in fig. 6.34 and 6.35.

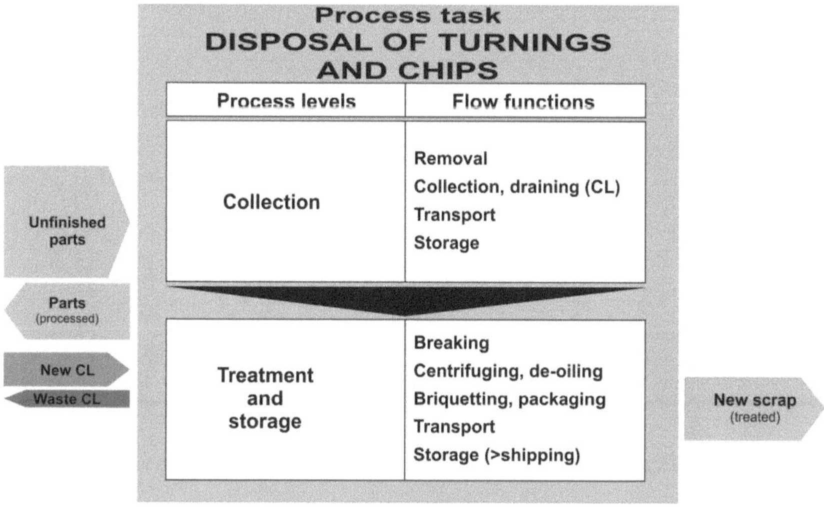

Fig. 6.34 General disposal process for turnings and chips

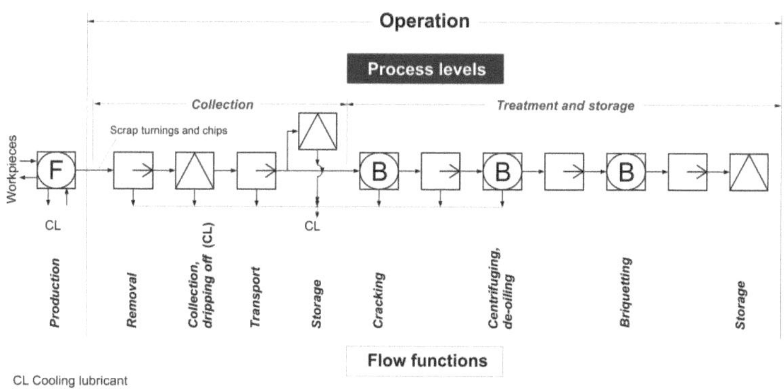

Fig. 6.35 General functional chain for disposal of turnings and chips

Technical ways of disposing of turnings and chips are depicted in the following figures.

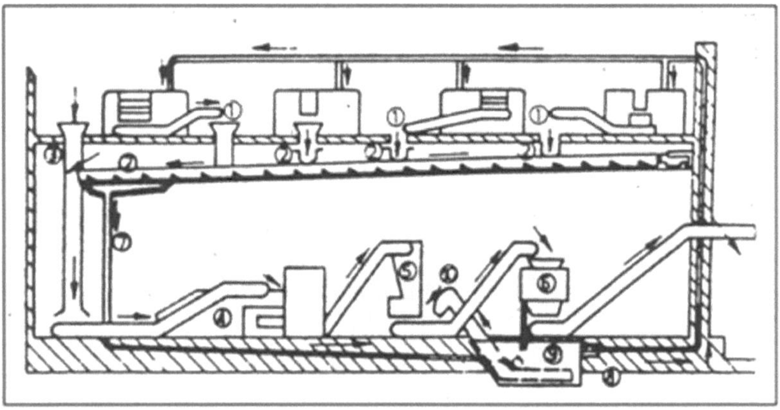

Fig. 6.36 Installation layout for centralized and peripheral disposal of turnings and chips

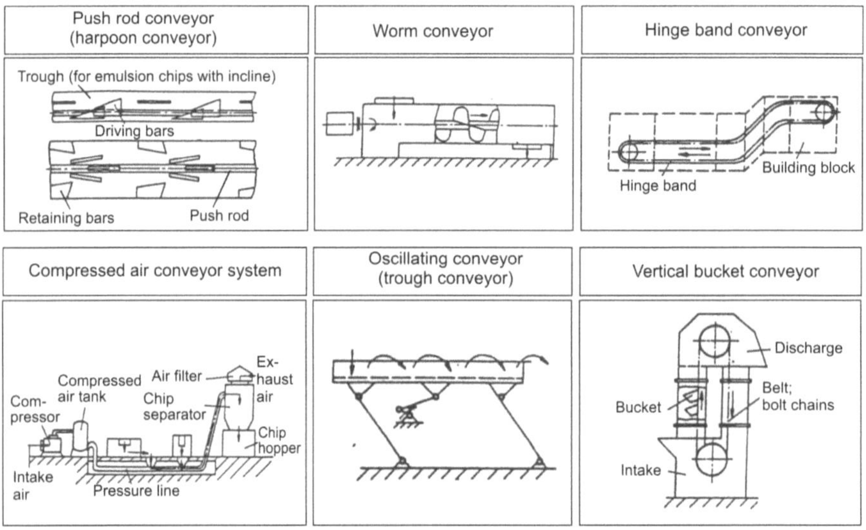

Fig. 6.37 Operating principles of flow system elements for the disposal of emulsion chips

Performance program

The disposal of turnings and chips in production/manufacturing systems is planned and executed, subject to the size and structure of production as well as to the type (material, condition) and quantity of the turnings and chips generated, either together with the detailed planning of the individual manufacturing equipment and THS processes (peripheral disposal integrated in the production area) or separately for centralized disposal of turnings and chips for the entire factory.

In the case of centralized disposal of turnings and chips, the tasks listed below were formulated as a task definition for a specialist supplier:

- the waste chips and turnings occurring in the production system for central disposal according to type (woolly, non-woolly), and form (screw, spiral etc.), condition (dry, wet) and quantity
- where possible, the quantities should be determined separately according to type/shape and condition on the basis of the quantity to be anticipated per machine [kgh-1]. depending on the availability or level of detail of the corresponding technological data for the parts program and the manufacturing procedure to be applied, this can be done on the basis of
  - the difference between quantity used and finished quantity or
  - the material usage factor or
  - the connected load of the machine tool.
- the detailed layout for the manufacturing system(s) where disposal is required (higher level);
- the required area and space for chip/turnings treatment, transport and storage
- the structural conditions for chip/turnings collection, treatment systems and storage.

For peripheral chip/turnings collection using containers, it makes sense to determine in advance the required number of transport containers per shift from information regarding

- container useable volume,
- chip and turning quantity accumulated during the period,
- average bulk density per chip/turning shape and
- filling factor of the type of transport container.

6.3.2.3 Energy flow

All high-voltage consumers in the object being planned must be supplied with electrical energy as required. This means that at every supply point between the operational high-voltage network and the individual consumer, the supply process must bring the (functional) consumer requirements into harmony with the different (functional and operational) connection conditions for the technical implementation of the corresponding process or functional chains (for the process task and general functional chain cf. fig. 6.38 and fig. 6.39).

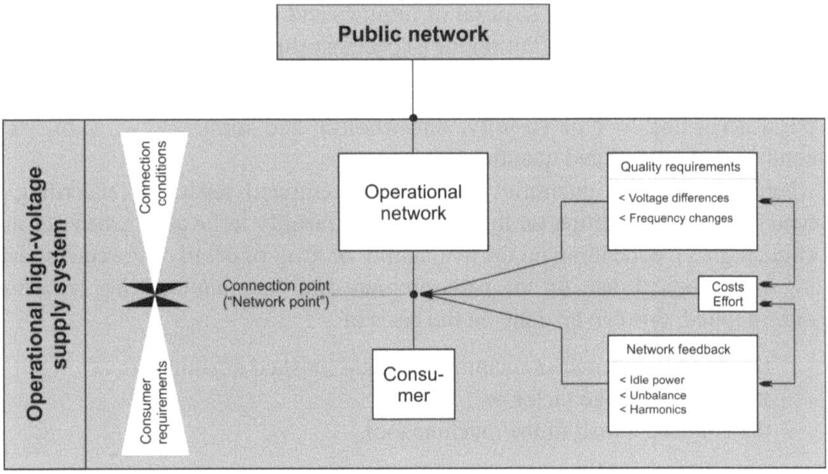

Fig. 6.38 General process for operational high-voltage supply

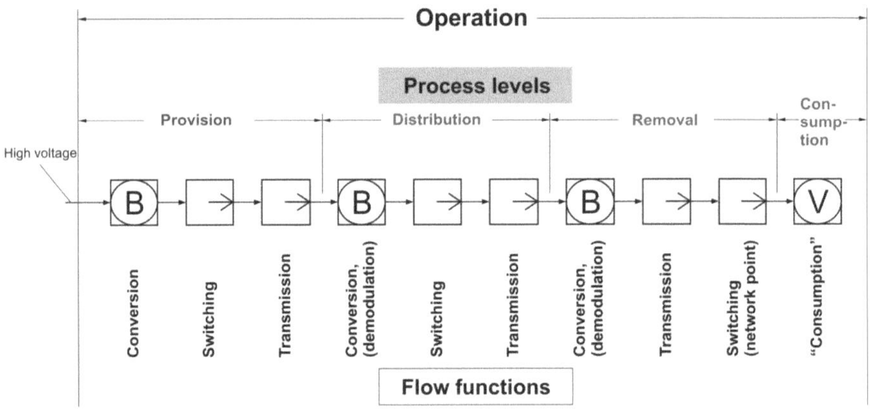

Fig. 6.39 General functional chain for operational high-voltage supply

Performance program

For the planning and execution of the high-voltage supply of production/manufacturing systems, the following tasks were formulated as a task definition for a specialist supplier.

- plants/equipment to be supplied (consumer connections) and lighting systems with the required network systems and network connection points;
- areas for distribution, cable ducts, load stations, emergency power systems, etc.;
- operating voltages (nominal voltages and their permissible tolerances for consumers, power points, lighting and emergency lighting), the voltage quality in

the connection point and the maximum permissible voltage loss up to the connecting terminals;

- required minimum protection class (against excessive contact voltages);
- installation types to be provided (e.g. duct, bracket, platform installation);
- supply reliability (constant availability of network voltage for automated process control);
- power requirements of the high-voltage consumers (energy demand list) as well as the total connected load and total power requirement for the planning object according to the rough calculation below;
- the total connected load according to which the entire high-voltage supply system must be dimensioned is calculated from the connected loads of the individual consumers.

$$P_{Cl\,tot} = \Sigma P_{Cl\,i} \tag{6.12}$$

I = 1 - n Consumers in the object $P_{Cl\,i}$ = Connected load of the individual consumer i

$$P_{An\,i} = P_{th\,i} = \frac{P_{Mot\,i}}{\eta_{Mot\,i}} \tag{6.13}$$

The total power requirement is calculated from the chronologically more or less overlapping demands of the groups of consumers

$$P_{tot} = \Sigma\,[f_{s\,j} \cdot \Sigma P_{Cl\,ij} \cdot f_{u\,ij}] \tag{6.14}$$

j = 1 - . . n groups of consumers

$f_{s\,j}$ Simultaneity factor of the consumers in consumer group j

(ratio of the highest simultaneous power requirement to the actual total power requirement of the group)

$f_{u\,i}$ Utilization factor of individual consumer i

(ratio of the effective power requirement to the connected load)

- compensation requirements for efficient energy utilization (reduction in idle power caused by inductance through an increase in the power factor cos φ);
- special requirements of the electrical installation in areas subject to the risk of fire and/or explosion.

At the same time, specifications are set out for the transmission systems (wiring, cables, ducts; distributors, transformer stations – power/transmission grids) including their installation (including on-floor, over-floor, under-floor, wall installation) for power and utilities in the form of pipeline, cable and route plans. Fig.

6.40 shows the types of connection and installation as well as the duct configuration for power and utility mains.

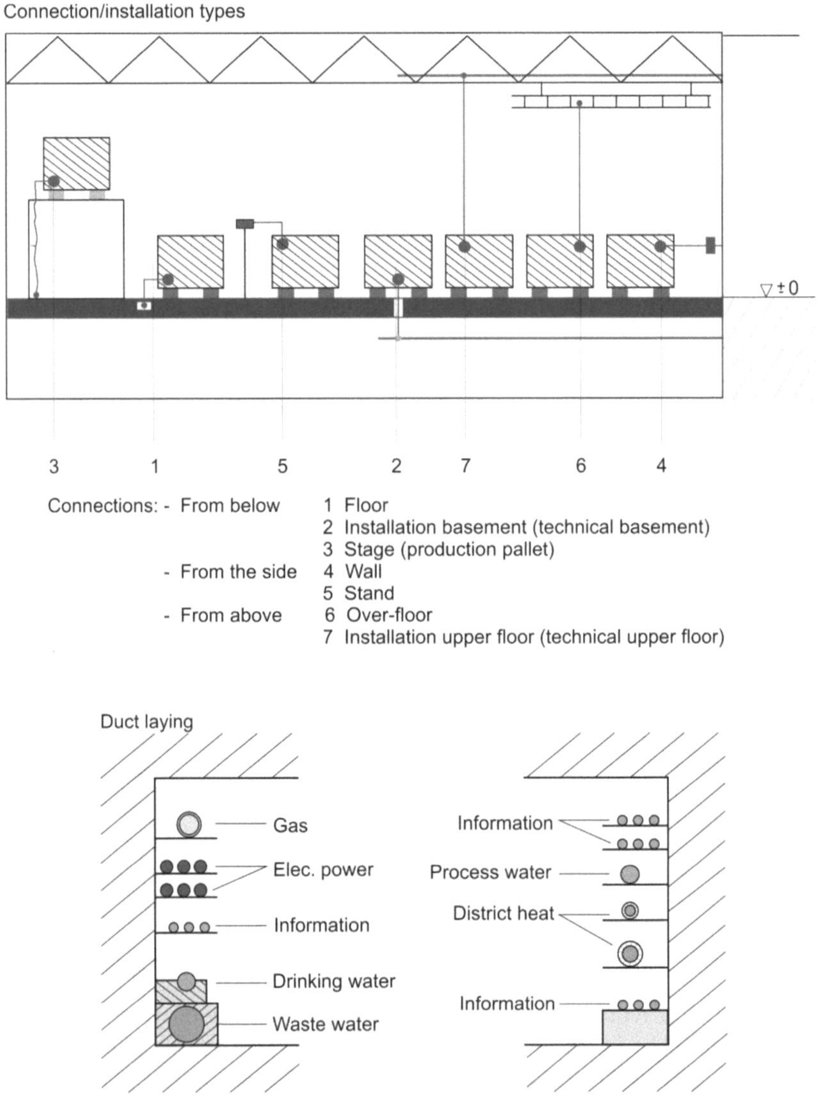

Fig. 6.40 Connection types and duct configuration for power and utility mains

Energy efficiency

Social development is characterized, among other things, by demographic development, individualization of need, intensification of purchasing power and expansion of environmental protection with increasing demand for energy and decreasing energy resources (e.g. fossil fuels). This requires energy efficiency in all areas, especially production (BREF 2008). Energy efficiency means: rational energy demand and high energy use capacity factor (VDI guideline 4661) along the value added chain. Energy management (VDI 4602) (Leven 2005) takes responsibility for ascertaining the actual level of energy procurement, transformation, distribution and use.

Energy management is becoming the most important focus of production facility design. (Neugebauer 2008). Specific areas of emphasis are achieving energy and material efficiency by increasing process stability, improving technological processes, introducing closed resource loops and networks and loss-free infrastructure management of systems.

a) General potential for savings

There is great savings potential, for example, in automobile manufacture, machine construction and metal production. Intelligent products with a high degree of energy efficiency are required in connection with CO_2 emissions reduction. They reduce energy costs in the face of constantly increasing energy prices. Production facilities transform energy and materials (VDI 4661) and should be used in an energy-efficient manner (Müller 2008).

Fig. 6.41 shows the energy control loop for production facilities (equipment and buildings).

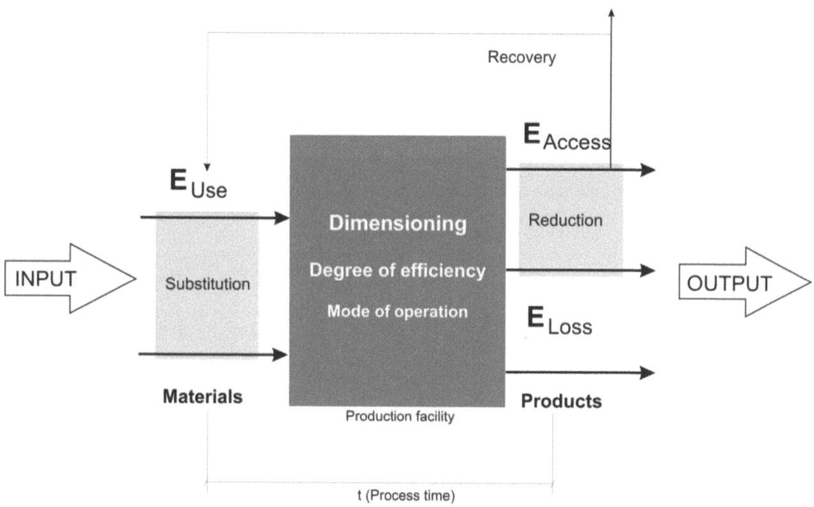

Fig. 6.41 Energy control and usage circuit (based on Engelmann 2008)

According to this, energy efficiency (energy savings) can be achieved in different ways through the:

- substitution of products/materials (e.g. lightweight construction), replacement of less efficient energy carriers (e.g. compressed air with electricity, thermoforming with cold forming);
- dimensioning of product, process and production facilities (miniaturization, correct arrangement of power generation, distribution and usage equipment)
- degree of efficiency (improvement of the quotient of the usable power that is emitted and supplied);
- mode of operation (energy-saving operation and organization of process flows, energy-optimized production and performance programs, reduction of stand-by times, avoidance of energy load peaks);
- reduction of losses (portions of supplied energy not used for the purposes of the process, reduction of process chains, process temperatures and times through maintenance and servicing);
- recovery, making waste energy useable for other production processes and systems.

b) Specific energy-saving potential

Via the energy flow, it is possible to derive the consumption using the energy supply types (provision), conversion and usage according to fig. 6.42 (Müller 2008).

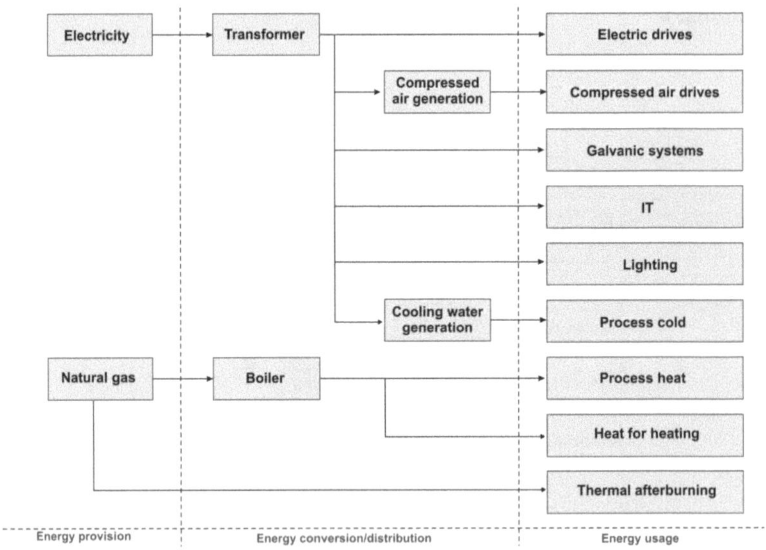

Fig. 6.42 Diagram of energy flow in a production facility

In particular, there are the following potential energy savings:

- Product-related
  Products determine the energy content for their production, usage and disposal. The energy demand is determined primarily by the material and its machining

and processing. Energy-efficient materials, processes and equipment must be selected (e.g. original form equal to end form).

- Process and equipment-related
  - energy-efficient process and equipment design,
  - process heat reuse, energy storage,
  - efficient energy flow circuit design (e.g. supply and disposal),
  - creation of integrated design solutions (integrate movements).
- Production and building-related
  - integration of production and technical building systems,
  - use of process heat for heating, ventilation, air conditioning,
  - regenerative energies and energy storage,
  - unite production and energy control center.
- Production building-related
  - energy-saving, insulated building (cladding, roof), e.g. using technical textiles,
  - operate heating, ventilation and air conditioning with regenerative energy (photovoltaic, solar, geothermal),
  - lighting through balancing of natural and artificial light,
  - minimizing spatial dimensions,
  - automating building control for heat and light via control center.

6.4 Information Flow

The information flow of a production facility includes all informational flows and functions for the preparation and execution of production with the goal of producing a product with optimal resource usage according to the customer requirements across the entire added value process.

It must be designed to accompany the product and process in connection with the material flow by linking operational functions via data along the value added chain.

The information flow is supported by information systems. The object of operational information systems for production is the functional, chronological and quantity-related planning, control and monitoring of all processes that are required for the production of goods. Examples of this are modular solutions for production planning and control (PPC) and for production control (production control stations) as instruments for the efficient execution of order processing.

Information systems for production include methods, models and software for production planning, control, monitoring and accounting, plus the necessary equipment for information/data acquisition, transmission, back-up, processing, storage and output.

Since in the literature there are different types of information systems, e.g. data processing, information technology (IT), information and communication technology (ICT), the generic term "information technology systems" will be used for information and data processing. (Krampe and Lucke 2007)

6.4.1 Functions of the information flow

The task of the planning process for the preparation and execution of production is to plan production on schedule with optimal resource usage eliminating all transition, changeover and waiting times for limited resources (capacity goals, time goals, inventory goals). The main task of planning is to solve the conflict of objectives that arises between optimal capacity utilization and simultaneously high flexibility with low costs. The seamless linking of the sales, purchasing/procurement, production and distribution processes is an important prerequisite for this. Therefore, for the preparation and execution of production, the focus should not just be on the smooth design of the material flow; instead the information flow should be considered to the same extent.

While during production preparation (planning) only information flows exist, during production execution (and control), the information and material flow must be regarded as a unit. For production therefore, the entire operational / production facility-relevant information flow must be fixed, and the performance program plus the flow function of the computer-aided information process must be dimensioned, structured and designed with the required equipment. The information flow, i.e. the preparation (acquisition and generation), processing, exchange and storage of information in coded form as data all takes place with computer assistance.

Technical equipment (instruments, systems) for safeguarding the data flow includes:

- data acquisition and preparation equipment (and systems), e.g. PDA devices,
- data transfer equipment (and systems), e.g. communication equipment, local networks,
- data backup equipment, e.g. access control, IDENT technology,
- data processing equipment (and systems), e.g. computers, control systems,
- data storage equipment (and systems), e.g. data storage media, databases,
- data output equipment, e.g. monitors, displays, signals.

These information systems are complex structures that map the concepts for planning and control according to the functions and selected organizational form for the operational production. Frequently, they can only be controlled by means of a modular design and breakdown into functional areas or organizational levels. In the process, data and information security becomes increasingly important.

Information security is not just a question of business ethics and, in terms of protection of expertise, not just a task that is in the interests of the company; rather it is also legally prescribed, e.g. for the protection of customer data. Information security (data security) serves to protect against dangers or threats, to prevent damage and minimize the risks of information processing and information storing systems. In practice, information security is usually oriented to the ISO standards family 27000 ff. Table 6.23 provides an overview of possible data security measures and how these measures must be considered as early as during the planning phase.

Table 6.23 Data and information security hazards and their consideration in the planning process

Hazard	Possible protective measures	Planning task affected in the business enterprise
Underestimation of the danger	Trained administrators Trained personnel and sensitizing of all employees	Personnel selection Workforce training
System failure	Create redundancies - cold redundancy (e.g. substitute computer not in operation) - warm redundancy (substitute system is already running, but is only used if the system fails, e.g. 1st disk is full, data is stored on 2nd disk) - hot redundancy (two elements always in operation)	Hardware procurement and dimensioning - amount of equipment Software procurement and dimensioning - number of licenses
	Duplication/multiplication of important data inventories through e.g. cloning of disks, back-up copies, back-up, storage according to the grandfather-father-son principle	Hardware procurement (computer configuration) - number of data storage units - required storage capacity
	Emergency (uninterruptable) power supplies (UPS)	Planning and layout Power supply concept
Acts of God (lightning strike, fire, flood)	Manual security concept	Establish emergency concept
Viruses, worms, spyware	Use anti-virus software/alternative: Linux	Software procurement
	Reliance on products from smaller and less well-known companies or on open source software, because these are subject to fewer attacks	Software procurement
Unwanted access from outside - Spying - Sabotage - Theft	Configure firewall	Software procurement and software set-up
	Limit and control access to the business enterprise	Hardware procurement (e.g. digital IDs, monitoring systems)
	Isolated hardware systems (stand-alone computers) for security-relevant data	Hardware structure
	Few and highly-qualified people with access authorization (access limitation)	Personnel selection
	Limit e-mail access and Internet access to necessary external contacts Do not send any files	Computer network
Unwanted access from inside	Do not digitize personnel files, lock them instead	Hardware procurement Software procurement
	No interface to data storage on the PCs	Hardware procurement

	(no drives, no USB interfaces, etc.)	(computer configuration)
System misuse (e.g. illegitimate resource usage; changing of contents)	Distribution and use of passwords, access codes, digital IDs, fingerprints, iris scans, etc.	Hardware procurement (hardware equipment)
	Encrypt sensitive data	Software environment
Feigning a fake identity (spoofing, phishing, pharming, vishing)	Integrate additional random security measures on a prophylactic basis	Software environment Organization
Faulty operation	Determine risks on basis of an FMEA, evaluate and remedy these	Hardware procurement Software procurement Software configuration
Deception Theft	Dual control principle as organizational principle	Organization (flow organization)
Outdated security system	Audits/inspection = guaranteeing of the IT security standard	Organization IT quality management

Fig. 6.43 provides a general depiction of the flow functions of the information flow that accompanies the processes throughout the order workflow.

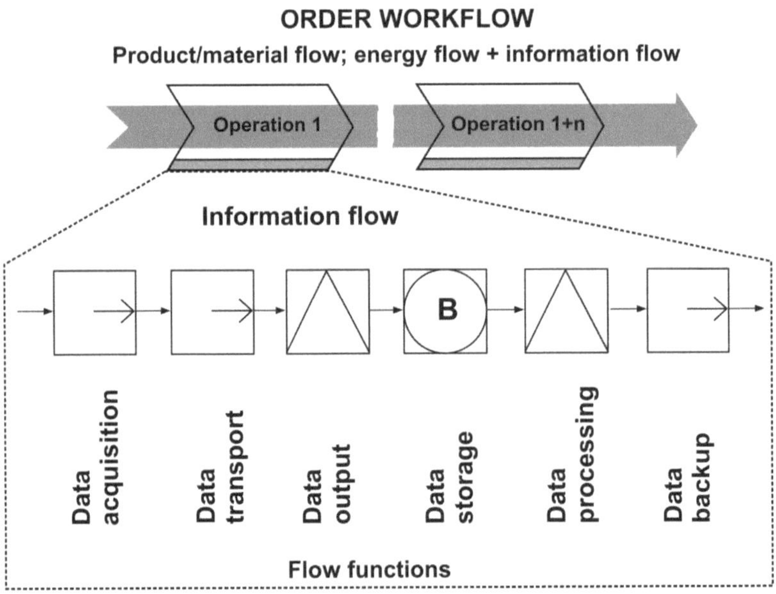

Fig. 6.43 Information flow functions throughout the order workflow

The core function of the information flow is the planning and control of the order workflow. It is governed by the business enterprise type and its specific production organization. The data and information must – subject to the situation - flow in accordance with the control task through the specific depiction of the flow functions on the basis of an organizational design solution.

In Eversheim and Schuh 1996, p. 14 – 60 ff various PPC design solutions are specified for various types of business enterprise. They depend, among other things, on the order trigger type, the spectrum and structure of products, demand for products/components, procurement type, stocking, production type, type of workflow in parts production and assembly, production structure and need for change. There are specific design solutions for: contract manufacturers, program manufacturers, system manufacturers, machine tool manufacturers, system suppliers and series manufacturers.

In addition, there are numerous hybrid forms. Two extreme examples show the different ways of linking functions for different organizational design solutions.

Example 1: "Contract manufacturer" PPC design

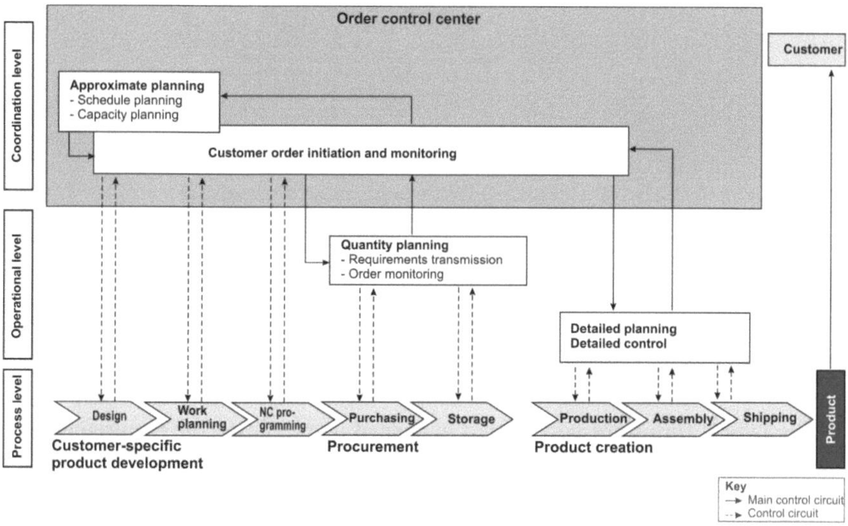

Fig. 6.44 "Contract manufacturer" PPC design fields (Eversheim, Schuh 1996, p. 14-48)

Example 2: "Series manufacturer" PPC design

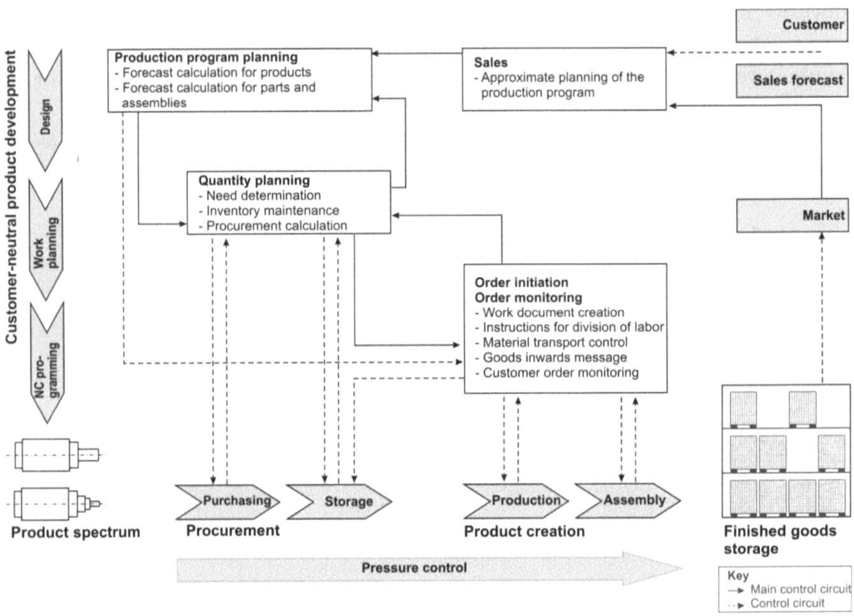

Fig. 6.45 PPC series manufacturers (Eversheim, Schuh 1996, p. 14-78)

The hybrid forms are very varied in shaped due to customer requirements and the organization of the business enterprise. Data flow and data processing must be adapted to the respective PPC solutions. Data flow plans must be created for the specific application depending on the situation.

The planning (and selection), design and operation of information systems for production planning and control (PPC) and for production control require a well-founded and goal-oriented conception whose strategic and operational orientation is developed and adjusted individually according to the needs of the business enterprise in question. Suitable planning and control systems can be determined on the basis of a variety of concepts, which have been formulated to represent the real production world. There are various PPC concepts in existence for production planning and control, as shown in fig. 6.46.

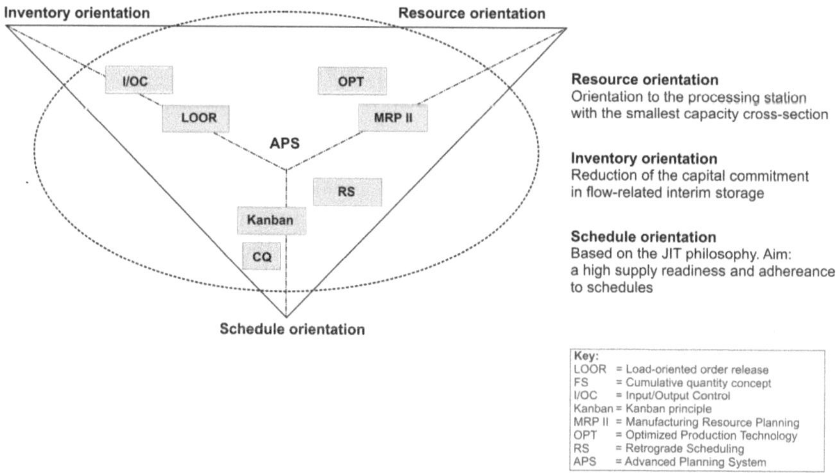

Fig. 6.46 Classification of PPC concepts

The material flow principle selected (push and/or pull) is decisive for the planning method used. If in the former case either total or successive planning approaches or combinations are used, then there are only two relevant implementation possibilities for the pull principle with the Kanban system or consumption control. Below is an overview of the procedures that are currently in use today, of which only MRP I/II, KANBAN and APS still have practical relevance.

Cumulative quantity concept

With the cumulative quantity concept, the requirements for a planning period are accumulated and depicted based on time. This results in a set curve with respect to time and quantity. The actual quantities are then entered on this chart in relationship to time so that the corresponding arrears or advance can be read off easily from the vertical distance to the set curve. This concept is primarily suited for the series production area.

KANBAN

The objective of the KANBAN principle is to increase reactivity and to reduce the time and effort required for repetitive processes. To this end the information flow required for production is decentralized and linked closely to the inverse flow of materials so that autonomous control loops are produced. By turning the bring (push) principle around to the collect (pull) principle, the production becomes consumption oriented. A production order is only initiated if a real need is indicated by the next stage in the value added chain.

Manufacturing Resource Planning

With this concept, the planning and control problems are considered throughout the entire logistics chain. For this purpose, material resource planning is expanded to include additional hierarchical elements from strategic planning and sales planning to rough planning on through to production planning. Many of the ERP programs used today are originally based on this approach.

Optimized Production Technology

With OPT, there is a practical implementation of bottleneck control. In the process, individual and sub-measures are not considered, but rather the overall operating performance. The level of material throughput within production is determined through bottleneck aggregates. This is based on the assumption that the bottleneck capacities are the essential factors for both throughput times and inventories.

Retrograde scheduling

The methods of retrograde scheduling are based on the network planning technique. Starting from this network plan, production orders are dispatched into production at the latest possible point in time. In the process various network plans for assemblies can be generated and launched according to the production flow. Retrograde scheduling enables low-stock, efficient production.

Advanced Planning and Scheduling (APS)

APS systems are software modules that can be used as intelligent planning and control systems across the entire supply chain and which replace conventional production planning modules in ERP and classic PPC systems. The APS system aims to efficiently manage the complexity of enterprise-wide processes by integrating information flows. The latest mathematical algorithms and technologies (e.g. genetic algorithms) are used in order to create enhanced plans. An important and essential characteristic in contrast to classic PPC systems is the speed of recalculations attained. APS systems are marketed by a multitude of different suppliers. When using them it is important to take into account that the performance of APS systems depends entirely on the quality of the individual configuration and the available data.

(Multi) Agent Systems

Control systems based on distributed, artificial intelligence (agent systems) represent a new approach, with which software modules (agents) as representatives of real resources/systems and their objectives and restrictions search for solutions according to particular principles (e.g. market principle). These systems are currently being used only experimentally or to solve specific transport problems.

In practice, it is often the case that a combination of several control concepts is used. To achieve efficient solutions, therefore, operational information systems must frequently also be adapted to the corresponding operating conditions and adjusted on an individual basis. Due to the dependency of the information needs and the control and planning needs on the existing material flow, in practice it is often necessary for different production areas to be subject to different control concepts.

6.4.2 Information system planning

Information systems are planned and designed on the basis of a concept that must take into account both the methodical solutions (models, processes, software) and the information (flow)-related (hardware) solutions.

The procedure for introducing information systems depicted in fig. 6.47 has proven useful in this respect.

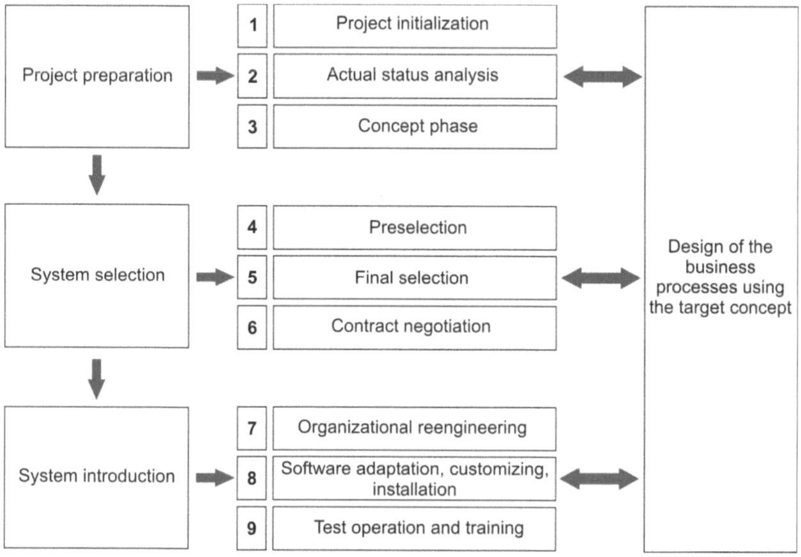

Fig. 6.47 Procedure for introducing information systems

When designing or introducing operational information systems, first of all organizational issues must be clarified in a project preparation phase. This includes specifying the area for consideration, the contact persons and the schedule. When outlining the area for consideration, special emphasis must be placed on identifying material and information flow interfaces to other organizational units along the supply chain so that they can be clearly defined.

This should be followed by a detailed examination (analysis) of the defined area for consideration. In addition to personnel-related and commercial aspects such as organization and enterprise orientation, special attention must be paid to technological considerations such as the existing or planned infrastructure for the secure planning and control of the material and information flows and the logistics processes.

Weak points, e.g. in the existing control concept, can be identified as part of this step and evaluated if necessary.

In the context of concept development, decisions must be made with respect to the planning and control systems to be implemented and also regarding a multitude of information-related, technical, organizational and personnel decisions. These must then be set out in an appropriate target concept.

By working through project planning steps, hardware and software solutions can be determined systematically from the perspectives of both engineering and information technology.

6.4.2.1 Hardware determination project planning steps

Performance program

The performance program includes the determination and selection of the data acquisition, transmission, back-up, processing, storage equipment and data output equipment required for the information flow in question.

It is based on the business and production process information flow analysis. This must be executed both for production preparation and production execution, on the basis of the functional areas and levels for the existing (actual) and planned (target) status. General criteria include:

1) an analysis of the task definition for the organization and information flow project

- customer requirements (e.g. identification, determination),
- business and organizational objectives (quantities, throughput times, capacity utilization),
- spatial and time-related structure of production (e.g. workshop, group, continuous flow production),
- hierarchy levels (production areas, groups, sections, divisions),
- organizational and control model for PPC solutions,
- automation level of the technological processes and systems (CNC, DNC, conventional control).

2) Analysis of the data flow

- type and structure of the data (numeric, alphanumeric, graphical data records),
- general data (master, inventory, transaction, logistics data),
- operation-specific data (production, time, volume, location data),
- type and form of data recording (paper, data storage media, screen),
- volume of information and data per type,
- frequency of the information per time unit (information frequency, intensity, speed, rate, e.g. information flow for the duration of a shift),
- range of the information transfer, acquisition location (source) and recipient location (sink),
- type of data transfer (e.g. telephone, radio, fax, WLAN, teleprinter, computer network),

- type of existing and planned information processing (data storage media, data processing, data transmission),
- automation level of data processing.

3) Preselection of model and method
- basic organizational solution,
- production, planning and control tasks,
- planning and control models (e.g. production and production logistics control systems),
- technological and organizational structures,
- information flows within and between the hierarchical levels including the reorganization of the information flow structures,
- type, volume and frequency of information.

Determination of functions
The following must be available in order to determine functions:

- the performance program,
- the basic organizational solution including the documentation for the actual status analysis,
- technological project including schemes,
- concepts for linking information systems (interfaces, databases, software),
- equipment inventory documentation, catalogs, bids and guideline values for the selection of system element equipment.

a) Information acquisition equipment

The functions are determined via the type of data, data quantity, time and location-related structure of the data incidence (data acquisition reaction time). Criteria for determining the technical level are:

- acquisition of the data at the point of origin,
- guaranteeing the acquisition of the data volume,
- possibilities for parallel data acquisition,
- possibilities for releasing workers from data acquisition tasks,
- type of transmission and processing of data.

Equipment: reading and recording devices for barcodes, transponders, biometrics, image processing etc.

b) and c) Information processing equipment

This includes computers and control equipment (including storage), which are also determined by the data type, data quantity, time and location-related data processing structure, operating and software systems. Criteria for the hardware (computers) include:

- existing IT infrastructure,

- coupling opportunities in the multi-level concept (defined interfaces to local networks (WLAN, intranet) and public wide-area transport networks (WAN, Internet, DSL, ISDN)),
- network structures, e.g. hierarchy levels (operation - operating computer, division – division computer, section/cell – process/cell computer, production area - machine, robot, instrument computers), in modern computers they are offered as complete solutions,
- quality of the user interface (short reaction times, adaptation to the operator's skill level, ergonomic design),
- area of use (office and process automation),
- processing power including storage capacity (transfer time, data quantity),
  - computer (processor, available storage for required software),
  - storage capacity (hard disk, CD-ROM, DVD, mobile storage, etc.) according to the areas of application,
  - input and output devices (screen, printer, plotter, digitizer etc.),
- operating systems including the software systems.

Modern computing systems incorporate nearly all of the named capabilities and must be specified according to catalog and bid.

d)Information transfer and communication equipment

In order to network computers at various operational management levels, production planning, control and monitoring, and communication systems in the form of local networks (LAN, intranet, industrial Ethernet) are required for the immediate area and wide-range networks (WAN, Internet, intranet). There is a range of standards for this, which must be adhered to. The networks are designated as special communication transmission networks that use a common transmission medium and establish data traffic between end user devices capable of communication such as computers and terminals as well as equipment for automated production (machines, robots and sensors) via distributed control. The following criteria must be taken into account when determining functions:

- network topology for the remote and immediate areas (network, star, ring, bus),
- transmission medium (wires, coaxial cable, fiber optic cable),
- access procedures (polling),
- ISO/OSI reference models,
- transmission protocols: TCP/IP (Transfer control protocol/ Internet protocol), e-mail, www (world wide web), PROFINET.

e) Other organizational aids, display and presentation of information

The organization, planning and control of production includes equipment that must also be determined and calculated, such as

- screen and visualization technology,
- office, writing and communication technology,

- telephone, voice, dictation, recording and copying devices,
- forms, documents, data storage media,
- scheduling, registration, viewing, recording and monitoring equipment,
- organizational, functional and structural plans including organizational instructions,
- presentation equipment.

Dimensioning

The information flow systems are dimensioned in terms of each of their elements (acquisition, processing, storage and communication equipment) on the basis of

- the performance program determination,
- conditions of use,
- the usage parameters.

For the information flow, the required channel capacity for the corresponding equipment (and data quantity, transfer speed etc.) must be compared with what is available, i.e. the required and available output. This task can only be performed in detail by special project engineers (information technology specialists).

Structuring and design

Here three complementary structuring problems must be planned. These are:

a) Structuring of the tasks on the planning and control level

In order to master the complexity and variability of the information processes and data quantities, the information flow systems are organized vertically and horizontally and linked with one another. A vertical information flow is necessary for transmitting tasks on various levels, with only immediately adjacent levels communicating with one another. A horizontal flow of information is for synchronization and the transmission of tasks within a level.

Generally there are three separate levels, which are, however, becoming ever more closely linked.

Corporate management level, also known as the administrative level.

Tasks: order content and schedule specification from the production program derived from the supply program and shipping inquiry, resource management according to commercial aspects, material requirements determination, ordering and inventory-keeping along the logistics chain.

Information system: ERP (Enterprise Resource Planning)

Production management level, also known as the scheduling level.

Tasks: production planning and control including data acquisition, quality, personnel, equipment and material management. Situation-driven order processing (time, resource assignment, management) performance analysis, material provision and use for the material flow logistics chain

Information system: MES (Manufacturing Execution System)

Production level: also known as the operational control level (actuators, sensors).

Task: physical control of the processes and realization of the order process steps and tasks according to the work schedules, technical control of the specified resources (machines, plant, transport, storage and handling equipment), quality assurance, maintenance and material machining and processing in specified process steps.

Information system: protocol systems

- MAP (Manufacturing Application Protocol) for transmission to the management and control levels,
- TOP (Technical and Office Protocol) for application in the management and planning area,
- TCP/IP (Transmission Control Protocol/Internet Protocol) for controlling the data flow in LANs (intranet) and WANs (Internet).

On the control and actuator/sensor level, a development in networked automation based on industrial Ethernet standards and PROFIBUS / PROFINET (Siemens) is emerging.

The *process and control level* works in real time operation and controls machines, robots, transport, storage and transfer equipment.

Modern computers integrate the individual levels and control the suppliers' logistics chains to the end producer through to the control of the individual drives (axes). For data communication (networking) of the various information systems, a 7-layer model, "Open Systems Interconnections" (or OSI), has been developed by the international standards organization (ISO). Various standardized transmission protocols are based on it.

b) Structuring of production and organizational level tasks

This is done as already described in section 3.2.4.2, tab. 3.27 and by fixing the organizational solution according to structural units. The "segmentation" of production provides information about how the information flow can be implemented within and between a number segments and PPC levels. An example is the subdivision of parts production and assembly. For example, "mechanical production" is organized into 5 production sections (master production areas, profit centers). These sections consist of 2 process-oriented sections (e.g. cutting, hardening shop), 1 machine system, 1 production/assembly cell with mobile robots and 3 machines, and 1 special purpose section.

The production structure determines the structure of the information flow and communication between the computers.

c) Structuring of the tasks for the computing and communication level

This is done on the basis of the levels fixed for a) control and b) organization. The computer and communication levels can correspond to these (e.g. computer pool and networks) and/or be planned as an integrated solution that links all levels with one another. Modern computer technology is developing towards integrated solutions that unite tasks and levels in a single computer.

Fig. 6.48 shows a general computer and communication structure.

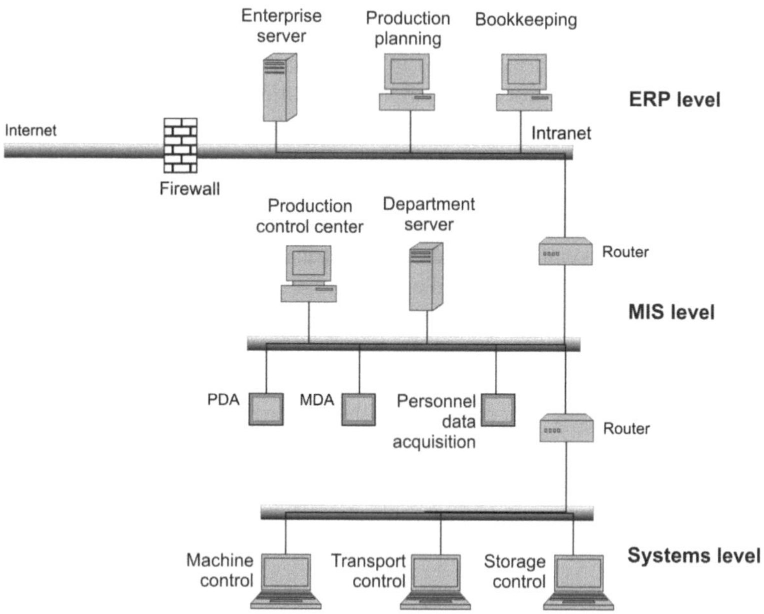

Fig. 6.48 General computer and communication structure

Example: Control centers and MES

Modern control centers, known as Manufacturing Execution Systems (MES), exist for the interaction of information flow and data processing. They are assigned under the PPC (production planning and control) and ERP (Enterprise Resource Planning) systems (Kletti 2006, Schuh 2006, Pawellek 2008).

On the one hand, the control centers are structured as pure *workshop control centers*, which primarily process technical data; on the other hand, they are used as *production logistics control center systems,* which optimize the material flow and organizational structures as part of an overall organizational solution.

The use of the MES (Manufacturing Execution System) according to VDI guideline 5600 (2007) supports the production-related business processes of work preparation, production, transport, materials management, quality assurance, personnel management, traceability, maintenance, continuous improvement (CIP) and controlling.

Note: IT solutions must always be selected in connection with the selected methodological solutions that are set out below.

6.4.2.2 Method and software determination project planning steps

The following remarks are intended to illustrate the planning of a factory's IT infrastructure. As part of process-oriented information management, the information management support functions must be oriented according to the business processes. (Schwarzer/Krcmar 1995, p. 33)

Business processes have been mapped and supported by IT at least since the triumphant introduction of Enterprise Resource Planning systems (ERP systems) and the euphoria over the Internet at the turn of the millennium. Up until now, however, in many cases the business processes have been oriented towards the limits and possibilities of monolithic IT systems. The necessary incorporation of flexibility into the processes must be reflected by the flexibility of the IT infrastructure.

If a particular process needs to be fully or partly supported by IT, then all information about the process that is necessary for the IT implementation must be available in documentary form in order for the IT to be developed. Possible steps for determining method and software are identified below:

(a) Process mapping

This information can be collected either in a structured interview or in group work as part of a workshop. The workshop variant has the advantage that several opinions about a process can be heard in a relatively short time, recorded with little effort and compared directly. For this reason, the following remarks focus on the workshop variant.

So that the discussion during the workshop focuses on the correct level of detail regarding the process, this level must first of all be defined. One way of doing this is to subdivide the process hierarchically into main and sub-processes and process steps and work steps. To record the relevant process information for IT-based support of the process, it must be collected on the lowest level of abstraction, either at process step level or at the most basic level of elementary work steps. Only in this way is it possible to ensure that the process flow can be integrated correctly into the algorithm so that the software and its components can be developed.

The first information collected relates to the activities that take place within the process step. The process step is written down on the Metaplan chart and attached to the pinboard on packing paper. By attaching it to the board, the sequence and parallelism of the process steps can be illustrated. Loops can also be charted on the paper using a marker pen. Additional information is not recorded on the Metaplan, but rather noted on the appropriate form. The moderator lets the participants discuss the process freely among themselves, asking targeted questions if necessary. The documenter is responsible for ensuring the completeness of the information collected. If necessary, the documenter may interrupt the discussions.

The input and output information and the areas of responsibility are the minimum requirements that must be recorded in order to implement a process. A process description is only complete when it also includes information regarding the process manager's support staff (where required) and what auxiliary equipment will be needed to execute the process (e.g. templates, existing IT).

The next thing to be recorded is where the information and input objects originate (e.g. storage location, department) and where the output objects and information belong. The last information specified is the starting point of the step. Generally, this is initiated after the foregoing step is completed; exceptions will be noted accordingly.

In order to convey the flow logic of a process to the IT, additional answers to the following questions are required, which can be asked once the individual steps have been modeled:

- what decisions are made when? (e.g. stop criteria)
- where do loops exist? (i.e. repeat steps)
- what steps occur in parallel?

After the process mapping workshop has been held, the process is modeled using a suitable process management tool and a modeling technique that is standardized throughout the company. The process model should then be returned to the workshop participants and adjusted if necessary. Producing an audio recording of the workshop together with the filled-out forms usually provides a sufficient basis for modeling the process, which is explained in more detail below.

(b) Process modeling

There are different process modeling techniques; these can be categorized as either diagram-based process modeling techniques and scripting language-based techniques due to their stark differences.

At the time of writing, scripting and diagram languages are slowly growing together. Diagram languages, for example, are usually used by the operational departments (i.e. those responsible for the process) in order to describe the processes. The level of detail is usually not less than the level of self-contained work steps or tasks. Scripting languages were developed from the IT environment in order to depict the flow logic of software programs and to underpin a business process in such detail that the algorithm for IT processing exists in pseudocode in a quasi-executable form. Therefore, scripting languages usually operate on a much more detailed level than the processes described by the operational departments using diagram languages. And this is where the two modeling approaches are linked.

Accordingly, there must also be tools that make it possible for the operational departments to model the processes while ensuring that the modeled processes can be translated into executable code. A multitude of software tools relating to the topic of business process management came onto the market as process orientation became more prevalent. At the same time the different software manufacturers have focused on different areas of emphasis and implemented different functionalities.

This development, whereby process model and software design are growing together is reflected in the IT tools: the current trend away from monolithic IT systems towards process-oriented IT systems that can be adapted dynamically to the requirements of the processes has been mentioned several times already. Irrespective of the technological perspective, it is possible to say that in the future, systems must be underpinned by a flow logic that is derived from business processes.

c) Business process modeling

Three representatives of diagram-based process modeling techniques will be discussed in brief below.

The event-driven process chain was developed in 1991 at the Institut für Wirtschaftsinformatik (Institute of Information Management, Iwi) for modeling business processes as part of the architecture of integrated information systems (ARIS) and implemented in 1994 in the ARIS Toolset (Scheer 1994).

It maps the functional process in terms of both time and logic. The dynamic process basis can be mapped using the concept of event control. EPCs are directed graphs that use function, event and logical connector model elements to model the control flow. Functions are time-consuming processes that transfer input into output data. Graphically, functions are indicated by round-edged rectangles. They are triggered and scheduled by events. An event is the occurrence of a defined state and in itself does not take any time. It is achieved only after the function has been fully executed and triggers one or several follow-up functions. In EPC models, they are depicted as hexagons (Keller, Nüttgens and Scheer 1992, p. 9ff.).

An EPC model maps processes as transitions between events and functions that are connected by directed lines (arrows). Processes are generally not linear, i.e. there are branches in the function sequence. With the help of logical connectors, it is possible to specify switch rules. The possible logical connectors are:

- conjunctive link AND (“and”),
- disjunctive link XOR (“either-or”) and
- adjunctive link OR (“and/or”).

Connectors are branch nodes between events and functions that are depicted as circles. Inside each of them is one operation as an input or output link. They can divide the flow up into several sub-flows (split, with several outputs) or combine several sub-flows into a whole (join, with several inputs). (Nüttgens 1997, p. 2 ff.)

If for an EPC the process flow is demonstrated only using events and functions, an eEPC (expanded event-driven process chain) represents an EPC expanded to include additional information such as output, organizational units or data.

EEPCs are depicted vertically, i.e. exclusively from top to bottom. The control flow is depicted using dashed lines and the information flow with continuous lines. The "goods have arrived" event triggers the "inspect goods" function. Thus an EPC begins and ends with an event or with a process marker that points to a further process. Functions and events, in other words the entire process, must be arranged alternately. The inspection of the goods occurs in the example with the help of the order and delivery slip and is executed by the goods inwards department. The outcome of the inspection is recorded in an inspection report. It can be seen that organizational units are always arranged to the right of the function and all other objects to the left next to the associated functions. After the inspection, all of the delivered goods can either be released for the production execution subprocess or blocked or rejected for the quality inspection sub-process. It is important to note that due to the XOR connector only one of the three events may take place (Nüttgens and Rump 2002).

UML (Unified Modeling Language) is an object-oriented modeling language for software and other systems. In UML there are 13 different diagram types and display options.

The emphasis of business process modeling is on the activity diagrams, which are considered here. They combine the status diagrams with flow diagrams and Petri nets and thus offer the best opportunity for depicting business processes since their focus is on describing activity sequences. They are differentiated from other modeling methods through the ability to depict parallel processes.

Activities are the individual steps in a processing sequence. They are linked automatically through transitions, which are shown as arrows. If several transitions are combined with one another, then Boolean conditions are used. In order to better depict the organizational responsibilities, swim lanes are used. There are two different ways of mapping activity diagrams: with UML2 and UML 1.x. In UML2, asynchronous communication mechanisms can be incorporated and they are very similar to the Petri nets. Disadvantages of activity diagrams are the limited mapping of organizational structures for complex business processes and the modeling of role concepts. (Bullinger and Schreiner 2001, p. 47ff.)

(d) IT-based process modeling

If the specification for IT components is to be developed from a process description, then it is important to model the flow logic of the process too. In order to generate this flow logic, various roles within the process must be identifiable. For example, it is important to show who has to perform a task and who is responsible for it. The representation of decisions is important in order to check where a change of responsibilities, departments, etc. must be considered. Such a change is integrated into the new system as an interface. Loops including their termination conditions are important in order to identify fallback points in the process and consider them accordingly in the IT system.

In order to implement a process in IT systems, in addition to the aspects mentioned above, data is also required about which IT systems are already involved in the process, which IT resources are required and which information objects need to be processed. Data, information and resources can be modeled in various ways. Process modeling techniques must be in a position to map IT requirements (IT mapability).

The criteria listed here are strongly oriented towards the software engineering process, i.e. the future creation of software. A process depiction resulting from a well-engineered process modeling technique should therefore be able to answer the following questions for the software developer:

- which functions / services must the software that is being developed provide?
- what input and output documents must be used/generated?
- what kind of documents are these (format), and from what data and information can they be made up?
- what roles should the software take into account?
- who may do what (rights)?
- what is the flow logic (when does what occur)?

If a process depiction is able to answer these questions, it can be used as a specification for the software development process.

e) IT requirements/infrastructure

Today there are many IT systems that can map a factory over the entire (product) life cycle. Examples of systems that are widely used internationally are SAP, Oracle and Microsoft Navision. In addition there are a multitude of specific systems often limited to a single industry or country such as IFAK, infor, PSI and proAlpha.

On the business enterprise level, standardized ERP systems are used, which are supported by specific solutions in the production facilities.

For example supply chain management information systems therefore include logistics processing of the value added chain with the functional areas of Enterprise Resource Planning (ERP) systems (procurement, materials management, production and sales). When planning information systems, norms and standards such as the OSI model (Open Systems Interconnection Reference Model) in the form of an open 7-layer model or, for electronic data exchange, UN/EDIFACT (United Nations Electronic Data Interchange for Administrations, Commerce and Transport) should be considered. The same applies for the use of data warehouses for data collection, analysis and presentation in a homogenous dataset. Fig. 6.49 shows the information systems that are used in connection with production planning and control.

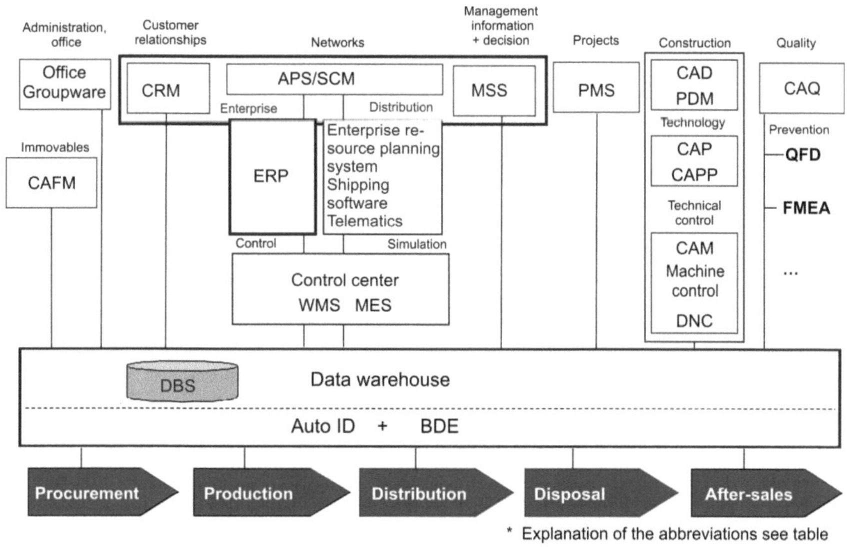

Fig. 6.49 Information systems (source: Glistau and Wirth 2008)

In table 6.24 they are classified by task, software range and main functions with software production examples, and explained in more detail.

In production planning and production itself, complete transparency of data exchange is the objective. This is achieved on the basis of the problem and task by combining and networking the information systems listed via their software (Spath 2007).

For example, using SAP-certified Coscom software modules, the process chains from control center detailed planning, MDA/PDA (machine/production data acquisition), PDM (production data management), CAM/simulation, DNC (Direct Numerical Control) and tool management are linked into a complete process chain (Coscom 2008).

Since various software components are available for information systems, a comparative examination of the software variants is necessary. Table 6.23 offers a checklist to assist in the selection of software.

Table 6.24 Overview of selected enterprise software for planning and control

Name	Task (T) + Software range (S) + Main functions (M)	Product examples
Office Systems + Groupware	(T) For the support of office work in administration and for cooperation in teams	• MS Office • OpenOffice
	(S) Licensed and open-source software	
	(M) Word processing, spreadsheets, databases, presentation programs, e-mail programs, file sharing	
CAFM Computer Aided	(T) For the planning, administration and management of buildings, systems and facilities with the objectives of improved usage flexibility, labor productivity and return on capital	• Planon Facility

Facility **M**anage-ment Sys-tem	(S) As a stand-alone system or a module in ERP systems	Solution • Speedikon FM
	(M) Building planning; floor space management; commercial building management; technical building management; FM service; real estate management; across-the-board functions	
CRM **C**ustomer **R**elation-ship **M**an-agement systems	(T) Customer relationship management = managing and documenting relationships with customers and tailoring business processes directly to customers	• SAP CRM • Microsoft Dynamics CRM • Adito online • Siebel CRM
	(S) As a stand-alone system, as an add-on or as an integrated module of ERP systems	
	(M) The task of analytic CRM is to analyze customer data, filter out important information and visualize it if necessary. Operational CRM enables as smooth a dialogue as possible with customers and supports all transactions relating to customers. Communicative CRM provides the interface to customers.	
APS / SCM-System **A**dvanced **P**lanning **S**ystem; **S**upply **C**hain **M**anage-ment Sys-tem	(T) Integrated planning of the entire supply chain = enterprise-wide coordination of all goods and information flows along a value added chain (supply chain), i.e. from the raw material supplier across all levels of the value added process to the end customer.	• SAP APO • Oracle E-Business Suite • Infor APS
	(S) As a stand-alone system, as an add-on or as an integrated module of ERP systems	
	(M) Network design; network planning; sales planning; procurement planning; production planning and production flow planning; distribution planning and transport planning; availability planning	
ERP-System **E**nterprise **R**esource **P**lanning **S**ystem	(T) Fully integrated software solutions for production, finance, logistics, human resources, project, sales and other enterprise areas; offers functionalities for all areas of a business enterprise	• SAP • Infor • Pro Alpha • Business Solutions • Microsoft Dynamics Nav
	(S) As a stand-alone system, today usually designed in modular fashion; Frequently there are special industry solutions	
	(M) Procurement/purchasing; storage; production; maintenance; sales; shipping; customer service; accounting and controlling; financial management; human resources; research and development	
Shipping software	(T) Distribution support for enterprises with their own fleet of distribution vehicles	• Spedifix • soloplan
	(S) Special software tools where necessary with link to or integration into telematics solutions	
	(M) Master data management; tender preparation/freight cost calculation; order entry; journey and route planning; human resources, vehicle fleet and storage equipment management; display of organizationally-relevant vehicle data (e.g. maintenance intervals, tank and repair data as well as storage equipment inventories); controlling/statistics; document management system for receipts, freight documents, etc.); open interfaces to other software (e.g. for CEP service providers, telematic storage management, movement, accounting); financial accounting	
Telematics systems	(T) "Telematics" is a term that combines "telecommunications" and "informatics" - with the implementation of telematics systems in shipping, the objective is to connect all participants in an enterprise's freight activities via a network and thus to achieve greater transparency along the transport chain and constant information availability by digitizing the data. Telematics enables the current optimization of transport utilization with respect to performance, security, environmental friendliness, economy and management effort.	• Trafficmaster PLC • Fleetboard • Comlog. Com

	(S) Linking of identification, positioning, mobile communication, data exchange and driver assistance; frequently offered in individual modules: www.euro-telematik.de (M) Journey planning; vehicle navigation; fleet monitoring; fleet control; driver assistance; message transmission; order management and accounting; cost management; activity and working time recording; shipment tracking (tracking and tracing); supply chain integration	
PCC Production control center **MES** Manufacturing Execution System	(T) Control center is the link between planning and operational execution (S) Often individual solutions with interfaces to ERP systems (M) Checking of planning measures and detailed planning; scheduling and breakdown management; controlling, monitoring and visualization of process flows; evaluation of data; process progress reporting; simulation of scenarios for alternative planning; time management	• INTEGRA Fertigung • AHP-Leitstand • PSImes • PLL
Identification systems (Auto-ID)	(T) Identification of non-material and material objects (S) Hardware and software for 1D, 2D, 3D barcodes, transponder, biometrics, image processing systems, chip cards and OCR (M) Identification of objects, goods, people, animals, etc.; access and authorization control, generation of planning and control information for logistics processes, tracing of goods or flows of goods, implementation of the I-point, in addition some data storage	• Hardware and software for 1D, 2D, 3D barcodes, transponder, biometrics, image processing systems, chip cards and OCR
PDA systems Production data acquisition systems	(T) Production data acquisition (PDA) is a collective term for the recording of actual production status and process data (S) Hardware and software solutions; often in combination with identification systems and technical controllers; Usually offered as specialized individual solutions (M) Can be distinguished functionally according to • order-related production data such as times, volumes, weights and qualities • machine-related data such as run times, interruption times, units produced per machine, messages and breakdowns, operator interventions, consumption of materials, power and auxiliary materials • material-related data, e.g. influx and outflow of materials • employee-related data such as posting data for wage calculation, booking of arrival, departure and breaks. Meaningful, current and correct production data is best recorded at the point of origin by the employees.	• Scales • Time management systems • Sensors • Video systems
MSS Management Support System	(T) Provision and preparation of aggregated information for the analysis and decision-making in management for different tasks and hierarchy levels (S) As a stand-alone system or as an add-on or as an integrated module of ERP systems Various names for the software, e.g. Business Information System (BIS); Computer Information System (CIS) Decision support system (DSS) Management Support System (MSS), Executive Support Systems (ESS) Management Information System (MIS), Executive Information System (EIS)	• Strategic Planner • MIC Business Intelligence

	(M) Recording of all relevant information and data; analysis, planning, simulation and presentation of the information; powerful graphics for visualization	
PMS **P**roject **M**anagement **S**ystems	(T) For the support of the planning, control, documentation and cost accounting of projects (S) As a module in ERP systems or special software tools (M) Activity planning, flow planning, schedule planning, resource planning, schedule monitoring, performance monitoring, cost monitoring, if-then analyses, reports and documentation	• MS Project 2007 • Projekt Kickstart • Milestones • Project Scheduler 7 • PSI PM
WMS **W**orkflow **M**anagement-**S**ystem	(T) The task of the workflow management system is to ensure the execution of workflows on the basis of a specification with the help of IT systems. It assists with the active control of processes based on the division of labor. (S) Individual software solutions or integration into business enterprise software (M) Support of processes based on the division of labor along the workflow with the objective of improving the quality of the overall processes. Furthermore, it serves to unify the processes, reduce the processing times and thus costs, increase the availability of information, avoid media disruption and increase the transparency of the processes (status determination, documentation of decisions).	• IBM Lotus Notes/ Domino
CAD **C**omputer **A**ided **D**esign	(T) Conception, design and detailing of products, assemblies and individual parts. CAD requires data from ERP systems (customer orders, work schedules, materials and parts list data, information about equipment) and supplies data to ERP systems, e.g. drawings, parts lists, costings, technical documentation. (S) Individual solutions with interfaces to ERP systems (M) Design engineering support (preparation of drawings, parts lists, costings, etc.)	• AutoCAD • SolidWorks • Pro/ ENGINEER • Catia • Solid Edge
CAP **C**omputer **A**ided **P**lanning Systems **CAPP** **C**omputer **A**ided **P**rocess **P**lanning **S**ystems	(T) Builds on CAD and other design engineering data in order to create work schedules and data for parts production and assembly (S) Individual systems, but also implemented in many ERP systems (M) Formulation, optimization and management of work schedules, specification of • raw materials • processing steps (type, number and sequence) • tools (type and material) • technology data (feed rate, number of revolutions) • machinery • jigs and fixtures • times	• HSi
QMS **Q**uality **m**anagement systems	(T) System for the specification of quality policy and objectives as well as the achievement of these objectives. The QMS includes the totality of the construction and process organization-related design and linking the quality-related activities to one another as well as with respect to a uniform, purposeful planning, implementation and control of quality management measures in the business enterprise. (S) Integrated into ERP systems or offered as an add-on (M) In addition to the standard ERP functions and flows, also inspection planning; control planning; goods inwards inspection;	• IBS • CRM • SINIC-CAQ • QSys QM

	goods outwards inspection; supplier assessment; specimen inspection; in-process production inspection; complaints management; testing equipment management; analysis of main sources of errors; action management	
Individual QM systems such as **FMEA** **F**ailure **M**ode and **E**ffect **A**nalysis	(T) Prevention of errors at the earliest possible stage of development of products or processes (S) Software applications exclusively for editing forms; software applications as independent, specialized FMEA modules; software applications in the framework of a CAQ system (M) Systems elements and structures; function lists; cause analysis; malfunction structures; measures; risk assessment	• Byteworx FMEA • FMEA Executive V4
Individual QM systems such as **QFD** **Q**uality **F**unction **D**eployment	(T) Quality Function Deployment (QFD) is a quality assurance methodology for the customer and market-oriented development of products and services. (S) In more complex QM tools or as a specific module (M) Analysis and preparation of the various requirements of customers, developers, producers, and salespeople; definition of the competitive advantages; creation of the assessment profile and interrelationships between the individual requirements and solution characteristics; specification of the product characteristics that are important for the quality of the product and clinch the sale from the customer's point of view; determination of the cost-value relationships; documentation of the entire process	• Decision CAPTURE • QFD/Capture V. 4.0 • QFD Designer V4.0 • Qualica QFD 2.8 • Logic MindGuide • QFD-Scope
DBS **D**ata**b**ase **S**ystem	(T) A database system is a system for electronic data management consisting of database, database management system and communication interface. The essential task of a DBS is to store large quantities of data efficiently, without discrepancies and permanently, and to provide any sub-sets that are required in various forms for users and user programs. (S) Database with management system for the administration and user interface for queries (M) A database is used • to store data • to ensure data integrity • to manage large data inventories • to manage data access rights • to guarantee data security • to secure multi-user operations via transactions • to optimize queries (quick data access)	• Oracle • Microsoft SQL Server • DB/2 • Sybase • MySQL
Data warehouse	(T) Hold, analyze and present data (S) Database with algorithms for extracting, compressing and displaying (M) Data warehouse in the larger sense: • integration of information into a homogenous data inventory • ETL process (extract, transform, load) • data analysis: - data mining: filter datasets and prepare for wide-ranging analysis - OLAP (Online Analytical Processing): data display in particular dimensions • data presentation (data visualization, reporting) = intuitive, easy-to-understand presentation and animations for end user	• Oracle BI • Data Warehousing Suite • Microsoft SQL 2008 • My SAP Business Intelligence • DB2 Enterprise 9 • HP Neoview Enterprise Data • Warehouse

Table 6.25 Checklist for selecting software

❒		**Supplier**		
	❍	Expertise		
	❍	Size		
	❍	Supplier security		
	❍	References		
		How satisfied are users with the system?		
		What experiences have users had with the implementation process?		
		Individual impression after visit on-site		
❒		**Functionality and scope of services**		
	❍	Areas of application		
	❍	Standards		
	❍	Master data management		
	❍	Resource management		
	❍	Client capability		
	❍	Incorporation of business processes		
	❍	Help function		
	❍	Services required		
	❍	Services desired		
	❍	Services offered as standard		
	❍	Additional (optional) services offered		
	❍	License type		
	❍	License scope		
❒		**Special user requirements**		
	❍	Operating system		
	❍	Language		
	❍	Hotline		
	❍	User guidance		
	❍	Environmental conditions		
❒		**Software type and structure**		
	❍	**Standard software**	❍	**Individual software**
		Is standard software available on the market for the desired functions?		How much time does development take up?
		How easy is the system to configure?		How experienced is the supplier in my industry?
		How good (individually) is the support, especially in the course of usage?		How great will the dependence on the supplier be?
		How error robust is the system?		What additional effort is required of the user?
	❍	**Open**	❍	**Closed**
	❍	**Modular**	❍	**Non-modular**
	❍	Module overview		
	❍	Module size		
	❍	Performance description of the modules		
	❍	Combinability of the modules		
	❍	Extendability of the software		
	❍	Customizing of the modules possible/necessary		
❒		**Hardware**		
	❍	Compatibility with the existing system		
	❍	Storage space		
	❍	Server		
	❍	Power		
❒		**Adaptability**		
	❍	List generator		

	❍	Document creation
	❍	Graphical analysis
	❍	Adaptability to the company
	❍	Individual programming
	❍	Comment functions
❒		**Integration**
	❍	Internal integration
	❍	Implementation in own system desired/required?
	❍	Interoperational integration
	❍	Connection to the suppliers possible/desired
	❍	Connection to the customers possible/desired
❒		**Interfaces**
	❍	Hardware-side
	❍	Software-side
	❍	Future
❒		**Service**
	❍	Training offered
	❍	Hotline
	❍	Emergency service
	❍	Online help/info pages on the Web
	❍	Analysis of system errors
❒		**Implementation**
	❍	How is the configuration designed?
	❍	To what extent is the implementation/configuration documented?
	❍	Are test runs made?
	❍	Will employee training be offered? To what extent?
❒		**Ease of use**
	❍	How self-explanatory is the software?
	❍	How is the user interface designed?
	❍	Is adaptation to individual users possible?
	❍	Will employee training be offered? To what extent?
❒		**Data security**
	❍	Is access control easy to guarantee?
	❍	Is data protection a given?
	❍	Is data protection ensured?
	❍	Automated back-ups?
	❍	Restore after system crash?
❒		**Maintenance/release/update/upgrades**
	❍	How is the system administration organized?
	❍	How are the responsibilities divided up?
	❍	How time-consuming is the daily/weekly/monthly maintenance?
	❍	What processes/system parts can be monitored?
	❍	What processes/system parts must be monitored?
	❍	Can the system be checked sufficiently?
	❍	How does the system react in case of breakdowns?
	❍	Is error correction possible?
	❍	Is a release change/update/upgrade possible?
	❍	What does a release change/update/upgrade cost?
	❍	What effort does a release change/update/upgrade involve?
❒		**Costs**
		Acquisition costs
	❍	What does the software cost?
	❍	What do the required hardware components cost?
	❍	What does the other equipment cost?

		Implementation costs
	❍	What does the implementation cost?
	❍	What additional costs arise?
		Training costs
		What are the costs for training the "pros"?
		What does "normal" employee training cost?
		What does subsequent additional training cost?
		Maintenance costs
		What ongoing costs does maintenance cause?
		Do maintenance costs arise only at longer intervals? If so, how often and how much?

6.4.3 Service-oriented architectures as flexible IT infrastructure

A service-oriented architecture can create a flexible connection between business processes on the one hand and modular IT components on the other. The processes form the upper layer, which accesses the IT beneath via a service layer. The IT applications provide their functionality in the form of services, which are coupled dynamically to the processes.

Service-oriented architecture (SoA) refers to an architectural concept of a system. The provision of services is a central element of it. In the context of SoA these services are limited exclusively to the technical context; accordingly, these are not services based on the business term "service."

Essentially, the principle of service orientation builds on the concept of distributed components and objects and adapts this approach. As is the case in the classic model of distributed components, there is an attempt to encapsulate functionalities and data in independent units – the services – and to make these accessible to the network using a standardized interface. However, the SoA has one characteristic that distinguishes it from other architecture models. The characteristics of the services are (Ort 2005, Richter 2005, Erl 2005, Hashimi 2004):

- transparency,
- flexibility and scalability,
- interoperability, loose interconnections and composability,
- platform independence,
- statelessness,
- reusability,
- encapsulation / autonomy.

For the SoA services, the principle of transparency applies; the function of the service and its location as well as the possibility of access should be made apparent to everybody in the network. In the process it is important to categorically separate the service description and service usage. Additional important characteristics are those of encapsulation and autonomy. Each service is responsible for its own task area in isolation – so that the service user does not have to know about its dependencies - and should be in a position to deal with it autonomously. The criterion of loose interconnections is essential for a service; it must be ensured that it can be bound dynamically at run time and that it is possible to combine it with other services. For this, unlimited interoperability in interplay with other services is required. A basic requirement of a service in SoA is that it is not bound to a particular platform. The paradigm of the architecture is often brought into connection with the business process orientation since the flexibility and the reusability of distributed resources can be implemented thus.

In a service-oriented architecture, three roles emerge that shape this system architecture. These are service provider, service requestor and service registry (Carmona et al. 2003). It is not obligatory that every role be executed by a different program. Particular interactions between these roles are specified that enable service usage.

The *Service Provider* is the supplier of a service. Its responsibility is to create the service and the service description as well as to provide the service. It ensures that the service is registered with at least one *Service Registry* together with its description. The service requestor is the service user. It engages the service and sends the service call. According to the principle of loose interrelationships, the address of the service must first be established through a search query with the service registry. The service engagement embodies the client-server relationship between service user and provider.

The service registry is a central directory that mediates between the two other parties. It provides the service descriptions and publishes them on the network. On the one hand, it enables the service providers to enter themselves in this directory; on the other hand, it provides the service requester with the opportunity to search in the collection of service descriptions. By centrally listing the services, it is possible for the service user to assess and compare them.

(a) Design of the infrastructure as SoA

In order to build an SoA, various technical and organizational adaptations must be made. On the organizational side, there must be a process-oriented installation of the business enterprise since an SoA connects the business processes dynamically with the IT infrastructure. A process orientation requires the execution of the approach discussed above on the one hand, and a release from the traditional function-oriented structural organization in favor of a process organization on the other hand. On the technical level, various tools must be provided. Thus on the topmost level, tools are required that provide the different user interfaces for the various application equipment. For the development of portal-based information systems, for example, Microsoft provides the SharePoint Portal Server. Users quickly find relevant information here since portal content and portal layout can be adapted, personalized and tailored to user groups.

On the level below, tools for the coupling of Web service and process are required. These must be in a position to model processes and to execute them. An example of these is the Microsoft BizTalk Server, which can be used with a BPEL attachment for Microsoft Visio. The processes can be modeled in Visio; these are then coupled with Web services using the process engine of the BizTalk server and they can be executed. An extensive modeling, simulation or even controlling of the processes is not possible, however. For this, there are specialized providers of process management tools such as IDS Scheer (Aris) or EMPRISE Process Management GmbH (Bonapart).

In order to provide the Web services, development tools are also required in order to be able to incorporate old applications into the SoA. An example of this is Microsoft Visual Studio Professional 2008. Using this tool, it is possible to develop Web services in many different languages.

On the lowest level, integration tools are required for the various applications to interact. The topic of integration tools will be discussed below. For this, an application and Web server are required, such as the ones in the Microsoft Windows Server 2008 family. For the incorporation of Web services, for example, there is Internet Information Services (IIS) 7 as a component of Windows Server 2003.

(b) Integration approaches

The integration of software systems in information processing can be examined on the levels of the data, applications and processes.

Data integration:

The data integration is established on the level of the data inventories / on the basis of the databases that underlie the schemes. In most cases the objective is ensuring application-wide data consistency. Data access middleware is frequently used in heterogeneous system environments for data integration. The schemes are thus integrated solely on the logical level.

Application integration:

In addition to the integration of schemes and data for various applications, the integration of the applications themselves is also important. On the one hand, the objective is the abovementioned application-wide data consistency. On the other hand, the aim is the greatest possible reuse and usage of existing functionality from the integrated system (functional integration) as well as a uniform user interface (GUI integration). Integration platforms serve as the basis for system coupling; they provide the required services. Frequently, message-oriented middleware is used for the application integration. The architectures that can be realized with it combine functional integration with the advantages of asynchronous communication and enable loose interrelationships between the participating applications. The message format is typically XML-based in order to make the incorporation of additional systems easier.

Process integration:

Data and application integration are not sufficient if complex business processes must be supported which are also subject to constant change. In this case it must be possible to easily extend or adapt the business processes to include additional activities. This requires a higher control level with which application-wide business processes can be realized. Concepts from the areas of workflow management (WFM) and enterprise application integration (EAI) are used to implement this. The focus of process integration is on the business management and customer oriented business processes. The objective of process integration is the complete electronic processing of all customer-oriented processes within the business enterprise.

If the integration levels are examined more closely, it quickly becomes clear that real integration requirements often cannot be assigned exactly to an integration level. Instead, these grow together so that integration takes place on several levels.

(c) Integration technologies

There are four distinct classic integration technologies, which over the course of time have grown ever closer together. Enterprise application integration systems in particular have been elevated to a higher and more dynamic level by the new Web services technology. The service oriented architectures that combine workflow management systems with web service-based EAI technologies constitute the highest level.

Point-to-point integration:

Point-to-point integration describes the development of individual interfaces between two applications, which can then be connected to one another permanently. If there is a multitude of different applications to be integrated this results in what is known as "spaghetti architecture." Point-to-point connections can also be implemented using Web services. However, the difficult problem of maintaining spaghetti architecture remains (Klöckner 2002, p. 18).

Middleware:

Application-independent technologies that provide communication services between applications are called middleware. Middleware conceals the complexity of the underlying operating systems and networks in order to make the integration of various applications easier (Ruh et al. 2001, p. 2).

Middleware extracts information from applications using adapters and forwards it to others (Winkeler et al. 2000). It is used to connect two or more applications and guarantee their connectivity and interoperability, thus simplifying communication between different applications. Middleware is a type of software that lies between the operating system and the applications as an additional abstraction layer.

Workflow Management System:

A workflow management system (WfMS) handles the active control and automation of processes that are based on the division of labor. It concentrates on processes that require user interactions and frequently works with unstructured data. In order to be able to interpret and process this data practically, manual intervention is usually necessary. WfMS map a usually Web-based presentation layer over the actual applications and use adapters to access the actual applications or to integrate Office applications, for example, via Object Request Broker.

Enterprise Application Integration:

Enterprise application integration (EAI) refers to the creation of operational application systems by combining individual applications jointly using middleware (Ruh et al. 2001, p. 2). EAI tools enable the integration of existing applications on the data, application and process level. Starting from a central server, an adapter must be created for the integration using an EAI tool for each application that is to be integrated. The central server thus represents the communication hub for the processes in the business enterprise. The adapters execute the transformation of the data and the remote system calls. They are generally based on proprietary class libraries. EAI tools are suitable first and foremost for fully-automated workflows that work on structured data.

With the growing popularity of Web services, EAI providers have also recognized the potential of this technology. Thus, today many EAI tool providers offer integration not just using proprietary adapters, but they also allow integrators to incorporate applications into the EAI tool via Web services. The Web services standards offer standardized protocols and interface descriptions for finding and using remote services. Thus a provider can make application functionality available as a service via a Web service; it can be used by any user (people or software systems). Web services are thus suitable for loosely coupling software systems.

Web services technology on its own does not allow various successive service calls to be related to one another (unless the participating applications enable it) or a sequence of different service calls to be specified. It only ever concerns itself with precisely one usage of a service. In turn the EAI system can take on this role; using Web services (as an adapter) it is able to establish connections between the individual systems (Klöckner 2002, p. 18). Because of the standardization of the Web services its advantage consists in its platform independence. For example, it is not necessary to develop an individual Web service for each EAI system to enable access to a particular application.

The individual components of an EAI solution are adapters (Kaib 2002) - which recently are ever more frequently Web services – middleware and the physical network, which enable the connectivity of the applications. Message management includes transformation services, synchronization services and transactionality. Process management covers process modeling, control and checking. In addition, there is a metadatabase and additional services such as system management.

Service-oriented architectures:

Service-oriented architectures go a step further than EAI tools. Ideally, in an SoA all applications are set up so that they provide their functions as services. WS-BPEL (and earlier also PBEL4WS) is an XML-based (Extensible Markup Language) description language for the specification of business processes. WS-BPEL (Web Services Business Process Execution Language) is specially tailored to the coordination of Web services within an SoA. The execution of a business process specified in WS-BPEL is controlled and monitored by what is known as a "BPEL engine." For the future, the combination of WS-BPEL and Web services will offer an alternative to traditional EAI tools. In principle, with Web services and BPEL, integration scenarios of any type can be realized (Boles et al. 2004, p. 10).

Overall it can be said that the various integration technologies are growing ever closer together. Thus today a series of EAI tools already support Web services and integrate BPEL engines to an appreciable extent (Boles et al. 2004). Thus it is also no wonder that the service-oriented architectures represent the most appropriate integration technology, since they combine the advantages of Web service-based EAI with the automation of processes in a manner similar to a WfMS.

Once the integration problem has been solved, planning the operation of the infrastructure can begin.

6.4.4 Operation of the IT infrastructure

Application Service Provider

There is agreement that core competencies should remain in the business enterprise and non-core competencies should be examined to see if it is possible to outsource them. Questions in IT are becoming ever more complex and the corresponding expertise must be built up, maintained, and continuously enhanced. A specialized IT service provider and outsourcing partner is in a better position to expand this expertise because their experience is spread across a number of different customers. It is possible to outsource all IT services or just a portion of them. Application service providing is a special form of outsourcing relationship. The IT service provider is called the "Application Service Provider" (ASP). A service level agreement is implemented to fix the quality of the service. In case of non-adherence, contractual penalties are specified thus increasing the security of the service provision with the ASP model. The advantages of the ASP model are discussed in brief below.

Cost advantages:

- lower (investment) costs and TCO,
- lower capital commitment and higher liquidity,
- better cost control and more planning security through transparent cost structures,
- optimized resource usage and concentration on core competencies possible.

Technological advantages:

- always up-to-date due to the use of the most current software and hardware,
- always the latest security standards,

- fast adoption of applications due to quick and easy implementation,
- high degree of flexibility, easier and quicker migration,
- low hardware requirements,
- freedom from ever-shorter software life cycles,
- quicker and easier scalability,
- improved administration.

Service & performance advantages:

- legal certainty due to service level agreements (SLA),
- clearly-defined performance and responsibilities,
- guarantee of high level of availability, system stability and data security,
- first level support: strong skills and expertise at the ASP,
- central data storage with worldwide and time-independent access.

Further added value:

- concentration on core competencies,
- process optimization,
- decentralized corporate structure possible: incorporation of mobile employees or home offices.

Once the cooperation with the IT service provider has been defined, the actual IT services can be planned, either in-house or externally. The ITIL is a useful tool for this.

ITIL – IT Infrastructure Library

The Infrastructure Library (trademark of the OGC) is a set of rules consisting of modules and disciplines. With consistent implementation, these guidelines ensure a more effective usage of IT resources and guarantee a better quality of service from the IT service provider.

ITIL is not a standard, but rather a type of toolbox that was gained from actual experience and should be regarded as "best practice" in the area of IT service management. At its heart is the Configuration Management Database, a practice-oriented database system for managing process dependencies. IT service management plans, monitors and controls the quality and quantity of IT services. ITIL is a generic model and is thus generally applicable. However, it must be adapted to the enterprise's particular circumstances. The model has a role concept and is organization-neutral, extensive and complete. The ITIL modules put the IT service management disciplines into practice. They are formatted according to the following levels:

- role definition,
- marketing for required change measures,
- planning,
- implementation,
- inspection,

- auditing.

ITIL puts the (IT) service manager in a position to incorporate information into the product development and the product life cycles since it provides a holistic view of the IT services. ITIL provides the basic conditions and objectives for achieving optimal support of the entrepreneurial core processes in the production process through established and functional tools (cf. KESS DV-Beratung GmbH (2002)). The limit of ITIL is reached with the description of the body of rules. The concrete implementation of these formalities occurs in the company and is supported by tools in order to adjust the IT area for this new paradigm change (FNT-GmbH 2004).

In times of rapid technological and therefore mental developments, in the future there will be new challenges for the information management of a factory. This chapter looks ahead to a few of them below.

6.4.5 Development perspectives

The digitization of product, process and control data was introduced using the CIM philosophy of the 1980s and 1990s. In recent years, wider-reaching information systems have come into being that take management and logistical functions into account. Existing information systems and those still in development are increasingly becoming building blocks (modules) of integral production planning and control. They can be used in relation to a particular problem or order by linking suitable models for different use cases. At the same time, they also form the basis for the "digital" production facility/factory. In that digital product and production are integrated with digital production planning and control, a new quality for holistic production facility planning has emerged.

The digital factory networks essential models in production facility planning as shown in fig. 6.50.

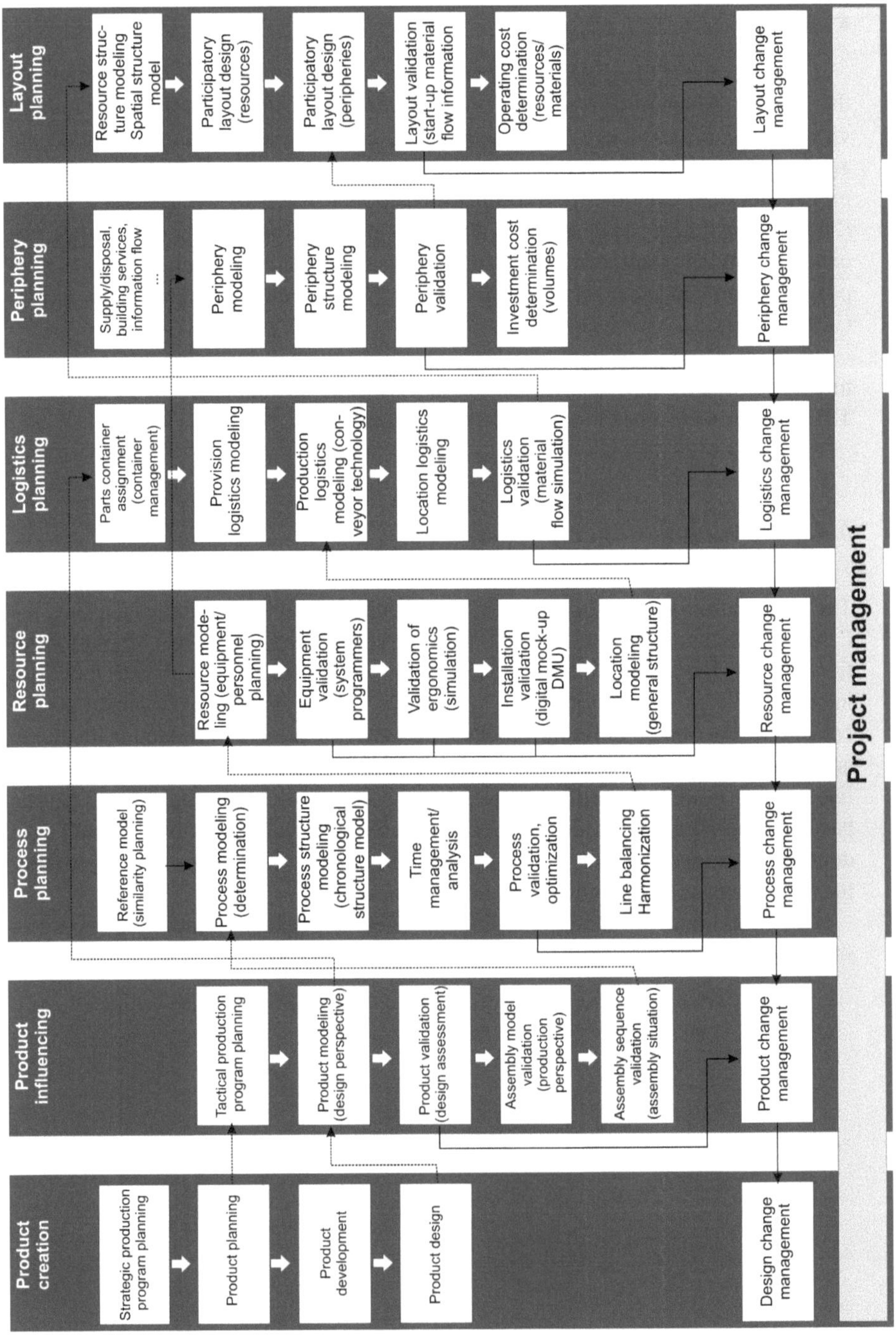

Fig. 6.50 System for linking integrated production facility planning models (based on Lehmann 2008, p. 66)

The digital factory is thus also becoming the data supplier for the operation of production facilities. Fig. 6.51 demonstrates the approach to integration and digitization of various functions in a business enterprise (Aldinger 2007; Lehmann 2008).

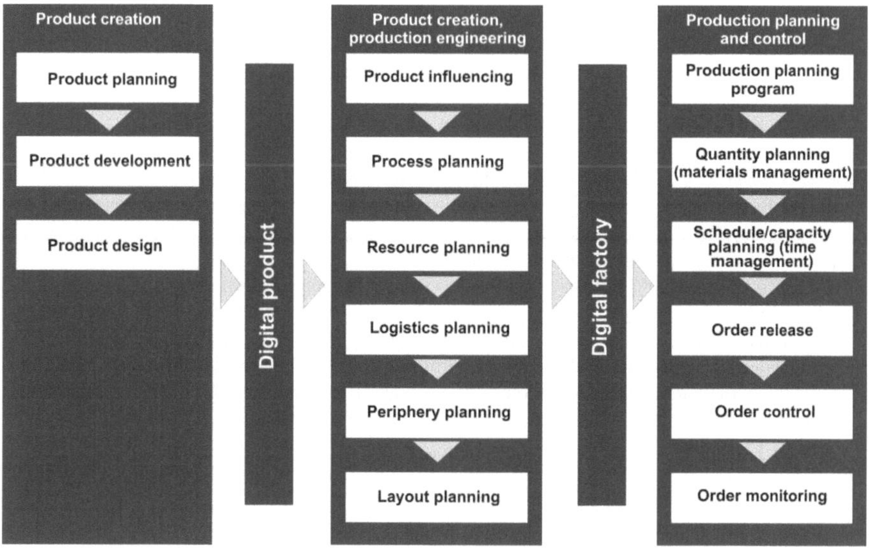

Fig. 6.51 The digital factory as data supplier to PPC (Lehmann 2008, p. 157)

Manufacturing systems engineering (MSE) methods continue to be enhanced and tailored to meet customer needs.

In the future, it will be possible to gradually transfer adapted forms of digital components developed for large business enterprises (e.g. automobile manufacturers) to small and medium-sized enterprises (SMEs) (Dombrowski 2005).

Digitization is and will remain just a tool for the efficient design of real production facilities with real production of real products.

This development is being encouraged by Internet-based modes of operation. It enables regional and inter-regional networking with and access by medium-sized companies to the customer.

Thanks to service-oriented software architecture, all computers in the network can be used and program functions are publicly available and can be further adapted.

References

Aldinger L, Kapp R, Westkämper E (2007) Echtzeitfähiges Fabrik-Cockpit-System. Wt Werkstattstechnik online 97H5, S 358 – 362

Arbeitsschutzgesetz, Arbeitssicherheitsgesetz, Betriebssicherheitsgesetz, Brandschutzverordnung, Immissionsschutzgesetz

Arbeitsstättenverordnung (Zulässige/empfohlene Klimawerte)

Arbeitsstättenrichtlinien ASR5 (Lüftung) und ASR 6/1,3 (Raumtemperaturen)

ASR 7/3: Künstliche Beleuchtung von Arbeitsstätten in Gebäuden

ASR 7/4: Sicherheitsbeleuchtung

ASR 41/3: Künstliche Beleuchtung für Arbeitsplätze und Verkehrswege im Freien

Arbeitsstoffverordnung

Baer R (1990) (Hrsg.) Beleuchtungstechnik, Technik, Berlin

BGR 163 Regeln für Sicherheit und Gesundheitsschutz beim Umgang mit krebserzeugenden und erbgutverändernden Gefahrstoffen

BGR 217 Umgang mit mineralischem Staub

BREF (2008) European Commission on Best Available Techniques for Energie Efficiency

Boles C, Friebe J, Luhmann T (2004) Typische Integrationsszenarien und deren Unterstützung durch Web Services und andere Technologien. In: Hasselbring W, Reichert M (Hrsg.) Tagungsband des GI-/GMDS-Workshops „Enterprise Architecture", Fachgruppe 2.5.2 'Entwicklungsmethoden für Informationssysteme und deren Anwendung (EMISA), Oldenburg. Download from: http://akea.iwi.unisg.ch/downloads/eai2004-paper6.pdf. most recently visited on 15.04.2005.

Bürck W (1981) Lärm – Messung, Bewertung und Wirkungen auf den Menschen. In: Schmidtke H (Hrsg) Lehrbuch der Ergonomie. Wien, München, Hanser

Bullinger HJ, Schreiner P (2001) Business Process Management Tools – Eine evaluierende Marktstudie über aktuelle Werkzeuge. Fraunhofer IRB, Stuttgart.

Bundesimmissionsschutzgesetz

Carmona DS, Injey F, Hore D, Lastra JLF (2003) XML on z/OS and OS/390: Introduction to a Service-Oriented Architecture. IBM redbooks. Download from: http://www.redbooks.ibm.com/abstracts/sg246826.html. most recently visited on 20.07.2005.

Coscom (2008) www coscom.en2008: Vollständige Transparenz in der Fertigung, VDI-Z Integrierte Produktion, H9, S 86

Deutsche Forschungsgemeinschaft: Senatskommission zur Prüfung gesundheitsschädlicher Arbeitsstoffe: MAK- und BAT-Werte-Liste in der jährlich aktualisierten Ausgabe

DIN 1946: Raumlufttechnik

DIN 5034: Tageslicht in Innenräumen

DIN 33403: Klima am Arbeitsplatz und in der Arbeitsumgebung

DIN 45630: Grundlagen der Schallmessung

DIN 45641: Mittelung von Schallpegeln

DIN 45645: Ermittlung des Beurteilungspegels aus Messungen

DIN 50011: Klimaprüfeinrichtungen

DIN 52900: DIN-Sicherheitsdatenblatt für chemische Stoffe und Zubereitungen; Formblatt und Hinweise zum Ausfüllen

DIN EN 12464-1: Beleuchtung von Arbeitsstätten, Teil 1 Arbeitsstätten in Innenräumen

DIN EN 12665: Grundlegende Begriffe und Kriterien für die Festlegung von Anforderungen an die Beleuchtung

DIN ISO 7730: Gemäßigtes Umgebungsklima (PMV, PPD, therm. Behaglichkeit)

Dombrowski U, Tiedemann H, Quack S (2005) Einführung der Digitalen Fabrik bei KMU, ZWF 1-2, S 10-13

Engelmann J (2008) Methoden und Werkzeuge zur Planung und Gestaltung energieeffizienter Fabriken, Diss, TU Chemnitz

Erl T (2005) Service-Oriented Architecture - Concepts, Technology, and Design. Schriftenreihe: Prentice Hall, New Jersey
Eissing G (1988) Klima am Arbeitsplatz, Messung und Bewertung. Köln
Fichter K (2007) Zukunftsmarkt energieeffizienter Rechenzentren. Studie im Auftrag des Bundesministeriums für Umwelt, Natur und Reaktorsicherheit (BMU); Borderstep Institut
FNT-GmbH (2004) ITIL - von der Theorie zur Praxis. Download from: http://www.fnt.de/itil/index.html most recently visited on 23.03.2005
Fröhlich J (2004) Fabrikökologie/Entsorgungslogistik, Studienbrief 11, TU Dresden
Gefahrstoffverordnung
Gottschalk K, Graham S, Kreger H, Snell J (2002) Introduction to Web services architecture. In: IBM Systems Journal, Issue 41-2. Download from: http://www.research.ibm.com/journal/sj/412/gottschalk.html. most recently visited on 17.08.2005.
Hanhart D (2008) Mobile Computing und RFID im Facility Management: Anwendungen, Nutzen und serviceorientierter Architekturvorschlag, Springer Berlin
Hashimi S (2004) Service-Oriented Architecture Explained. Download from: http://dev2dev.bea.com/pub/a/2004/05/soa_hashimi.html. most recently visited on 02.06.2004.
Hintermann R, Pfahl S (2007) BITKOM, Bundesministerium für Umwelt, Naturschutz und Reaktorsicherheit; Energieeffizienz im Rechenzentrum. Download from: http://www.bitkom.org/files/documents/Leitfaden_Energieeffizienz_in_RZ_final_31072008.pdf. most recently visited on 23.10.2008
Hoffmann H, Lübke A v (1990) 0 Dezibel + 0 Dezibel = 3 Dezibel. Berlin, Erich Schmitt Verlag
Hülsenbeck F (2005) RFID 2005 – Logistiktrends in Industrie, Dienstleistung und Handel. WWW. 2006. – URL http://www.iwi.uni-hannover.de/trim/huelsenbeck.pdf. – Zugriffsdatum: 2006-08-29
ISO-DIS 7933: Bekleidungsisolation
Kaib M (2002) Enterprise Application Integration. Wiesbaden
Keller G, Nüttgens M, Scheer AW (1992) Semantische Prozessmodellierung auf der Grundlage Ereignisgesteuerter Prozessketten (EPK). In: Veröffentlichungen des Instituts für Wirtschaftsinformatik, Heft 89, Saarbrücken
Kletti J (2006) MES – Manufacturing Execution Systems, Moderne Informationstechnologie zur Prozessfähigkeit und Wertschöpfung, Berlin, Springer
KESS DV-Beratung GmbH (2002) IT Infrastructure Library (ITIL) Basis und Standard für ein erfolgreiches IT Service-Management. Download from: http://www.kess-dv.de/WirUeberUns/Beschreibungen/ITIL.pdf. most recently visited on 01.04.2005.
Klöckner M (2002) Web-Services: Die Lösung aller Integrationsprobleme? In: Objektskektrum, Heft 5/2002, S 16-19
Konietzko D (1989) (Hrsg.) Handbuch der Arbeitsmedizin. Ecomed, Landsberg
Krampe H, Lucke H-J (2007) Grundlagen der Logistik, Huss, München
Kubert K, Teuteberg F, Hilker J (2008) Mobile Business-Anwendungen im Enterprise Ressource Planning: Mobilitätspotentiale entlang der ERP-Funktionskreise; Europa-Universität Viadrina Frankfurt (Oder). Download from: http://www.vg-u.de/euv-new-site/doc/mobcom/mob-erp.pdf. most recently visited on 23.10.2008
Lehmann J, Müller E (2006) Mit Methode zum Erfolg – Ansatz zur Entwicklung von Methoden zur ganzheitlichen integrierten Planung kundenorientierter wandlungsfähiger Fabriken mit Hilfe der Digitalen Fabrik. Diplomarbeit, TU Chemnitz
Lehmann J (2008) Entwicklung von Methoden zur ganzheitlichen integrierten Planung wandlungsfähiger Fabriken unter Nutzung der digitalen Fabrik, Dissertation, Fakultät Maschinenbau, TU Chemnitz
Leven B (2005) Energiemanagement in der Investitionsgüterindustrie, Diss, Universität Stuttgart
Lindemuth F, Krochmann J Empfehlungen zur Messung von Beleuchtungsanlagen. Schriftenreihe der Bundesanstalt für Arbeitsschutz. Fb 567, Wirtschaftverlag NW

Luczak H (1998) Arbeitswissenschaft. Berlin, Heidelberg, New York, Springer
Müller E, Engelmann J, Strauch J (2008) Energieeffizienz als Planungsprämisse, Industrie Management, GITO, Berlin
Müller E, Engelmann J, Strauch J (2008b) Energieeffizienz als Zielgröße in der Fabrikplanung. Wt Werkstatttechnik online, H 7/8, S 634 – 639
Neugebauer R (Hrsg) (2008) Energieeffizienz in der Produktion, FhG-Institute, BMBF, FZ Karlsruhe, FKZ 02PV1000
Nüttgens M (1997) Ereignisgesteuerte Prozessketten (EPK) – Forschungsansätze in der wissenschaftlichen Literatur und Praxis. In: Desel J, Reichel H (Hrsg): Grundlagen der Parallelität. Dresden
Nüttgens M, Rump FJ (2002) Syntax und Semantik Ereignisgesteuerter Prozessketten (EPK). Promise, S 64-77
o.V.: BITKOM: Vorteile des ASP-Modells. Download from: http://www.bitkom.org/files/documents/Vorteile_ASPModell.PDF. most recently visited on 23.10.2008
o.V.: Integration von RFID-Systemen. http://www.rfid-ready.de/298-0-rfid-integration.html. most recently visited on 23.10.2008
Ort E (2005) Service-Oriented Architecture and Web Services: Concepts, Technologies, and Tools, Sun Microsystems. Download from: http://java.sun.com/developer/technicalArticles/WebServices/soa2/. most recently visited on 02.06.2005
REFA (1989) Klimagestaltung (Lehrbuch), Darmstadt
Richter J (2005) Wann liefert eine Serviceorientierte Architektur echten Nutzen? In: Software Engineering 2005, Fachtagung des GI-Fachbereichs Softwaretechnik. Essen
Ruh WR, Maginnis FX, Brown, WJ (2001) Enterprise Application Integration. New York
Scheer AW (1994) ARIS-Toolset: Die Geburt eines Softwareproduktes. In: Scheer AW (Hrsg): Veröffentlichungen des Instituts für Wirtschaftsinformatik, Heft 111, Saarbrücken
Scheer AW (1994) Wirtschaftsinformatik – Referenzmodelle für industrielle Geschäftsprozesse. 5. Auflage, Berlin
Schuh G (2006) Produktionsplanung und -steuerung, Springer, Berlin
Schwarzer B, Krcmar H (1995) Zur Prozeßorientierung des Informationsmanagements. In: Wirtschaftsinformatik, Jg. 37 (1995) Nr. 1, S. 33-39
Spat D, Lentes J, Richter M (2007) Neue Ansätze für die Integration von Konstruktion und Planung im Rahmen der digitalen Produktion, ZWF, H 1-2, D 73-77
Staud JL (2006) Geschäftsprozessanalyse – Ereignisgesteuerte Prozessketten und objektorientierte Geschäftsprozessmodellierung für betriebswirtschaftliche Standardsoftware, Springer
TA Lärm (1998) - Technische Anleitung zum Schutz gegen Lärm. GMBl. S 503
TA Luft
TRGS 100: Auslöseschwelle für gefährliche Stoffe
TRGS 102: Technische Richtkonzentration (TRK) für gefährliche Stoffe
TRGS 402: Ermittlung und Beurteilung der Konzentration gefährlicher Stoffe in der Luft in Arbeitsbereichen
TRGS 900: Luftgrenzwerte am Arbeitsplatz - MAK und TRK
TRGS 903: Biologische Arbeitsplatztoleranzwerte – BAT-Werte
TRGS 905: Verzeichnis krebserzeugender, erbgutverändernder oder fortpflanzungsgefährdender Stoffe
Unger H (2008) Arbeitsumwelt, Vorlesungsskript, TU, Prof. Arbeitswissenschaft, Chemnitz
UVV Lärm (BGV B3) der Gewerblichen Berufsgenossenschaften
VDI 4499 (2006) (Hrsg) Richtline VDI 4499 Digitale Fabrik-Grundlagen, Düsseldorf
VDI 5600 (2007) Fertigungsmanagementsysteme, Manufacturing Executions System, VDI Handbuch, Informationstechnik, Düsseldorf
VDI 2058: Beurteilung von Lärm
VDI 2711: Schallschutz durch Kapselung

VDI 2720: Blatt 2 Schallschutz durch Abschirmung in Räumen
VDI 2720: Blatt 3 Schallschutz durch Abschirmung im Nahfeld
VDI 4602 (2006) Energiemanagement, Begriffe und Definitionen
VDI 4661 (2003) Energiekenngrößen, Begriffe – Definitionen - Methodik
VEM-Handbuch Beleuchtungstechnik. Verlag Technik Berlin 1975
Weis SA (2005) Security Parallels between People and Pervasive Devices. In: PERCOMW '05: Proceedings of the Third IEEE International Conference on Pervasive Computing and Communications Workshops (PERCOMW'05). Washington, DC, USA: IEEE Computer Society, S 105–109
Winkeler T, Raupach E, Westphal L (2000) EAI – Enterprise Application Integration. In: PriceWaterhouseCoopers (Hrsg.) Leitfaden E-Business. (o.O.)

Subject index

M. Schenk et al., *Factory Planning Manual*, DOI 10.1007/978-3-642-03635-4,

MIX
Papier aus verantwortungsvollen Quellen
Paper from responsible sources
FSC® C105338

If you have any concerns about our products,
you can contact us on
ProductSafety@springernature.com

In case Publisher is established outside the EU,
the EU authorized representative is:
Springer Nature Customer Service Center GmbH
Europaplatz 3, 69115 Heidelberg, Germany

Printed by Libri Plureos GmbH
in Hamburg, Germany